Essential Physics of Medical X-ray Sources and Detectors

Ke Li

Essential Physics of Medical X-ray Sources and Detectors

Lecture Notes on X-ray Tubes, Linacs, Photon-Counting Detectors, and Energy-Integrating Detectors

Springer

Ke Li
Department of Imaging Physics
UT MD Anderson Cancer Center
Houston, TX, USA

ISBN 978-3-032-27799-2   ISBN 978-3-032-27800-5 (eBook)
https://doi.org/10.1007/978-3-032-27800-5

This Springer imprint is published by the registered company Springer Nature Switzerland AG
The registered company address is: Gewerbestrasse 11, 6330 Cham, Switzerland

In Memory of Dr. Edward F. Jackson

Preface

"This isn't an ordinary flashlight. It lights up when there's light," said Da Wenxi proudly.
"What about when there's no light?" asked the General.
"It absolutely won't light up," replied Da Wenxi proudly.

The dialog above originates from the 1994 Hong Kong comedy spy film *From Beijing with Love*. Da Wenxi, a fictitious character in the film, is an inventor of nonsensical spy gadgets. In one scene, he demonstrates his latest ridiculous invention—a solar-powered flashlight—to his boss, the General. The flashlight, powered by a solar panel, only works when there is already external light, creating a comedic effect because of the invention's complete uselessness.

I was amused by this nonsensical scene as well—until one day I realized that x-ray imaging devices are, in a sense, analogous to Da Wenxi's flashlight. When light (x-rays) is incident on a detector, the detector "lights up" and produces an output signal; when there are no x-rays, it is desirable that the detector absolutely does not light up. In essence, the idea is quite simple: we need an external light (x-ray) source and a detector that converts the incoming x-rays into lights or other types of signal and immediately switch off when the external illumination disappears.

This book is about x-ray sources and detectors, focusing specifically on the essential physics underlying x-ray production and detection. If you are looking for a review of cutting-edge research on x-ray sources and detectors, this is not the book for you. It is also not intended for high-energy or condensed matter physicists seeking a comprehensive textbook on semiconductor devices or linear accelerator systems—numerous specialized textbooks cover those topics. Instead, this book is aimed at new learners of medical physics, medical imaging professionals, and radiotherapy practitioners who want to gain a deeper understanding of the physical principles behind x-ray imaging devices and linear accelerators used in medical practice.

I was once such a learner during my time in a medical physics graduate program. While there were excellent textbooks covering the engineering, quality evaluation and assurance, and clinical operation aspects of x-ray sources and detectors, I often struggled to grasp the physics behind some of the governing equations. For example, I was told that the peak instantaneous power of an x-ray tube is proportional to the $3/2$-th power of the focal spot size, and that the tube current may be limited by the tube voltage according to the same $3/2$-th power. I was also told that, in a cold cathode, the tube current is related to the work function of the cathode material—magically and coincidentally—by that very same $3/2$-th power. I became fascinated by the origin of all these $3/2$ power laws and wanted to understand the underlying physics.

After becoming a medical physics practitioner and researcher, the gaps in my fundamental physics knowledge became increasingly apparent. Concepts such as those underlying the so-called Shockley-Ramo theorem are often well understood by nuclear physicists and engineers who originally developed semiconductor detector technologies, yet this knowledge is rarely passed along to the end clinical users of the detectors. As medical physicists, we have our own areas of expertise, such as detector quality assessment in contexts directly relevant to clinical imaging tasks. However, these strengths do not diminish the need to engage deeply with the underlying physics—especially if we aim to contribute to technical advancements in the field.

I am not alone in recognizing the pressing need to strengthen the physics foundation of medical physicists and other medical practitioners. Around 2016, the Chair of the Medical Physics Department at the University of Wisconsin (UW)–Madison, Dr. Edward F. Jackson, initiated a transformation of the department's graduate curriculum. At that time, I was a newly appointed Assistant Professor in the department and had the privilege of witnessing his vision firsthand: to integrate more fundamental physics, mathematics, biology, and chemistry into the traditional medical physics curriculum. His goal was to help students gain a deeper understanding of core principles, better prepare for research challenges, and advance not only the field of medical physics but also the broader domains of medical imaging and radiotherapy.

Dr. Guang-Hong Chen was one of the faculty members who played a leading role in implementing the curriculum transformation. He completely redesigned a core medical physics course, *Radiation Physics and Dosimetry*, so that the key formulas involved in stopping power, cavity theory, and interaction cross-sections were derived from first principles, yet presented in a way that students without extensive graduate training in nuclear physics could understand. Remarkably, he developed all the lecture notes without any prior formal training in radiation dosimetry or medical physics. I can imagine that, trained as a theoretical physicist, he could intuitively derive everything directly from fundamental principles.

Since its implementation in 2017, the restructured *Radiation Physics and Dosimetry* course has had a profound impact on many students. For me personally, it provided a solid physics foundation behind radiation detectors, inspired me to explore the physics of x-ray tubes, linear accelerators, and x-ray detectors, and greatly benefited my research, ultimately leading to multiple patents and a series of publications.

Recognizing my potential as an effective teacher, Drs. Chen and Jackson encouraged me to participate in the development of a new, advanced course on the physics of radiation detectors. In 2019, I prepared a set of lecture notes covering the physics of semiconductor and scintillator detectors, which laid the groundwork for this book. Together with two other faculty members, Drs. Wes Culberson and Jon Engle, we began teaching a new course, *Radiation Production and Detection*. Initially, I was responsible for the lectures on scintillator and semiconductor detectors, but over the years my teaching duties expanded to include x-ray tube and linear accelerator (Linac) physics, covering a total of 18 lectures. Many of the detailed derivations and underlying physics principles could not be fully taught during the limited lecture hours but are now included in this book.

Dr. Jackson unfortunately passed away in 2020. His loss was deeply felt, and it strengthened my resolve to carry on his vision and passion for medical physics education. I dedicate this book to his memory, in honor of his enduring commitment to teaching and advancing the fields of diagnostic and therapeutic medical physics.

From 2020 to 2023, I spent a substantial part of my time to the teaching of *Radiation Production and Detection*, as well as several other medical physics courses. During this time, I received numerous valuable suggestions from UW colleagues such as Drs. Guang-Hong Chen, Tim Hall, Michael Speidel, Ran Zhang, Yijing Wu, Wes Culberson, and Jon Engle. Jeff Radtke from the UW Medical Physics Department also offered important insights into the experimental aspects of semiconductor detectors and helped design a laboratory exercise that uses a simple LED flashlight to trigger another reverse-biased flashlight to produce an electrical signal—essentially Dan Wenxi's "flashlight experiment."

I am grateful to the former Chair of the Department of Radiology at UW–Madison, Dr. Thomas Grist, who was the first radiologist to ask me about the concept of holes in semiconductors and their connection to the operating principles of semiconductor photon counting detectors. His leadership, scientific curiosity, and mentorship have been invaluable to me.

I have also received much helpful feedback and encouragement from UW-Madison medical physics graduate students, which has inspired me to revise and improve my lecture notes over the years. In particular, I would like to thank my PhD students Kevin Treb, Xu Ji, and Mang Feng, who provided the first proofreading of the initial draft of the lecture notes. They have since become experts in their respective fields such as photon counting detector CT imaging. Former UW students Chengzhu Zhang and Linying Zhan also offered valuable feedback that helped enhance the lecture notes. I am also grateful to the students in the 2017 Sino-U.S. Advanced Medical Physics Training Program, who were the victims of my first medical imaging teaching experience and nevertheless offered me tremendous encouragement.

In 2024, I assumed a new faculty position in the Department of Imaging Physics at the University of Texas MD Anderson Cancer Center, where Dr. Jackson worked prior to becoming Chair of the Department of Medical Physics at UW–Madison. I continue to teach graduate-level medical physics courses while also providing clinical imaging physics services and conducting research. The teaching style brought from Wisconsin seems to be well received by students at MD Anderson. In my spare time, I began expanding my lecture notes into a book—one of the first to present both Linacs and x-ray tubes within a unified framework of x-ray production.

My efforts have been greatly encouraged by the MD Anderson Imaging Physics Department Chair, Dr. John Hazle; the Chief of the Radiological Physics Section, Dr. Frank Dong; my faculty mentors, Dr. Jeff Siewerdsen and Tinsu Pan; and many other colleagues in the Department of Imaging Physics. I have also had the pleasure of collaborating with radiation therapy physicists at MD Anderson, including Drs. Jingwei Duan, Ming Yang, and Ronald Zhu, who have provided numerous insightful suggestions on Linacs and image-guided radiation therapy.

I am grateful to the following individuals from the medical physics community, whose encouragement has strengthened my determination to complete this book: the Jackson family, Cynthia McCollough, Shuai Leng, Lifeng Yu, Nobert Pelc, Zhihua Qi, Andrew Karellas, Srinivasan Vedantham, Wei Zhao, Michael F. McNitt-Gray, Marc Kachelrieß, Ken Taguchi, Scott Hsieh, Ehsan Samei, York Haemisch, and Hao Gao. I would also like to acknowledge two mentors from outside the medical physics community: Dr. Zhenlin Wang of the Physics Department at Nanjing University, my undergraduate research advisor who taught me electrodynamics, and Dr. Ezekiel Johnston-Halperin of the Physics Department at The Ohio State University, my master's research advisor who taught me semiconductor physics.

I am grateful to my friends Dongxu Wang, Xuran Zhao, Zhiqiang Liu, and Yangguang Ma for their encouragement and support over the years.

I would like to thank Margaret Moore, Swathiga Karthikeyan, Shanjini Rajasekaran, and Patricia Gutierrez from Springer Nature for their trust, patience, and outstanding editorial assistance.

Last but not least, I extend my heartfelt thanks to my wife and daughter for their unwavering support for this book project. They also contributed to the design of some of the artwork in this book. Thank you very much.

Houston, TX, USA Ke Li

Contents

Chapter 1
Basic Semiconductor Physics

Contents

K. Li, *Essential Physics of Medical X-ray Sources and Detectors*,
https://doi.org/10.1007/978-3-032-27800-5_1

1.1 Introduction

We are often introduced to the band structures that distinguish conductors, semiconductors, and insulators through an illustration like Fig. 1.1. When I first encountered it, I found myself wondering: Why has a material that sits between a conductor and an insulator become so central to modern technology—powering so many electronic devices from LEDs to the microprocessors in smartphones and computers?

After all, when high conductivity is required, such as in power lines, we use conductors. When high resistivity is desired, as in heating elements, we use insulators. So what practical advantage does a material that sits somewhere in between offer?

The answer lies in the unique tunability of semiconductor conductivity. Unlike conductors and insulators, semiconductors allow their electrical properties to be precisely engineered. By introducing small amounts of impurity atoms—a process known as doping—the conductivity of a semiconductor can change by several orders of magnitude, sometimes over just a few microns. This abrupt, controllable change in conductivity enables the creation of devices, such as diodes and transistors, with behaviors that are simply not possible in other materials.

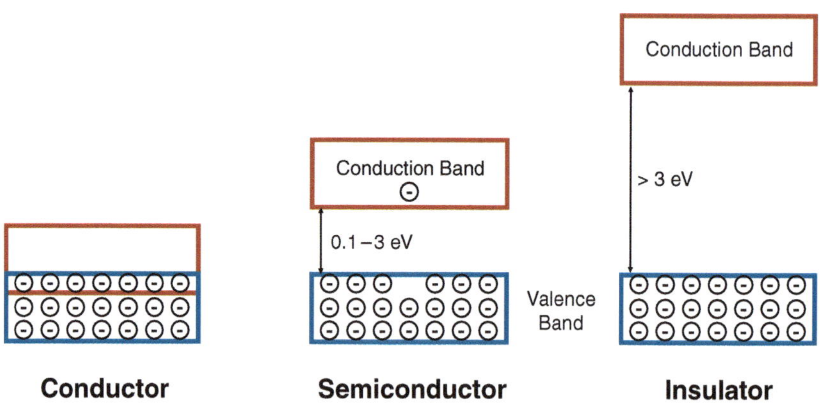

Fig. 1.1 A "layman" version of the energy band diagrams for conductors, semiconductors, and insulators

This chapter, along with the next two, aims to provide examples and physics insight into why semiconductors are uniquely suited for X-ray detection applications and how their intermediate nature is not a limitation, but a powerful advantage.

In X-ray detection applications, semiconductors may serve as X-ray sensors themselves, or as components of electronic devices that measure the electrical or optical signals generated by X-rays in another material (e.g., a scintillator). When a semiconductor is used directly as an X-ray sensor, the working principle is as follows: The energy of an incoming X-ray photon is partially or completely transferred to the semiconductor, generating a certain number of charge carriers. Driven by an applied electric field, these carriers move through the material, producing an electric current. This current induces a signal in the readout circuit, which is then amplified and recorded to extract information about the incident X-ray.

Several factors directly influence this current: the speed of charge carriers, the conductivity of the semiconductor, the loss of charge carriers during their transport, and the number of charge carriers generated per unit of absorbed energy. One challenge in using semiconductors as radiation sensors is their strong temperature dependence. To understand this behavior, we need to examine the energy band structure of semiconductors—a feature fundamentally determined by their crystal structure.

So let us begin there.

1.2 Crystal Structure of Semiconductor Materials

1.2.1 The Formation of Solids: Atomic Bonding

Atoms in a solid remain together only through bonding forces. Generally, there are four main types of bonding in solids:

Ionic Bonding Atoms located at opposite ends of the periodic table (excluding inert gases) tend to lose or gain valence electrons, forming ions. For example, Group I elements typically lose one electron to become positively charged ions, while Group VII elements tend to gain an electron to become negatively charged ions. These oppositely charged ions experience Coulomb attraction, resulting in an ionic bond. A classic example of ionic bonding is found in sodium chloride (NaCl), as illustrated in Fig. 1.2a.

Covalent Bonding Covalent bonds form when atoms share valence electrons to complete their outer electron shells. For instance, silicon (Si) and other Group IV elements have four valence electrons and need four additional electrons to fill their outer shells. If a Si atom has four nearest neighbors, and each neighbor shares one valence electron, the central atom effectively achieves a full octet of eight electrons in its outer shell (Fig. 1.2b). Since each Si atom can form multiple covalent bonds with its neighbors, they collectively create a stable, densely packed, and periodic crystal structure, as further described in Sect. 1.2.2.

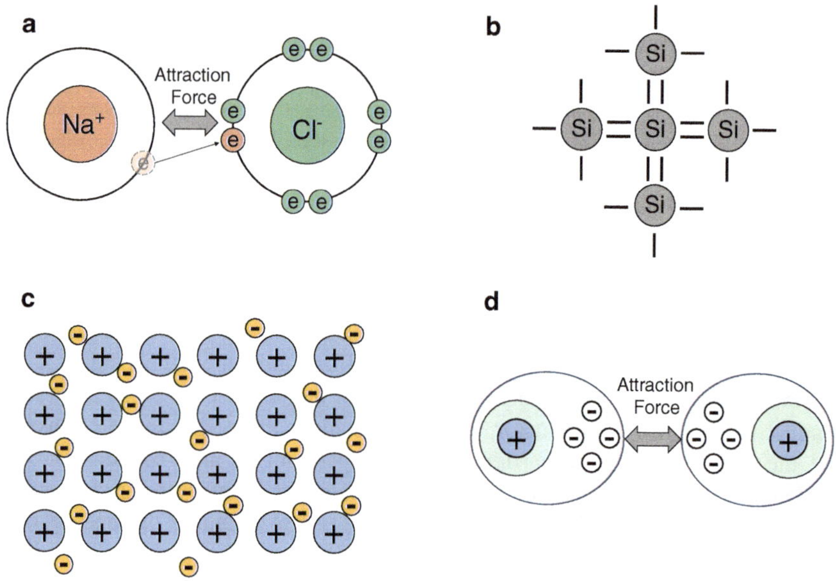

Fig. 1.2 Schematic illustration of ionic, covalent, metallic, and van der Waals bondings. (**a**) Ionic bonding. (**b**) Covalent bonding. (**c**) Metallic bonding. (**d**) Van der Waals bonding

Metallic Bonding When two metallic atoms are brought into close proximity, the valence electrons interact in a way similar to that in covalent bonding. However, the valence electron can also interact with many other delocalized conduction electrons in the metal. In the end, the solid is being held together by the electrostatic attractive force between conduction electrons (in the form of an electron cloud of delocalized electrons) and positively charged metal ions. It can be considered as the sharing of free electrons among a structure of positively charged ions (Fig. 1.2c).

Van der Waals Bonding These bonds arise from temporary or permanent distortions in the electron distribution *within* molecules, creating transient or permanent dipoles that induce electrostatic forces between neighboring particles (Fig. 1.2d). Van der Waals bonds are significantly weaker than covalent or ionic bonds. For example, two-dimensional layers of carbon atoms, known as graphene, are stacked and held together by van der Waals forces to form graphite. The bonding is so weak that a simple piece of scotch tape can be used to peel off a single layer of graphene from graphite.

1.2.2 Lattice Structure

Figure 1.3 shows three basic cubic crystal structures: **simple cubic (SC)** that has an atom located at each corner of each unit cell, **body-centered cubic (BCC)** that has

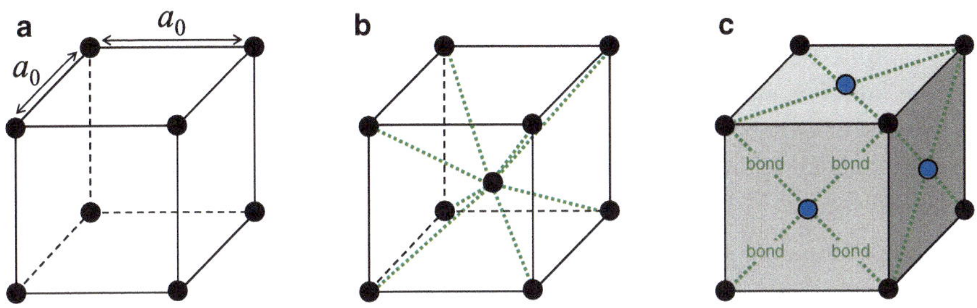

Fig. 1.3 Unit cells of three crystal lattice structures. The concept of the lattice constant (a_0) is illustrated in the figure. a_0 of a typical semiconductor is in the range of [3, 7] Å (e.g., a_0 of Si and Ge are 5.43 and 5.66 Å, respectively). (**a**) Simple cubic (SC). (**b**) Body-centered cubic (BCC). (**c**) Face-centered cubic (FCC)

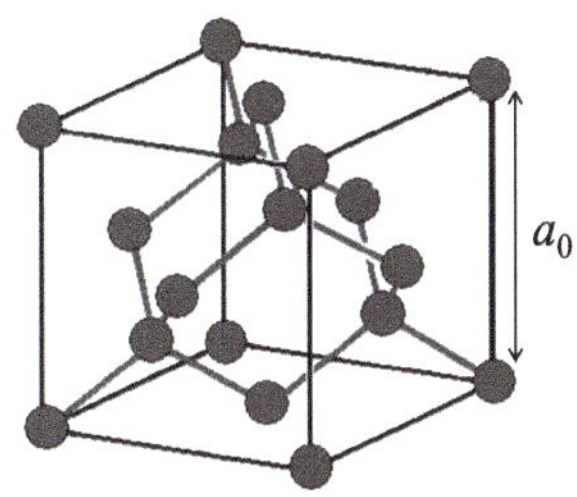

Fig. 1.4 A unit cell of the so-called diamond crystal structure

an additional atom at the center of the cube, and **face-centered cubic (FCC)** that has an additional atom at the center of each face plane (but not at the center of the cube).

For the SC structure, each atom in the 8 corners of a unit cell is shared by 8 neighboring unit cells. Therefore, each unit cell in effect contains $\frac{1}{8} \times 8 = 1$ atoms.

For the BCC structure, each corner atom contributes $\frac{1}{8}$ to a unit cell; the one atom at the center of the cell is not shared by other unit cells. Therefore, each unit cell in effect contains $\frac{1}{8} \times 8 + 1 = 2$ atoms.

For the FCC structure, each of the 8 corner atoms contributes $\frac{1}{8}$ to a unit cell; each of the 6 face atoms is shared by two unit cells. Therefore, each unit cell in effect contains $\frac{1}{8} \times 8 + \frac{1}{2} \times 6 = 4$ atoms.

Figure 1.4 shows the fourth crystal structure: the **diamond structure** in Si and germanium (Ge). Each of the 8 corner atoms is shared by 8 unit cells; each of the 6 face atoms is shared by 2 unit cells; the 4 internal atoms (in green color) are not shared by other unit cells. Therefore, each unit cell in effect contains $\frac{1}{8} \times 8 + \frac{1}{2} \times 6 + 4 = 8$ atoms.

By knowing the basic crystal structure of a material and its lattice constant, we can determine the volume density of atoms and electrons for the material.

Example: Calculation of volume densities of atoms and electrons in Si
The lattice constant of Si is $a_0 = 5.43$ Å $= 5.43 \times 10^{-8}$ cm. The volume density of atoms is given by dividing the effective number of atoms in a unit diamond cell by the volume of the unit cell:

$$\rho_{\text{atom}} = \frac{8}{a_0^3} = \frac{8}{(5.43 \times 10^{-8})^3} = 5.0 \times 10^{22} \text{ (atoms per cm}^3\text{)}. \tag{1.1}$$

Since each Si atom has 14 electrons, the volume electron density in Si is

$$\rho_{\text{e}} = 14 \times \rho_{\text{atom}} = 7.0 \times 10^{23} \text{ (electrons per cm}^3\text{)}. \tag{1.2}$$

Even if we only consider valence electrons (4 valence electrons in each Si atom), the volume density of valence electrons in Si is

$$\rho_{\text{e}_{\text{valence}}} = 4 \times \rho_{\text{atom}} = 2.0 \times 10^{23} \text{ (valence electrons per cm}^3\text{)}. \tag{1.3}$$

In reality, only about 1.5×10^{10} electrons per cm^3—less than 0.000, 000, 000, 01% of the total electrons—contribute to the formation of electrical signals at room temperature. A key objective of Sects. 1.3–1.6 is to understand why this fraction is so remarkably small.

1.2.3 Types of Solids

Solids can be classified in several ways. One common method categorizes them into **amorphous materials**, **polycrystalline materials (or polycrystals)**, and **single crystals**, based on the size and extent of atomic ordering within the material. An *ordered region* refers to a spatial volume in which atoms or molecules exhibit a regular geometric arrangement or periodicity.

As illustrated in Fig. 1.5, amorphous materials exhibit only short-range order—limited to a few atomic or molecular dimensions. In contrast, polycrystalline materials consist of many ordered regions, known as **grains**, each with its own crystal orientation. These grains are separated by **grain boundaries**, which disrupt the continuity of the crystal lattice. Single-crystal materials, on the other hand, maintain long-range order across the entire volume of the material.

A key advantage of single crystals is their superior electrical properties, as grain boundaries in polycrystalline or amorphous materials often degrade carrier mobility and other desirable electrical characteristics.

Single-crystal materials can be grown from a seed crystal using the Czochralski method [7]. In this process, the seed crystal is brought into contact with the surface

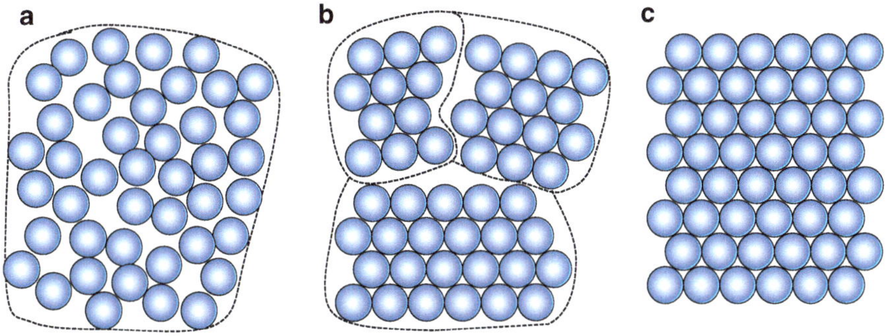

Fig. 1.5 Three types of solid structures. (**a**) Amorphous. (**b**) Polycrystalline. (**c**) Single crystal

of the same material in its liquid phase and then slowly pulled upward from the melt. As the seed is withdrawn, solidification occurs along the plane between the solid–liquid interface, adding single-crystal volume to the seed.

Another widely used technique for single-crystal growth is epitaxial growth. Like the Czochralski method, it begins with a single-crystal seed substrate, but it operates at temperatures below the material's melting point. For example, in the epitaxial growth of Si, a single-crystal Si substrate is heated and exposed to a chemical vapor containing silicon tetrachloride ($SiCl_4$). At the substrate surface, heated hydrogen reacts with $SiCl_4$, depositing Si atoms onto the substrate while producing hydrogen chloride (HCl) gas, which is then removed from the reactor. Epitaxial growth enables the fabrication of both homogeneous layers and complex heterogeneous structures, such as GaAs–AlGaAs–GaAs sandwiches, by varying the chemical composition of the vapor.

Due to the stringent technical requirements involved in single-crystal fabrication, certain single-crystal materials can be costly. Moreover, producing large-area single-crystal substrates is often challenging. As a result, many large-area electronic devices—such as the thin-film transistor (TFT) arrays used in medical flat-panel X-ray detectors—employ amorphous semiconductors like amorphous silicon (a-Si). This material can be deposited over large-area substrates, such as glass or metal, using plasma-enhanced chemical vapor deposition (PECVD) at a significantly lower cost.

However, a-Si exhibits some undesirable electrical properties, including low charge carrier mobility and a high tendency for charge trapping. These issues are primarily due to the presence of numerous structural defects, such as **dangling bonds** (silicon with three bonds) or **floating bonds** (silicon atoms with five bonds). To mitigate these defects, hydrogen atoms are introduced during the PECVD process (Fig. 1.6), resulting in hydrogenated amorphous silicon (often denoted as a-Si:H in literature). The hydrogen atoms passivate the dangling bonds, significantly reducing the defect density—often by several orders of magnitude.

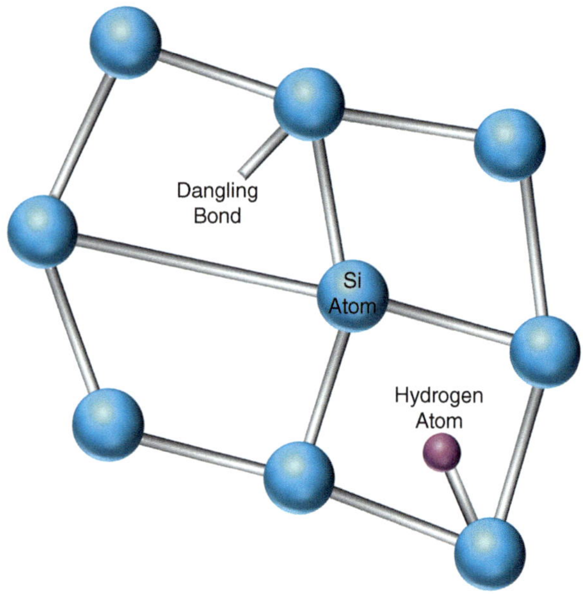

Fig. 1.6 Adding hydrogen atoms to amorphous Si helps in reducing the undesirable dangling bonds

1.2.4 Common Semiconductor Materials

In addition to the classification method presented in Fig. 1.5, solids can also be categorized as conductors, insulators, or semiconductors based on their energy band structures, as illustrated in Fig. 1.1. **A defining property of semiconductor materials is that their resistivity can be precisely controlled—often across several orders of magnitude—by introducing impurity atoms, a concept that will be explored later in this chapter. This tunability enables the development of a wide range of advanced electronic devices, including radiation detectors.**

Two general subcategories of semiconductors are:

- **Elemental semiconductor materials**: Consist of a single element, typically from Group IV of the periodic table, such as Si or germanium (Ge).
- **Compound semiconductor materials**: Formed from specific combinations of elements from Group III and Group V, or Group II and Group VI. Figure 1.7 shows a section of the periodic table highlighting the elements commonly used to form compound semiconductors.

A compound semiconductor composed of two elements is referred to as a **binary compound semiconductor**. For example, cadmium telluride (CdTe) is a binary II–VI compound semiconductor (Fig. 1.8). Other common binary semiconductor materials are listed in Fig. 1.7.

Fig. 1.7 The table in the top row shows a portion of the periodic table with elements that can form semiconductors. The table on the lower left provides a partial list of binary III–V compound semiconductor materials. The table on the lower right provides a partial list of binary II–VI compound semiconductor materials

	II	III	IV	V	VI
2		^{5}B	^{6}C	^{7}N	^{8}O
3		^{13}Al	^{14}Si	^{15}P	^{16}S
4	^{30}Zn	^{31}Ga	^{32}Ge	^{33}As	^{34}Se
5	^{48}Cd	^{49}In	^{50}Sn	^{51}Sb	^{52}Te
6	^{80}Hg	^{81}Tl			

Binary III-V Compounds	
GaAs	Gallium arsenide
InP	Indium phosphide
InAs	Indium arsenide
GaSb	Gallium antimonide
InN	Indium nitride

Binary II-VI Compounds	
CdTe	Cadmium telluride
CdS	Cadmium sulfide
HgS	Mercury sulfide
ZnS	Zinc sulfide
ZnTe	Zinc telluride

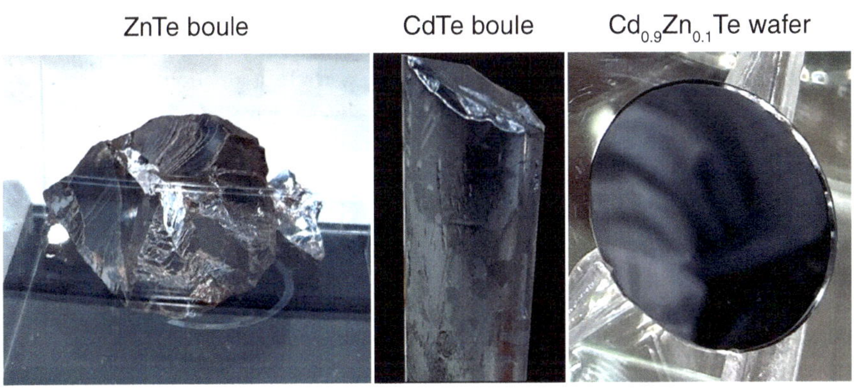

Fig. 1.8 Photos of two binary compound semiconductors (CdTe and ZnTe) and a ternary alloy semiconductor (CdZnTe)

When a compound semiconductor consists of three elements, it is classified as a **ternary alloy semiconductor**. The term alloy reflects the fact that it is typically a mixture of two binary semiconductors or, in some cases, a semiconductor and a metal. A well-known ternary semiconductor used in radiation detectors is cadmium zinc telluride (CdZnTe, or CZT). In CZT shown in Fig. 1.8, Cd and Zn are Group II elements, while Te is a Group VI element. CZT effectively combines the properties of CdTe and ZnTe, enabling a tunable bandgap and enhanced material characteristics. Its composition is often represented as $Cd_{1-x}Zn_xTe$, where x denotes the fraction of Zn substituting for Cd.

1.3 Energy Bands in Semiconductors

1.3.1 Energy Band, Bandgap, and E-k Diagram

The energy bandgap in a semiconductor originates from the quantum–mechanical interaction of electrons with the periodic atomic lattice of the crystal. This topic could by itself form an independent chapter, but it is beyond the scope of this book. Instead, we will focus on the essential physical principles. Let us begin with the time-independent Schrödinger equation [5]:

$$\left[-\frac{\hbar^2}{2m}\nabla^2 + U(\mathbf{x})\right]\varphi(\mathbf{x}) = E\varphi(\mathbf{x}), \tag{1.4}$$

where m is the mass of a particle such as an electron, $U(\mathbf{x})$ is the potential energy function representing the environment in which the particle exists, E is an allowed energy eigenvalue of the particle in that potential, $\hbar$ is the reduced Planck constant, and $\varphi(\mathbf{x})$ is the spatial component of the wave function. The squared modulus of $\varphi(\mathbf{x})$ gives the probability density of finding the particle at position $\mathbf{x}$.

1.3.1.1 Free Electron

To begin our analysis, let us consider a free electron (mass m_e) in a one-dimensional vacuum space of length L, where the potential energy is zero. Like traveling around the equator, moving a distance L brings you back to the origin. Therefore, we impose the boundary condition

$$\varphi(x = L) = \varphi(x = 0). \tag{1.5}$$

According to the Schrödinger equation, when $U(x) = 0$, we have

$$\frac{d^2}{dx^2}\varphi(x) + \frac{2m_e E}{\hbar^2}\varphi(x) = 0. \tag{1.6}$$

The general solution of this differential equation is

$$\varphi(x) = C\exp\left(\pm i\sqrt{\frac{2m_e E}{\hbar^2}}x\right). \tag{1.7}$$

Applying the boundary condition in Eq. (1.5) gives

$$\sqrt{\frac{2m_e E}{\hbar^2}}L = 2\pi n, \quad \text{where } n = 0, \pm 1, \pm 2, \cdots. \tag{1.8}$$

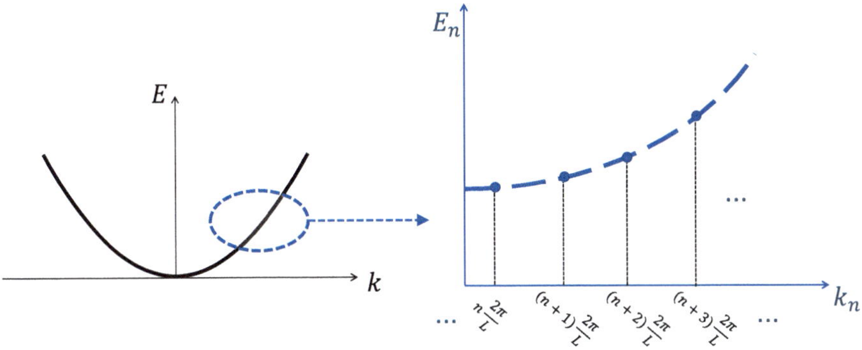

Fig. 1.9 $E - k$ diagram for free electron. While the diagram on the left appears to show a continuous curve, a zoomed-in view reveals that it actually consists of numerous closely spaced discrete $E - k$ points

Thus, the free electron is allowed only at quantized energy levels:

$$E_n = \frac{(2\pi n\hbar)^2}{2m_e L^2}, \quad (n = 0, \pm 1, \pm 2, \cdots). \tag{1.9}$$

According to de Broglie's hypothesis, the momentum of the electron is related to the wave vector k by de Broglie [4]

$$p = \hbar k. \tag{1.10}$$

For a free electron, the relation between energy and momentum is

$$E = \frac{p^2}{2m_e}. \tag{1.11}$$

Combining Eqs. (1.9)–(1.11) yields

$$E_n = \frac{(2\pi n\hbar)^2}{2m_e L^2} = \frac{p^2}{2m_e} = \frac{\hbar^2 k_n^2}{2m_e}. \tag{1.12}$$

Therefore, the wave vector is allowed only at the following quantized levels:

$$k_n = \frac{2\pi}{L} n. \tag{1.13}$$

In practice, the box size L is typically very large compared to the interatomic spacing a in solids. As a result, the discrete energy levels are very densely spaced, and the E-k relation of a free electron appears nearly continuous when plotted (Fig. 1.9).

1.3.1.2 Electron in a Crystal

For an electron in a crystal, the potential energy $U(x)$ arises from the periodic arrangement of atoms and is itself a periodic function with lattice constant a:

$$U(x) = U(x + ma), \quad m = 0, \pm 1, \pm 2, \ldots \tag{1.14}$$

Compared to the free-electron case, the presence of this periodic potential can be regarded as a perturbation of the free-electron wave function. According to Bloch's theorem, [3] the perturbed wave function can be written as

$$\varphi(x) = u(x)\,\mathrm{e}^{ikx}, \tag{1.15}$$

where e^{ikx} is a plane wave and $u(x)$ is the spatially varying amplitude of the wave, with the same periodicity as the lattice, $u(x + a) = u(x)$.

When the wave vector reaches the following certain critical values, [2]

$$k_m = \pm \frac{m\pi}{a}, \quad m = 1, 2, \ldots, \tag{1.16}$$

electron waves undergo Bragg reflection due to constructive interference with the periodic lattice. This effect opens forbidden energy gaps at k_m. **The width of each bandgap depends on the specific shape and strength of $U(x)$ and therefore varies from one material to another.**

Notice that the lattice constant a is on the order of a few angstroms (Å) and is much smaller than the total width of a solid material, denoted by L. As a result, between quantum numbers $m = -1$ and $m = +1$, there exist millions upon millions of discrete eigenstates. The spacing between adjacent quantum states, whether along the energy (E) or wave vector (k) axis, is so small that the E-k relationship can be treated as quasi-continuous.

1.3.1.3 *E*-*k* Diagram

Figure 1.10 illustrates a commonly used representation of the E-k diagram, where only the so-called reduced Brillouin zone is shown. As highlighted in the figure, the diagram actually consists of a dense collection of discrete points rather than a smooth curve. The total number of these available quantum states plays a critical role in determining the electrical properties of semiconductors—a topic we will revisit in more depth when we discuss the density of states in Sect. 1.4.

Since the periodicity a of the crystal potential varies depending on the crystal direction, the E-k diagram also depends on the orientation of the crystal plane. Figure 1.11 shows the E-k diagrams for Si and GaAs along two different crystal directions: [111] and [100]. Typically, these diagrams focus on the regions near the top of the filled energy band at absolute zero (called the **valence band**) and the

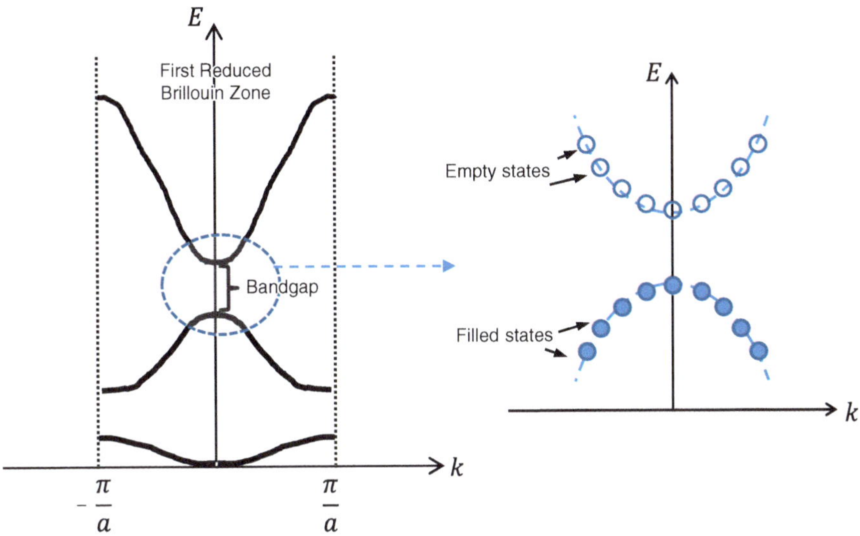

Fig. 1.10 (Left) The reduced Brillouin zone. (Right) A closeup of the top of the fully filled valence band and the bottom of the fully empty conduction band at 0 K

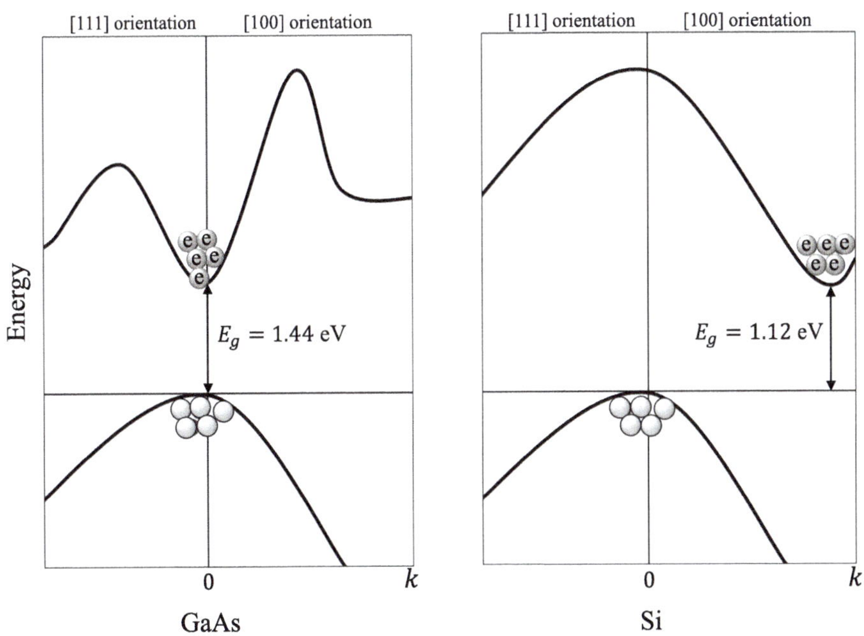

Fig. 1.11 *E*-*k* diagrams for GaAs (a direct bandgap semiconductor) and Si (an indirect bandgap semiconductor). The bandgap energy (E_g) is 1.44 eV for GaAs and 1.12 eV for Si. The diagrams are shown along two crystallographic directions: [111] and [100]

bottom of the empty band (called the **conduction band**). The maximum energy of the valence band is denoted E_v, while the minimum energy of the conduction band is denoted E_c. The energy difference between these two,

$$E_g = E_c - E_v, \tag{1.17}$$

is the **energy bandgap**.

In materials such as GaAs, both the conduction band minimum (E_c) and the valence band maximum (E_v) occur at $k = 0$. A semiconductor with this property is referred to as a **direct bandgap semiconductor**. Because these electronic transitions occur without a change in crystal momentum, direct bandgap materials exhibit strong interactions with light. This characteristic makes GaAs and similar materials well suited for use in light-emitting diodes (LEDs), lasers, and other optoelectronic devices.

In contrast, for Si, the valence band maximum E_v occurs at $k = 0$, but the conduction band minimum E_c is located at a different value of k. Such materials are known as **indirect bandgap semiconductors**. In these cases, electronic transitions between the valence and conduction bands require conservation of crystal momentum, which typically involves interaction with a phonon (a quantum of lattice vibration). This additional requirement makes indirect transitions less efficient for light emission.

In some literature (particularly in medical physics contexts), the energy band structure of a semiconductor is simplified as two horizontal lines representing E_v and E_c, as shown in Fig. 1.1. While this layman representation can be helpful for basic discussions, it omits critical information such as the curvature of the bands and whether the material has a direct or indirect bandgap. As a result, it provides only a limited understanding of the material's electronic and optical properties.

1.3.2 Charge Carriers in Semiconductors: Electron and Holes

At 0 K, a semiconductor's valence band is completely filled with electrons, while the conduction band is entirely empty. Although the material contains millions of millions of electrons, all are confined to the valence band and cannot contribute to electrical conduction. This is because, in a completely filled band, electrons cannot gain energy and change momentum in response to an external electric field—a requirement for generating electric current.

To see why, consider the following: Suppose an electron in the fully filled valence band is somehow accelerated. To do so, it would need to occupy a higher-energy quantum state. However, all such states in the valence band are already occupied due to the Pauli exclusion principle. With no available states to transition into, the electron cannot increase its energy and thus cannot be accelerated. This contradiction disproves our assumption.

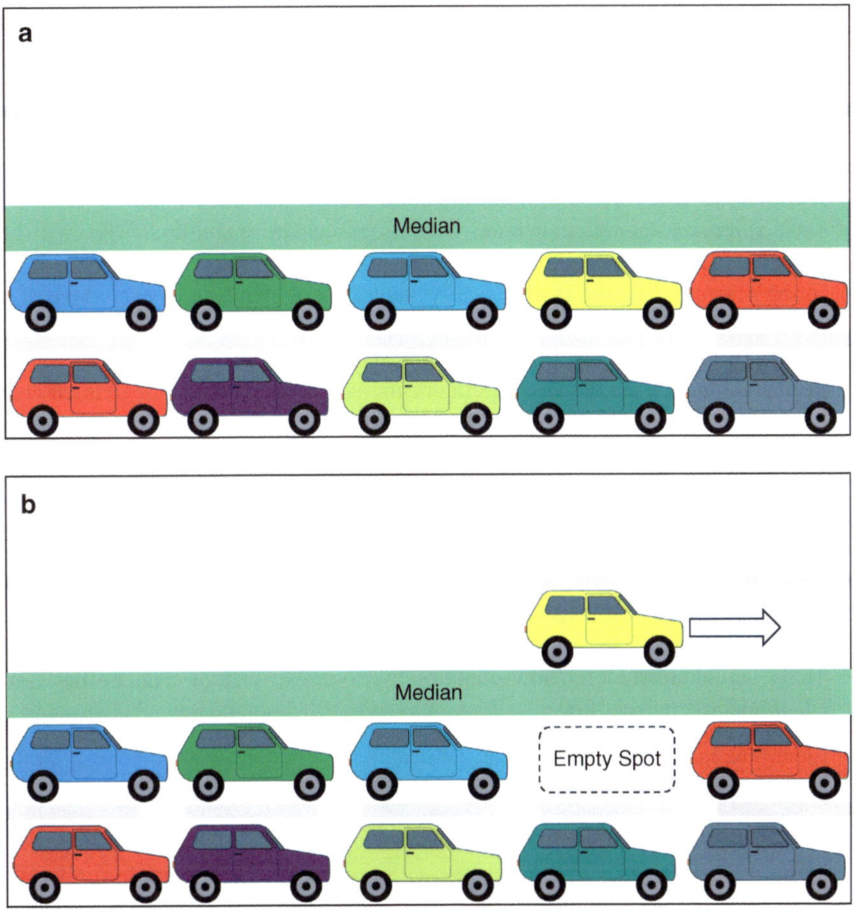

Fig. 1.12 Traffic-jam analogy for charge carriers in semiconductors. (**a**) When a highway is completely packed with vehicles, none can move—they are no longer functioning as transportation tools. (**b**) If one vehicle gains enough energy to jump across the median into an open lane, it can move freely. The empty space it leaves behind allows vehicles in the original lane to shift forward sequentially, making it appear as if the empty spot (a "hole") is moving in the opposite direction

To reinforce this idea, consider the following analogy: imagine one side of a divided highway in a complete traffic jam—every car represents an electron. Although each car has an engine (just as electrons have charge), none can move because there is no space ahead. As a result, the cars, despite being capable of motion, are effectively stuck—just like electrons in a filled band.

Now, if one car somehow gains enough energy to jump the median (analogous to crossing the bandgap into the conduction band), it can begin moving freely on the other side of the highway—becoming a "true" vehicle in motion (Fig. 1.12). In doing so, it leaves behind a gap in the jammed carriageway. This empty space—a "hole"—allows the cars behind it to shuffle forward, making it appear as if the gap is moving backward. This models how holes in the valence band can behave like positive charge carriers once an electron transitions to the conduction band.

At temperatures above 0 K, some electrons in the valence band may gain sufficient thermal energy to transition into the conduction band—analogous to a car jumping across the highway median in Fig. 1.12. While it is impossible to predict which specific electrons will make this transition, the macroscopic probability of such valence-to-conduction band excitations can be quantified. This will be addressed in detail in Sect. 1.5.

Since a semiconductor is electrically neutral as a whole, the excitation of a negatively charged electron from the valence band to the conduction band leaves behind a positively charged "empty state" at its original position within a covalent bond in the valence band. As temperature increases, more electrons gain enough thermal energy to "jump" into the conduction band, resulting in the creation of more such empty states in the valence band.

When an external electric field is applied, electrons in the conduction band can be accelerated, since there are plenty of available higher-energy states for them to occupy. Meanwhile, the presence of empty states in the valence band allows electrons there to also gain energy and move, contributing to electric current. This behavior is equivalent to treating the empty states in the valence band as positive charge carriers, known as holes.

Thus, the semiconductor now contains two types of charge carriers that contribute to electric current flow:

- Electrons, in the conduction band
- Holes, in the valence band

The total electric current density, J, is given by

$$\begin{aligned} J &= J_p + J_n \\ &= (+e)\sum_{i=1}^{N} |v_i^p| + (-e)\sum_{i=1}^{N}(-|v_i^n|) \\ &= +e\sum_{i=1}^{N}(|v_i^p| + |v_i^n|), \end{aligned} \tag{1.18}$$

where the velocity of the charge carriers, v, is defined to be positive in the direction of hole motion. In Eq. (1.18), the subscripts p and n denote positive (hole) and negative (electron) charge carriers, respectively.

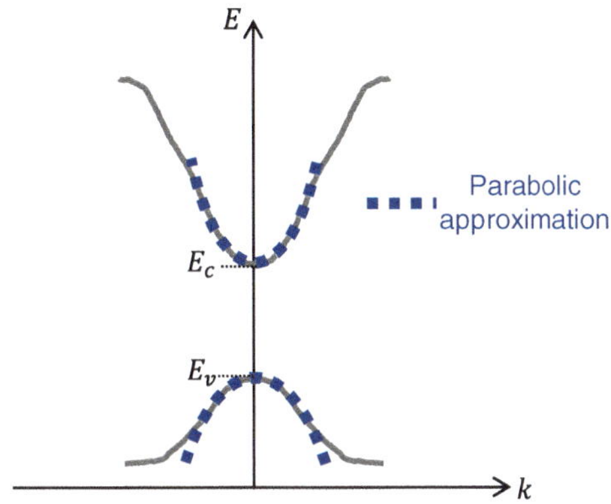

Fig. 1.13 Parabolic approximation near the conduction band minimum and the valence band maximum

1.3.3 *Effective Masses of Electrons and Holes*

Near the bottom of the conduction band in a semiconductor, the E-k relationship can be approximated using a Taylor expansion. For simplicity, assume the conduction band minimum occurs at $k = 0$. The Taylor expansion of $E(k)$ about $k = 0$ is

$$E(k) = E(0) + \left.\frac{\mathrm{d}E}{\mathrm{d}k}\right|_{k=0} k + \frac{1}{2}\left.\frac{\mathrm{d}^2E}{\mathrm{d}k^2}\right|_{k=0} k^2 + \cdots , \tag{1.19}$$

where $E(0)$ is the minimum energy of the conduction band and is typically denoted by E_c. Since we are expanding around the minimum point of the E-k curve, the first derivative vanishes. Neglecting higher-order terms, the Taylor series of $E(k)$ in Eq. (1.19) can be approximated by

$$E(k) \approx E_c + \frac{1}{2}\left.\frac{\mathrm{d}^2E}{\mathrm{d}k^2}\right|_{k=0} k^2, \tag{1.20}$$

which is known as the **parabolic approximation of the E-k relation near the conduction band minimum (Fig. 1.13).**

We now define the **effective mass of electrons** near the conduction band minimum, m_n^*, by

$$\boxed{\frac{1}{m_n^*} := \frac{1}{\hbar^2}\left.\frac{\mathrm{d}^2E}{\mathrm{d}k^2}\right|_{k=0}} . \tag{1.21}$$

Substituting into Eq. (1.20), we obtain

$$\boxed{E(k) - E_c \approx \frac{\hbar^2k^2}{2m_n^*}} . \tag{1.22}$$

This expression is analogous to the energy–momentum relation for a free electron, as seen in Eq. (1.12), with the key difference being that the free electron mass m_e is replaced by the effective mass m_n^*. This effective mass captures the influence of the periodic crystal potential on the electron's motion near the band minimum.

The effective mass is not the actual mass of an electron. Rather, it represents the mass an electron appears to have when responding to an external force $\mathbf{F}_{\text{ext}}$.

Why is this concept important? In a semiconductor crystal, electrons in the conduction band do not move like free particles. Instead, they interact with the periodic potential of the crystal lattice and experience internal forces, denoted by $\mathbf{F}_{\text{int}}$, arising from this potential. As a result, the total force acting on the electron is

$$\mathbf{F} = \mathbf{F}_{\text{ext}} + \mathbf{F}_{\text{int}}. \tag{1.23}$$

In principle, one could use Newton's second law with the total force $\mathbf{F}$ and the real electron mass m_e to calculate the electron's acceleration. However, determining the form of $\mathbf{F}_{\text{int}}$ is highly complex due to its dependence on the electron's position within the periodic lattice.

Fortunately, the concept of effective mass allows us to reduce this complexity. By incorporating the influence of the crystal potential into the effective mass m_n^*, we can model the electron **as if it were a free particle experiencing only the external force**. Its acceleration is simply [6]

$$\boxed{\mathbf{a}_n = \frac{\mathbf{F}_{\text{ext}}}{m_n^*}.} \tag{1.24}$$

Proof of Eq. (1.24)

To understand the notion of effective mass in a crystal, it is useful to begin from the electron's dynamical response to an external force. In a periodic lattice, the quantity that evolves under an applied force is not the classical momentum, but the **crystal momentum**, defined as [2]

$$\mathbf{P} = \hbar\mathbf{k}. \tag{1.25}$$

Under the action of an external force $\mathbf{F}_{\text{ext}}$, the crystal momentum follows an equation analogous to Newton's second law:

$$\mathbf{F}_{\text{ext}} = \frac{\mathrm{d}\mathbf{P}}{\mathrm{d}t} = \hbar\frac{\mathrm{d}\mathbf{k}}{\mathrm{d}t}. \tag{1.26}$$

While $\mathbf{k}$ governs the dynamics, the electron's velocity is determined by the band structure. From quantum mechanics, the velocity is given by the group velocity of the electron wave packet [6]:

(continued)

$$\mathbf{v} = \nabla_k \omega = \frac{1}{\hbar} \nabla_k E, \tag{1.27}$$

where $E = \hbar\omega$ has been used.

Near the bottom of the conduction band, the parabolic approximation gives

$$\nabla_k E = \frac{\hbar^2 \mathbf{k}}{m_n^*}. \tag{1.28}$$

Substituting this into Eq. (1.27), the velocity becomes

$$\mathbf{v} = \frac{\hbar \mathbf{k}}{m_n^*}. \tag{1.29}$$

We now differentiate the velocity with respect to time to obtain the acceleration:

$$\begin{aligned} \mathbf{a} &= \frac{\mathrm{d}\mathbf{v}}{\mathrm{d}t} \\ &= \frac{\hbar}{m_n^*} \frac{\mathrm{d}\mathbf{k}}{\mathrm{d}t}. \end{aligned} \tag{1.30}$$

Using Eq. (1.26) to replace $\mathrm{d}\mathbf{k}/\mathrm{d}t$, we arrive at

$$\mathbf{a} = \frac{\mathbf{F}_{\text{ext}}}{m_n^*}. \tag{1.31}$$

Similar to the derivation of Eq. (1.20), the E-k relation near the top of the valence band can be approximated using a Taylor expansion:

$$E(k) \approx E_v + \frac{1}{2} \frac{\mathrm{d}^2 E}{\mathrm{d}k^2}\bigg|_{k=0} k^2. \tag{1.32}$$

where E_v denotes the maximum energy of the valence band. We assume the maximum occurs at $k = 0$, and the curvature is negative. To account for this, we define the **effective mass of holes** as

$$\boxed{\frac{1}{m_p^*} := -\frac{1}{\hbar^2} \frac{\mathrm{d}^2 E}{\mathrm{d}k^2}\bigg|_{k=0}}. \tag{1.33}$$

Using this definition, Eq. (1.32) can be rewritten as

Table 1.1 Effective masses of electrons (m_n^*) and holes (m_p^*) for several common semiconductor materials: Si, Ge, CdTe, GaAs, and indium gallium arsenide (InGaAs). m_e denotes the electron rest mass

Material	m_n^*/m_e	m_p^*/m_e
Si	1.08	0.56
Ge	0.55	0.37
CdTe	0.11	0.4
GaAs	0.067	0.48
InGaAs	0.041	0.051

$$E(k) \approx E_v - \frac{\hbar^2 k^2}{2m_p^*}. \tag{1.34}$$

or, equivalently,

$$\boxed{E_v - E(k) \approx \frac{\hbar^2 k^2}{2m_p^*}}. \tag{1.35}$$

This is known as the **parabolic approximation of the E-k diagram near the valence band maximum.**

When an external electric field is applied, holes near the valence band maximum can be accelerated. Using the concept of effective mass, a hole behaves as if it were a *free* particle with mass m_p^*, and its acceleration is given by

$$\boxed{\mathbf{a}_p = \frac{\mathbf{F}_{\text{ext}}}{m_p^*}}. \tag{1.36}$$

Table 1.1 presents the effective masses of electrons and holes in various semiconductor materials. In the case of CdTe, a material commonly used in photon-counting detectors, the electron effective mass is only 28% that of the hole effective mass. This indicates that conduction electrons in CdTe respond significantly more rapidly to an applied electric field than holes in the valence band. This distinction is crucial in the design of CdTe-based X-ray detectors, where charge carrier transport velocity strongly influence performance. We will revisit this point in more detail in Chap. 3.

1.4 Density of States

The density of states describes the number of available quantum states at each energy level within a unit volume of a material. When combined with the occu-

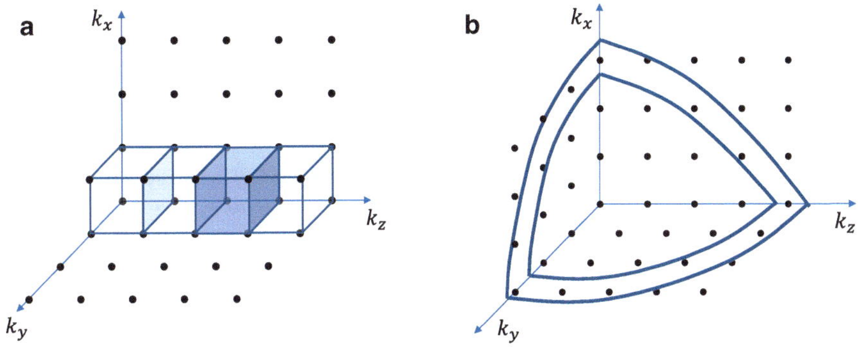

Fig. 1.14 (**a**) Discrete quantum states in three-dimensional (3D) k-space. (**b**) An octant of a spherical shell between k and $k + dk$ in 3D k-space, containing a finite number of discrete k-states

pation probability (discussed in Sect. 1.5), it enables the theoretical calculation of conduction-band electron and valence-band hole concentrations. These carrier concentrations directly affect a semiconductor's electrical conductivity—and, consequently, the output signal of a semiconductor detector.

1.4.1 Density of Quantum States in Momentum Space

According to quantum mechanics, the wave vector k (and thus the momentum) of electrons cannot take arbitrary values. In a one-dimensional material, the allowed values of k are quantized and determined by the total length L of the system, given by

$$k(n) = \frac{2\pi}{L} n \qquad (n = 0, \pm 1, \pm 2, \cdots). \tag{1.37}$$

Similarly, in a 3D material with dimensions $L_x \times L_y \times L_z$, the allowed values of the wave vector $\mathbf{k}$ are quantized as follows:

$$\begin{cases} k_x(m) = \frac{2\pi}{L_x} m & (m = 0, \pm 1, \pm 2, \cdots) \\ k_y(p) = \frac{2\pi}{L_y} p & (p = 0, \pm 1, \pm 2, \cdots) \\ k_z(q) = \frac{2\pi}{L_z} q & (q = 0, \pm 1, \pm 2, \cdots). \end{cases} \tag{1.38}$$

In other words, the wave vector $\mathbf{k}$ can only take discrete values in (k_x, k_y, k_z)-space. As illustrated in Fig. 1.14a, each cube of size $\left(\frac{2\pi}{L_x} \times \frac{2\pi}{L_y} \times \frac{2\pi}{L_z}\right)$ in $\mathbf{k}$-space contains eight allowed $\mathbf{k}$-states, one at each corner. However, since each corner is shared among eight adjacent cubes, each quantum state is effectively counted only once.

As a result, the average **k**-space volume occupied by a single allowed state is

$$\frac{(2\pi)^3}{L_x L_y L_z} \quad \text{per quantum state.} \tag{1.39}$$

In the Cartesian k-space coordinate system, the number of available quantum states within the volume element from (k_x, k_y, k_z) to $(k_x + \mathrm{d}k_x, k_y + \mathrm{d}k_y, k_z + \mathrm{d}k_z)$ is given by

$$\begin{aligned} N(k_x, k_y, k_z)\, \mathrm{d}k_x \mathrm{d}k_y \mathrm{d}k_z &= \frac{\mathrm{d}k_x \mathrm{d}k_y \mathrm{d}k_z}{\frac{(2\pi)^3}{L_x L_y L_z}} \\ &= \frac{V}{(2\pi)^3} \mathrm{d}k_x \mathrm{d}k_y \mathrm{d}k_z, \end{aligned} \tag{1.40}$$

where $V = L_x L_y L_z$ is the volume of the material.

Now let us switch from the Cartesian coordinate system to the spherical coordinate system in k-space. A given wave vector $\mathbf{k}$ can be described by the coordinates (k, θ, ϕ), where $k \triangleq |\mathbf{k}|$. We want to determine how many k-states lie within the spherical shell shown in Fig. 1.14b, with inner and outer radii k and $k + \mathrm{d}k$, respectively. The volume of this spherical shell in k-space is

$$V_k = 4\pi k^2 \mathrm{d}k. \tag{1.41}$$

Since each allowed k-state occupies a volume of $\frac{(2\pi)^3}{V}$ in k-space, the total number of k-states between k and $k + \mathrm{d}k$ is

$$\begin{aligned} N(k)\mathrm{d}k &= \frac{V_k}{\frac{(2\pi)^3}{V}} \\ &= \frac{4\pi k^2 \mathrm{d}k}{\frac{(2\pi)^3}{V}} \\ &= \frac{V}{2\pi^2} k^2 \mathrm{d}k. \end{aligned} \tag{1.42}$$

Up to Eq. (1.42), we have discussed the quantum mechanics of a general particle confined within a 3D material. From now on, we focus specifically on electrons, which are our primary interest in this chapter. For each k-state and energy level, the electron's spin provides two possible quantum states (spin-up and spin-down). Therefore, the total number of k-states in the spherical shell for **electrons** is

$$N_e(k)\mathrm{d}k = 2 \times \frac{V}{2\pi^2} k^2 \mathrm{d}k. \tag{1.43}$$

Dividing both sides of Eq. (1.43) by the total volume V of the 3D material gives the number of k-states *per unit volume* within the spherical shell:

$$g(k)\mathrm{d}k = \frac{N_e(k)\mathrm{d}k}{V} = \frac{k^2}{\pi^2}\,\mathrm{d}k. \tag{1.44}$$

1.4.2 Density of Quantum States in Energy Space

Instead of counting how many quantum states exist between k and $k+\mathrm{d}k$ in k-space, we are more interested in determining how many discrete energy states lie between E and $E+\mathrm{d}E$ in energy space. Here, the inner boundary E corresponds to the wave vector magnitude k, and the outer boundary $E+\mathrm{d}E$ corresponds to $k+\mathrm{d}k$. Since the magnitude of the wave vector $\mathbf{k}$, denoted k, has a one-to-one correspondence with energy E, the number of states satisfies

$$\begin{aligned} g(E)\mathrm{d}E &= g(k)\mathrm{d}k \\ &= \frac{k^2}{\pi^2}\,\mathrm{d}k. \end{aligned} \tag{1.45}$$

Next, we apply the parabolic approximation of the E-k relation near the conduction band minimum (Eq. (1.22)):

$$E \approx E_c + \frac{\hbar^2 k^2}{2m_n^*}, \tag{1.46}$$

which implies

$$k^2 = \frac{2m_n^*(E - E_c)}{\hbar^2}. \tag{1.47}$$

Therefore,

$$k = \left(\frac{2m_n^*(E - E_c)}{\hbar^2}\right)^{\frac{1}{2}}. \tag{1.48}$$

Differentiating, we get

$$\mathrm{d}k = \frac{1}{2}\left(\frac{2m_n^*}{\hbar^2}\right)^{\frac{1}{2}}(E - E_c)^{-\frac{1}{2}}\,\mathrm{d}E. \tag{1.49}$$

Based on Eqs. (1.47) and (1.49), we can substitute into the right-hand side of Eq. (1.45) to express the density of states in terms of energy:

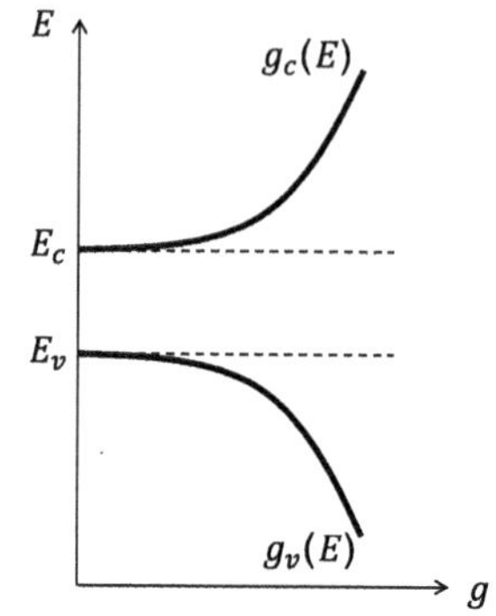

Fig. 1.15 The density of quantum states near the conduction band minimum and the valence band maximum, plotted as functions of energy

$$g(E)\mathrm{d}E = \frac{1}{\pi^2}\frac{2m_n^*(E-E_c)}{\hbar^2}\frac{1}{2}\left(\frac{2m_n^*}{\hbar^2}\right)^{\frac{1}{2}}(E-E_c)^{-\frac{1}{2}}\mathrm{d}E$$
$$= \frac{1}{2\pi^2}\left(\frac{2m_n^*}{\hbar^2}\right)^{\frac{3}{2}}(E-E_c)^{\frac{1}{2}}\mathrm{d}E. \quad (1.50)$$

Thus, the density of states in the conduction band as a function of energy is

$$\boxed{g_c(E) = \frac{1}{2\pi^2}\left(\frac{2m_n^*}{\hbar^2}\right)^{\frac{3}{2}}(E-E_c)^{\frac{1}{2}}}. \quad (1.51)$$

In Eq. (1.51), a subscript $_c$ is used to indicate that the expression applies to the conduction band. The function $g_c(E)$ is referred to as the **density of energy states for electrons in the conduction band**. It represents the number of available quantum states *per unit volume per unit energy*. It is important to note that this expression is strictly valid only near the bottom of the conduction band, where the parabolic approximation used in Eq. (1.46) holds true.

Using a similar method, we can derive the density of energy states for electrons in the valence band:

$$\boxed{g_v(E) = \frac{1}{2\pi^2}\left(\frac{2m_p^*}{\hbar^2}\right)^{\frac{3}{2}}(E_v-E)^{\frac{1}{2}}}. \quad (1.52)$$

As with the conduction band, the validity of Eq. (1.52) is limited to energies near the top of the valence band, since it is derived using the parabolic band approximation given in Eq. (1.35).

Figure 1.15 plots g_c and g_v as functions of energy. As E increases, the density of states in the conduction band increases, while it decreases in the valence band. There are no allowed quantum states within the bandgap; therefore, $g(E) = 0$ for $E_v < E < E_c$.

1.5 Occupancy Probability of Quantum States

In a semiconductor, the density of available charge carriers (electrons and holes) is significantly lower than the density of quantum states. This is because not all quantum states in the conduction band are occupied by electrons, and not all states in the valence band are unoccupied. To determine the charge carrier concentration in a semiconductor, one needs not only the density of states but also the **occupancy probability** of those states.

This probability depends on both the energy E of the state and the temperature T of the material. Intuitively, the higher the energy of a quantum state, the less likely it is to be occupied by an electron. Conversely, as the temperature increases, the probability of higher-energy states being occupied also increases.

At absolute zero temperature ($T = 0$ K), all electrons occupy the lowest available energy states, filling up to a certain energy level. Above this level, no states are occupied at $T = 0$ K. This energy threshold is called the **Fermi energy**, denoted by E_F.

To describe this behavior quantitatively, theoretical physicists have developed models that relate the occupancy probability to E and T. These models are reviewed below.

1.5.1 The Fermi–Dirac Function

While the Fermi–Dirac function is often referred to as the Fermi–Dirac distribution function, it is important to note that it is not a cumulative distribution function (CDF) in the conventional probabilistic sense. Instead, it describes the average number of fermions (e.g., electrons) occupying a quantum state with energy E at thermodynamic equilibrium. Alternatively, it can be interpreted as the probability that a quantum state of energy E is occupied by a fermion at thermodynamic equilibrium.

As shown in Appendix 1, the Fermi–Dirac distribution function is given by

$$\boxed{f_F(E) = \frac{1}{1 + \exp\left(\frac{E - E_F}{k_B T}\right)}} \tag{1.53}$$

where k_B denotes the Boltzmann constant.

1.5.2 The Boltzmann Function

For a system at thermodynamic equilibrium, the Boltzmann function gives the probability for an arbitrary quantum state to be occupied by distinguishable classical

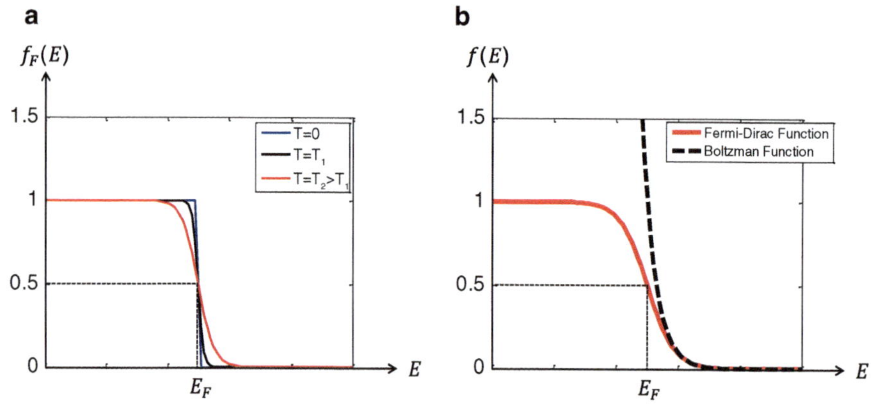

Fig. 1.16 (**a**) The Fermi–Dirac distribution as a function of energy at various temperatures. (**b**) The Boltzmann approximation of the Fermi–Dirac distribution

particles. There is no limit to the number of particles that can occupy a given sublevel (state). As shown in Appendix 1, the Boltzmann distribution function is given by

$$\boxed{f_B(E) = \exp\left(-\frac{E - E_F}{k_B T}\right)} \tag{1.54}$$

Electrons can be approximated as distinguishable classical particles when their occupancy probability becomes very small, i.e., when $f_F(E) \to 0$. As shown in Fig. 1.16a, this occurs under the condition

$$E - E_F \gg k_B T. \tag{1.55}$$

In this regime,

$$\exp\left(\frac{E - E_F}{k_B T}\right) \gg 1 \tag{1.56}$$

which simplifies the Fermi–Dirac distribution to

$$f_F(E) \approx \exp\left(-\frac{E - E_F}{k_B T}\right) = f_B(E). \tag{1.57}$$

In other words, the Fermi–Dirac function can be approximated by the Boltzmann distribution when $E - E_F \gg k_B T$. This simplification is known as the **Boltzmann approximation of the Fermi function**. As illustrated in Fig. 1.16b, this approximation holds well in limit given by Eq. (1.55).

1.6 Charge Carrier Concentrations of Intrinsic Semiconductors

After investing considerable effort in deriving the density of states and the occupation probability for semiconductor energy levels, you might reasonably ask: *Why are we learning all of this?* The following sections aim to answer that question—and, ideally, provide a sense of purpose and satisfaction for the groundwork we have laid in the previous sections.

The focus of this section is to determine the number of charge carriers in an intrinsic semiconductor that are available to contribute to electrical conduction. **An intrinsic semiconductor is a pure semiconductor with no significant dopant atoms.** Our primary goal is to derive expressions for the number of *accelerable* charge carriers per unit volume, specifically (1) the density of electrons in the conduction band and (2) the density of holes in the valence band. Why are these densities important? Because they directly determine the electrical properties of the semiconductor.

1.6.1 Electron Concentration in the Conduction Band

Mathematically, the density of electrons at an energy level E is given by the product of the density of available quantum states and the probability that each state is occupied:

$$n(E) = g(E)\, f(E). \tag{1.58}$$

The total electron density in the conduction band is obtained by integrating $n(E)$ over the range of energies within the band:

$$\begin{aligned} n_0 &= \int_{E_c}^{E_{\text{Top}}} n(E)\mathrm{d}E \\ &= \int_{E_c}^{E_{\text{Top}}} g_c(E)\, f(E)\mathrm{d}E, \end{aligned} \tag{1.59}$$

where E_c and E_{Top} denote the minimum and maximum energies of the conduction band, respectively.

Recall from Eq. (1.51) that the density of states near the conduction band minimum is given by

$$g_c(E) = \frac{1}{2\pi^2}\left(\frac{2m_n^*}{\hbar^2}\right)^{\frac{3}{2}} (E - E_c)^{\frac{1}{2}}. \tag{1.60}$$

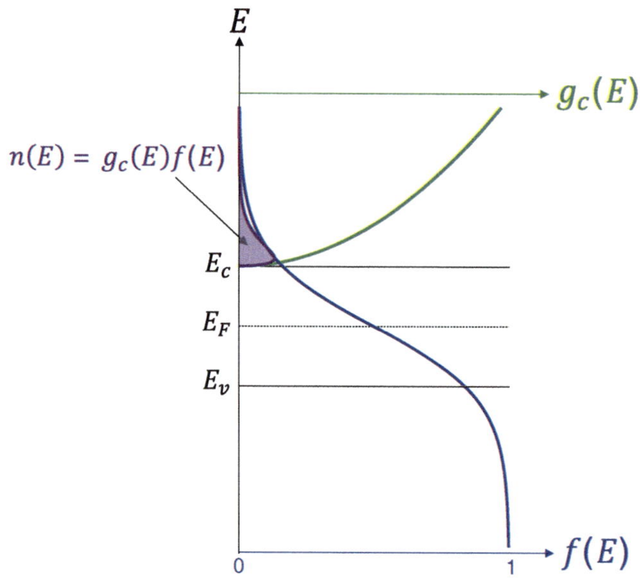

Fig. 1.17 The electron density in the conduction band, $n(E)$, is given by the product of the density of states in the conduction band, $g_c(E)$, and the occupation probability, $f(E)$

Under the Boltzmann approximation (Appendix 2), the occupation probability $f(E)$ simplifies to (Eq. (1.54))

$$f_B(E) = \exp\left(-\frac{E - E_F}{k_B T}\right). \tag{1.61}$$

Multiplying $g_c(E)$ and $f_B(E)$ gives the energy-dependent electron density $n(E)$, as described in Eq. (1.58):

$$n(E) = \frac{1}{2\pi^2}\left(\frac{2m_n^*}{\hbar^2}\right)^{\frac{3}{2}} (E - E_c)^{\frac{1}{2}} \exp\left(-\frac{E - E_F}{k_B T}\right). \tag{1.62}$$

The purple curve in Fig. 1.17 illustrates $n(E)$. Since the density of states $g(E)$ is zero for $E < E_c$, it follows that $n(E) = 0$ in that region. For values of E significantly greater than E_c, the occupation probability $f(E)$ decays rapidly due to its exponential form. As a result, the product $g_c(E) \cdot f(E)$ also decays rapidly, with the exponential term dominating the behavior at higher energies. Therefore, the upper limit of integration in Eq. (1.59) can be extended to $+\infty$, since $n(E)$ becomes negligibly small beyond E_{Top}:

$$
\begin{aligned}
n_0 &= \int_{E_c}^{E_{\text{Top}}} g_c(E)\, f(E) \mathrm{d}E \\
&\approx \int_{E_c}^{+\infty} g_c(E)\, f_B(E) \mathrm{d}E \\
&= \int_{E_c}^{+\infty} \frac{1}{2\pi^2} \left(\frac{2m_n^*}{\hbar^2}\right)^{\frac{3}{2}} (E - E_c)^{\frac{1}{2}} \exp\left(-\frac{E - E_F}{k_B T}\right) \mathrm{d}E \\
&= \frac{1}{2\pi^2} \left(\frac{2m_n^*}{\hbar^2}\right)^{\frac{3}{2}} \int_{E_c}^{+\infty} (E - E_c)^{\frac{1}{2}} \exp\left(-\frac{E - E_F}{k_B T}\right) \mathrm{d}E.
\end{aligned} \tag{1.63}
$$

To simplify the integral, we perform a change of variables:

$$
x = \frac{E - E_c}{k_B T}, \tag{1.64}
$$

which implies

$$
E = E_c + x\, k_B T, \qquad \mathrm{d}E = k_B T \mathrm{d}x, \qquad \frac{E - E_F}{k_B T} = x + \frac{E_c - E_F}{k_B T}. \tag{1.65}
$$

Substituting into Eq. (1.63), the integral becomes

$$
\begin{aligned}
n_0 &= \frac{1}{2\pi^2} \left(\frac{2m_n^*}{\hbar^2}\right)^{\frac{3}{2}} \int_0^{+\infty} (x\, k_B T)^{\frac{1}{2}} \exp\left(-x - \frac{E_c - E_F}{k_B T}\right) k_B T \mathrm{d}x \\
&= \frac{1}{2\pi^2} \left(\frac{2m_n^* k_B T}{\hbar^2}\right)^{\frac{3}{2}} \exp\left(-\frac{E_c - E_F}{k_B T}\right) \boxed{\int_0^{+\infty} x^{\frac{1}{2}} \exp(-x) \mathrm{d}x}.
\end{aligned} \tag{1.66}
$$

The boxed integral is a gamma function $\Gamma(z)$ evaluated at $z = \frac{3}{2}$. Its value is

$$
\Gamma\left(\frac{3}{2}\right) = \int_0^{+\infty} x^{\frac{3}{2}-1} \mathrm{e}^{-x} \mathrm{d}x = \frac{\sqrt{\pi}}{2}. \tag{1.67}
$$

Let us now revisit Eq. (1.66). The total electron density in the conduction band of a semiconductor can be expressed as

$$
\begin{aligned}
n_0 &= \frac{1}{2\pi^2} \left(\frac{2m_n^* k_B T}{\hbar^2}\right)^{\frac{3}{2}} \exp\left(-\frac{E_c - E_F}{k_B T}\right) \frac{\sqrt{\pi}}{2} \\
&= 2 \left(\frac{m_n^* k_B T}{2\pi\hbar^2}\right)^{\frac{3}{2}} \exp\left(-\frac{E_c - E_F}{k_B T}\right),
\end{aligned} \tag{1.68}
$$

which can be compactly written as

$$\boxed{n_0 = N_c \exp\left(-\frac{E_C - E_F}{k_B T}\right)}, \tag{1.69}$$

where

$$\boxed{N_c = 2\left(\frac{m_n^* k_B T}{2\pi\hbar^2}\right)^{\frac{3}{2}}}. \tag{1.70}$$

N_c is known as *the effective density of states function in the conduction band.* N_c characterizes the number of available electron states near the conduction band minimum and depends on both temperature (through the $T^{3/2}$ term) and the semiconductor material (via the effective mass m_n^*). A higher temperature or a smaller effective mass leads to a larger N_c, indicating more accessible quantum states for electrons in the conduction band.

Example: N_c of GaAs
The effective mass of electrons in GaAs is $m_n^* = 0.067 m_e$, where $m_e = 9.1 \times 10^{-31}$ kg is the rest mass of the electron. At room temperature ($T = 300$ K), N_c of GaAs is given by

$$\begin{aligned} N_c &= 2\left(\frac{m_n^* k_B T}{2\pi\hbar^2}\right)^{\frac{3}{2}} \\ &= 2 \times \left[\frac{0.067 \times 9.1 \times 10^{-31} \times 1.38 \times 10^{-23} \times 300}{2 \times 3.14 \times (1.05 \times 10^{-34})^2}\right]^{\frac{3}{2}} \\ &= 4.7 \times 10^{23} \ [\text{m}^{-3}] \\ &= 4.7 \times 10^{17} \ [\text{cm}^{-3}]. \end{aligned} \tag{1.71}$$

For common semiconductor materials, N_c is on the orders of 10^{17}–10^{19} cm^{-3}.

1.6.2 Hole Concentration in the Valence Band

The density of holes at a given energy level is determined by the density of electron states at that energy, multiplied by the probability that a state is **unoccupied** by an electron (i.e., vacant). Mathematically, this is expressed as

$$p(E) = g_v(E)\,[1 - f(E)]. \tag{1.72}$$

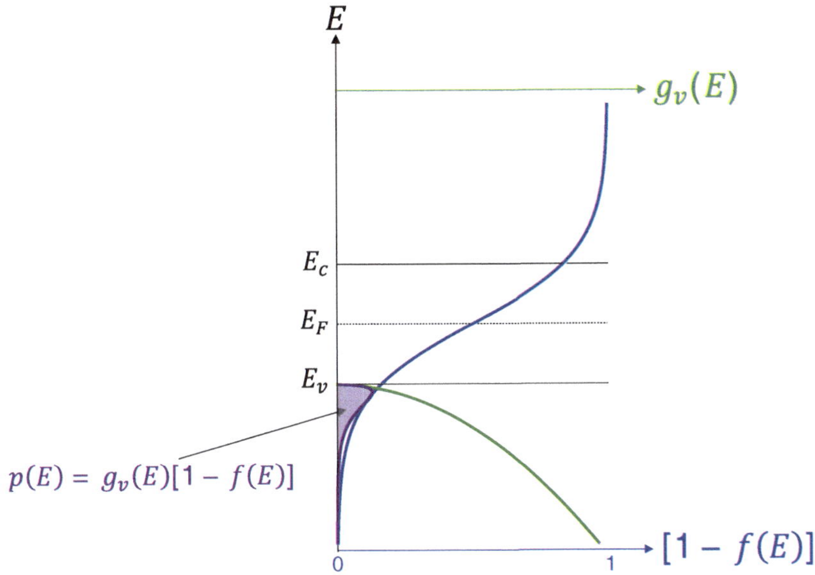

Fig. 1.18 The hole density in the valence band, $p(E)$, is given by the product of the density of states in the valence band, $g_v(E)$, and the pobability of a state not being occupied, namely, $1-f(E)$

The total hole density in the valence band is given by integrating $p(E)$ over the energy range of the band:

$$p_0 = \int_{E_{\text{Bottom}}}^{E_v} p(E)\text{d}E$$

$$= \int_{-\infty}^{E_c} g_v(E)\,[1 - f(E)]\text{d}E, \tag{1.73}$$

where E_v and E_{Bottom} denote the maximal (top) and minimal energies of the valence band, respectively. We use the approximation $p(E) \approx 0$ for $E < E_{\text{Bottom}}$, since the density of states is negligible in that region (Fig. 1.18).

Recall that the formula of $g_v(E)$ has been provided in Eq. (1.51):

$$g_v(E) = \frac{1}{2\pi^2}\left(\frac{2m_p^*}{\hbar^2}\right)^{\frac{3}{2}}(E_v - E)^{\frac{1}{2}}. \tag{1.74}$$

For $1 - f(E)$, we can use the following approximation:

$$1 - f_F(E) = 1 - \frac{1}{1 + \exp\left(\frac{E - E_F}{k_B T}\right)}$$

$$
\begin{aligned}
&= \frac{\exp\left(\frac{E-E_F}{k_B T}\right)}{1+\exp\left(\frac{E-E_F}{k_B T}\right)} \\
&= \frac{1}{\exp\left(\frac{E_F-E}{k_B T}\right)+1} \\
&\approx \exp\left(-\frac{E_F-E}{k_B T}\right) \quad \text{if } (E_F-E) \gg k_B T.
\end{aligned}
\tag{1.75}
$$

This exponential approximation holds under the condition $(E_F - E) \gg k_B T$, which is typically valid since $k_B T = 0.026\,\text{eV}$ at room temperature, and the Fermi level E_F is often higher than the top of the valence band by 0.3 eV or more in intrinsic or lightly doped semiconductors.

Substituting Eqs. (1.74)–(1.75) into the expression for p_0 in Eq. (1.73), we obtain

$$
\begin{aligned}
p_0 &\approx \int_{-\infty}^{E_c} \frac{1}{2\pi^2}\left(\frac{2m_p^*}{\hbar^2}\right)^{\frac{3}{2}} (E_v - E)^{\frac{1}{2}} \exp\left(-\frac{E_F-E}{k_B T}\right) \mathrm{d}E \\
&= \frac{1}{2\pi^2}\left(\frac{2m_p^*}{\hbar^2}\right)^{\frac{3}{2}} \int_{-\infty}^{E_c} (E_v - E)^{\frac{1}{2}} \exp\left(-\frac{E_F-E}{k_B T}\right) \mathrm{d}E.
\end{aligned}
\tag{1.76}
$$

Using a change of variables analogous to the one used in the derivation of the electron concentration n_0, the above integral evaluates to

$$
\boxed{p_0 = N_v \exp\left(-\frac{E_F - E_v}{k_B T}\right)}, \tag{1.77}
$$

where

$$
\boxed{N_v = 2\left(\frac{m_p^* k_B T}{2\pi\hbar^2}\right)^{\frac{3}{2}}} \tag{1.78}
$$

is *the effective density of states function in the valence band.* The value of N_v depends on the temperature (via term T) and the semiconductor material (via term m_p^*).

1.6.3 Intrinsic Carrier Concentration and Intrinsic Fermi Level

In an electrically neutral intrinsic semiconductor, every electron excited into the conduction band leaves behind a corresponding hole in the valence band. As a result, the concentration of electrons in the conduction band is equal to the concentration of holes in the valence band:

$$n_0 = p_0 = n_i, \tag{1.79}$$

where n_i denotes the concentration (density) of electrons or holes in an *intrinsic semiconductor* (i.e., without any doping). The Fermi energy level in an intrinsic semiconductor is referred to as the *intrinsic Fermi level*, denoted by E_{F_i}.

Based on the general expressions for n_0 and p_0 in Eqs. (1.69) and (1.77), n_i must simultaneously satisfy both of the following conditions:

$$n_i = n_0 = N_c \exp\left(-\frac{E_c - E_{F_i}}{k_B T}\right) \tag{1.80}$$

$$n_i = p_0 = N_v \exp\left(-\frac{E_{F_i} - E_v}{k_B T}\right). \tag{1.81}$$

Multiplying Eqs. (1.80) and (1.81) yields

$$\begin{aligned} n_i^2 &= N_c N_v \exp\left(-\frac{E_c - E_{F_i}}{k_B T}\right) \exp\left(-\frac{E_{F_i} - E_v}{k_B T}\right) \\ &= N_c N_v \exp\left(-\frac{E_c - E_v}{k_B T}\right), \end{aligned} \tag{1.82}$$

which simplifies to

$$\boxed{n_i^2 = N_c N_v \exp\left(-\frac{E_g}{k_B T}\right)}, \tag{1.83}$$

where $E_g = E_c - E_v$ is the bandgap energy. Equation (1.83) reveals that, **for a given semiconductor and a temperature, the intrinsic carrier concentration n_i is independent of the Fermi energy.**

In summary, the values of N_c, N_v, and n_i depend solely on temperature and the semiconductor material. These quantities are unaffected by doping. This is not generally true for n_0 and p_0 in doped semiconductors, a point we will explore in Sect. 1.7.

For reference, Table 1.2 lists commonly accepted values of N_c, N_v, and n_i at 300 K.

Using Eqs. (1.80) and (1.81), we can also solve for the intrinsic Fermi energy level E_{F_i}. Instead of multiplying the two equations, we equate their right-hand sides:

$$N_c \exp\left(-\frac{E_c - E_{F_i}}{k_B T}\right) = N_v \exp\left(-\frac{E_{F_i} - E_v}{k_B T}\right) \tag{1.84}$$

Table 1.2 Effective density of states, effective mass, intrinsic carrier concentration, and room-temperature bandgap for Si, GaAs, Ge, and CdTe. m_e denotes the rest mass of electron

Material	N_c (cm^{-3})	N_v (cm^{-3})	m_n^*/m_e	m_p^*/m_e	n_i (cm^{-3})	Bandgap (eV)
Si	2.8×10^{19}	1.0×10^{19}	1.08	0.56	1.5×10^{10}	1.12
GaAs	4.7×10^{17}	7.0×10^{18}	0.067	0.48	1.8×10^{6}	1.44
Ge	1.0×10^{19}	6.0×10^{18}	0.55	0.37	2.4×10^{13}	0.67
CdTe	7.9×10^{17}	1.8×10^{19}	0.10	0.35	7.6×10^{5}	1.52

Taking the natural logarithm of both sides and simplifying:

$$\exp\left(\frac{2E_{F_i} - (E_c + E_v)}{k_B T}\right) = \frac{N_v}{N_c} \tag{1.85}$$

$$\Rightarrow \quad \frac{2E_{F_i} - (E_c + E_v)}{k_B T} = \ln\left(\frac{N_v}{N_c}\right) \tag{1.86}$$

$$\Rightarrow \quad E_{F_i} = \frac{E_c + E_v}{2} + \frac{1}{2} k_B T \ln\left(\frac{N_v}{N_c}\right). \tag{1.87}$$

According to the definitions of N_c and N_v in Eqs. (1.70) and (1.78), the ratio of N_c and N_v is equal to $(m_p^*/m_n^*)^{3/2}$. Therefore, Eq. (1.87) can be written as

$$\boxed{E_{F_i} = \frac{E_c + E_v}{2} + \frac{3}{4} k_B T \ln\left(\frac{m_p^*}{m_n^*}\right)}. \tag{1.88}$$

As shown in Table 1.2, the ratio m_p^*/m_n^* is typically close to 1, except in materials such as GaAs. Therefore, the intrinsic Fermi level E_{F_i} generally lies very close to the *midgap energy*, $(E_c + E_v)/2$.

- If $m_p^* > m_n^*$ (e.g., in GaAs), then E_{F_i} lies slightly above the midgap energy.
- If $m_p^* < m_n^*$ (e.g., in Si), then E_{F_i} lies slightly below the midgap energy.

1.7 Effects of Doping on Charge Carrier Concentrations

1.7.1 Donors and Acceptors

Although intrinsic semiconductors, as discussed in Sect. 1.6, are of fundamental interest, the true utility of semiconductors lies in their ability to be doped—that is, modified by the intentional addition of small amounts of impurity atoms. As we will see later in this section, even a tiny concentration of dopants can dramatically alter the electrical properties of a semiconductor, enabling the creation of a wide range

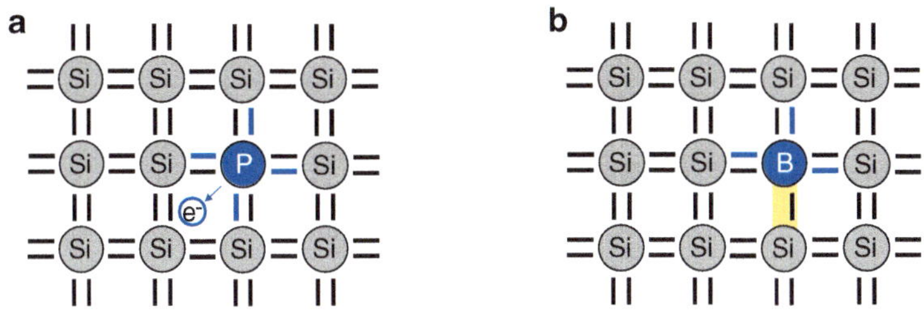

Fig. 1.19 (**a**) A Si crystal lattice is doped with a phosphorus atom. (**b**) A Si lattice crystal is doped with a boron atom

of optoelectronic devices, including X-ray detectors. A semiconductor that has been modified by such impurities is called an **extrinsic semiconductor**.

One common form of doping involves introducing a Group V element (such as phosphorus, P) into a Si crystal lattice. When a P atom substitutes for a Si atom, four of its five valence electrons form covalent bonds with neighboring Si atoms. The fifth electron, however, is only loosely bound to the phosphorus nucleus (see Fig. 1.19). This extra electron is known as a **donor electron**, and the phosphorus atom is referred to as a **donor impurity**.

To illustrate the concept of a donor impurity, consider the following analogy: Imagine four students sharing an apartment. Each has similar financial means and can contribute just one-quarter of the rent, forming a stable group. Occasionally, a student might earn a bit of extra money and contribute to group activities, but overall, the group's funds remain limited. Now, suppose one student leaves and is replaced by a wealthier student who is willing to contribute extra money to the group's activities. As a result, there is more free money circulating within the group—and potentially within the wider community. Similarly, a donor atom introduces extra charge carriers (electrons) into the conduction band of a semiconductor, increasing the overall conductivity.

The energy required to excite an orbiting electron of a donor atom into the conduction band of the semiconductor is much lower than that required for electrons of the intrinsic semiconductor. Once excited, the electron contributes to electrical conduction, while the remaining positively charged P ion remains fixed in the crystal lattice. Since these additional electrons come from the dopant atoms—not from the valence band—they do not create holes in the valence band. A semiconductor doped in this manner is called an **n-type semiconductor** (where n stands for "negative charge").

Conversely, doping with a Group III element (such as boron, B) introduces a different type of behavior. A boron atom has only three valence electrons, which are used to form covalent bonds with three neighboring Si atoms. The fourth neighboring Si atom is left with an incomplete bond, creating a "vacant" bonding site (see Fig. 1.19).

If a small amount of energy (such as thermal energy) is provided, a valence electron from a nearby Si atom can be excited to fill this vacancy. This process leaves behind a hole in the valence band, which can move through the lattice and contribute to electrical conduction. The B atom, now negatively charged after accepting the electron, remains fixed in the lattice. Since the hole originates from the valence band, no electrons are added to the conduction band. The boron atom is called an **acceptor impurity**, and the resulting material is referred to as a **p-type semiconductor** (where p stands for "positive").

1.7.2 Equilibrium Distribution of Electrons and Holes in Extrinsic Semiconductors

Adding donor or acceptor impurities alters the distribution of electrons and holes in a semiconductor. Since the Fermi energy determines this distribution, doping shifts the Fermi level from its intrinsic value E_{F_i} to a new value, denoted E_F. Based on Eq. (1.69), the electron concentration in the conduction band of a doped (extrinsic) semiconductor becomes

$$\begin{aligned} n_0 &= N_c \exp\left(-\frac{E_c - E_F}{k_B T}\right) \\ &= N_c \exp\left[\frac{-(E_c - E_{F_i}) + (E_F - E_{F_i})}{k_B T}\right] \\ &= \boxed{N_c \exp\left(-\frac{E_c - E_{F_i}}{k_B T}\right)} \exp\left(\frac{E_F - E_{F_i}}{k_B T}\right). \end{aligned} \tag{1.89}$$

According to Eq. (1.80), the boxed term above is simply the intrinsic carrier concentration n_i. Therefore, Eq. (1.89) simplifies to

$$n_0 = n_i \exp\left(\frac{E_F - E_{F_i}}{k_B T}\right). \tag{1.90}$$

Similarly, using Eq. (1.81), the hole concentration in the valence band becomes

$$p_0 = n_i \exp\left(-\frac{E_F - E_{F_i}}{k_B T}\right). \tag{1.91}$$

An important consequence of these expressions is that *the product of electron and hole concentrations is independent of the Fermi level*:

$$\begin{aligned} n_0\, p_0 &= n_i \exp\left(\frac{E_F - E_{F_i}}{k_B T}\right) n_i \exp\left(-\frac{E_F - E_{F_i}}{k_B T}\right) \\ &= n_i^2. \end{aligned} \tag{1.92}$$

Table 1.3 Impurity ionization energies in Si

	Donors		Acceptors	
Impurity	P	As	B	Al
Ionization energy (eV)	0.045	0.05	0.045	0.06

This result is identical to the electron–hole product in an undoped (intrinsic) semiconductor. In other words, *for a given semiconductor material and a temperature, the product* $n_0 p_0$ *is constant, regardless of the level or type of doping.* This relationship is known as the **fundamental semiconductor equation**, and its importance will become more evident in the sections that follow.

Note Equation (1.92) is derived under the Boltzmann approximation. When this approximation no longer holds (e.g., in cases of very high doping or low temperature), the equation may no longer be valid.

1.7.3 Ionization of Donors and Acceptors

In Sect. 1.5, we discussed how the probability of a specific energy state E being occupied by an electron in a semiconductor depends on the energy level itself, the temperature T, and the Fermi energy E_F. While similar principles apply to donor and acceptor energy states, the exact occupation probability expressions differ slightly and will not be detailed here.

Instead, Table 1.3 presents the energy differences between the conduction band minimum and donor levels ($E_c - E_d$) and between the acceptor levels and the valence band maximum ($E_a - E_v$) for Si. These values represent the **ionization energies** required to free an electron from a donor or to generate a hole via an acceptor.

As shown in the table, these ionization energies are significantly smaller than the bandgap energy of Si (1.12 eV). At room temperature, thermal energy is typically sufficient to ionize most donor atoms. Therefore, as an approximation, **donor atoms are assumed to be fully ionized at room temperature**—meaning nearly every donor atom contributes one electron to the conduction band.

Similarly, **acceptor atoms are also considered fully ionized at room temperature**. Each acceptor atom captures an electron from the valence band, thereby creating a hole. This ionization process—along with the generation of electrons in the conduction band and holes in the valence band—is illustrated in Fig. 1.20.

In practice, an n-type semiconductor may contain both donor and acceptor impurities. However, for it to behave as n-type, the donor concentration N_d must exceed the acceptor concentration N_a (i.e., $N_d > N_a$). Similarly, a p-type semiconductor typically has both types of dopants, but with $N_a > N_d$.

Due to the requirement of **charge neutrality**, the total positive charge in the material must equal the total negative charge. If all donor atoms are fully ionized,

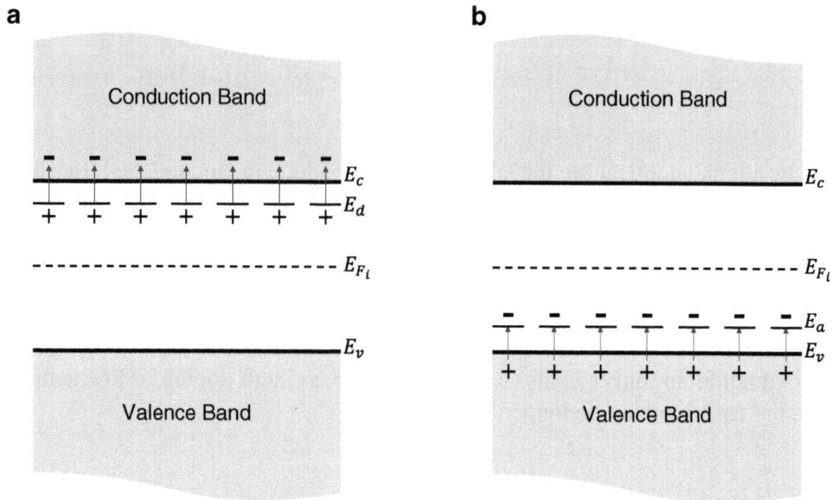

Fig. 1.20 (**a**) Energy band diagrams showing complete ionization of (**a**) donors and (**b**) acceptors

they each contribute one positive charge (from the ionized donor). Likewise, each ionized acceptor atom contributes one negative charge.

Let n_0 and p_0 denote the concentrations of free electrons in the conduction band and free holes in the valence band, respectively. Then, applying the principle of charge neutrality, we can write

$$n_0 + N_a = p_0 + N_d. \tag{1.93}$$

Using the fundamental semiconductor equation, we can express the hole concentration as $p_0 = \frac{n_i^2}{n_0}$. Substituting this into the charge neutrality condition, we obtain

$$n_0 + N_a = \frac{n_i^2}{n_0} + N_d. \tag{1.94}$$

Rearranging terms:

$$n_0^2 - (N_d - N_a)n_0 - n_i^2 = 0. \tag{1.95}$$

This is a quadratic equation in n_0, and its solution is

$$\boxed{n_0 = \frac{N_d - N_a}{2} + \sqrt{\left(\frac{N_d - N_a}{2}\right)^2 + n_i^2}}\,. \tag{1.96}$$

This expression gives the electron concentration n_0 in an n-type semiconductor where $N_d > N_a$.

Similarly, for a p-type semiconductor, we can write Eq. (1.93) as

$$\frac{n_i^2}{p_0} + N_a = p_0 + N_d. \tag{1.97}$$

The solution of p_0 is

$$\boxed{p_0 = \frac{N_a - N_d}{2} + \sqrt{\left(\frac{N_a - N_d}{2}\right)^2 + n_i^2}}\,. \tag{1.98}$$

This equation gives the electron concentration in a p-type semiconductor for which $N_a > N_d$.

Approximate Case: Heavily Doped n-Type Semiconductor If the donor concentration is much greater than the intrinsic carrier concentration ($N_d \gg n_i$) and there are no acceptor atoms ($N_a = 0$), then Eq. (1.96) simplifies significantly:

$$n_0 \approx N_d. \tag{1.99}$$

That is, **the majority carrier concentration is approximately equal to the donor concentration.**

For an n-type semiconductor, the minority carriers are holes, and their concentration can be found using the fundamental relation:

$$p_0 = \frac{n_i^2}{n_0} \approx \frac{n_i^2}{N_d}. \tag{1.100}$$

This result emphasizes that as donor doping increases, the minority carrier (hole) concentration decreases.

Approximate Case: Heavily Doped p-Type Semiconductor If $N_a \gg n_i$ and $N_d = 0$, according to Eq. (1.98), p_0 in a p-type semiconductor can be approximated by

$$p_0 \approx N_a. \tag{1.101}$$

Here, the holes are the majority carriers. The minority carriers are electrons, and their concentration can be determined using the fundamental semiconductor relation:

$$n_0 = \frac{n_i^2}{p_0} \approx \frac{n_i^2}{N_a}. \tag{1.102}$$

Equations (1.96) and (1.98) provide the majority carrier concentrations for n-type and p-type semiconductors, respectively. For minority carriers, the fundamental semiconductor equation $n_i^2 = n_0 p_0$ can be used to determine their concentration.

Finally, it is important to note that based on Eqs. (1.96) and (1.98), when $|N_d - N_a| \gg n_i$, the majority carrier concentration is approximately equal to $|N_d - N_a|$.

Example: Determine the room-temperature electron and hole concentrations in a Ge sample with a donor concentration of $N_d = 5 \times 10^{13}$ ***cm***$^{-3}$ ***and no acceptor*** ($N_a = 0$.)

First, we can determine this is an n-type semiconductor. The majority charge carriers are electrons.

Second, we can use Table 1.2 to determine the intrinsic carrier concentration of Ge to be $n_i = 2.4 \times 10^{13}$ cm^{-3}.

Third, we can use Eq. (1.96) to calculate the concentration of electrons:

$$\begin{aligned} n_0 &= \frac{N_d - N_a}{2} + \sqrt{\left(\frac{N_d - N_a}{2}\right)^2 + n_i^2} \\ &= \frac{5 \times 10^{13} - 0}{2} + \sqrt{\left(\frac{5 \times 10^{13} - 0}{2}\right)^2 + (2.4 \times 10^{13})^2} \\ &= 6 \times 10^{13}\ \text{cm}^{-3}. \end{aligned} \tag{1.103}$$

Depending on the donor concentration levels, n-type semiconductors can be further classified as n (moderately doped), $n+$ (heavily doped), and $n{+}{+}$ (very heavily doped). Similarly, p-type semiconductors can be classified as $p, p+, andp++$. An $n{+}{+}$ or $p{+}{+}$ semiconductor is also referred to a degenerately doped semiconductor that exhibits behavior similar to that of a metal. The specific doping concentration thresholds for these subcategories are listed in Table 1.4.

Table 1.4 Classification of n- or p-type Si based on the doping levels

Classification	Donor concentration (N_d)
n	10^{13}–10^{17} cm^{-3}
$n+$	10^{18}–10^{19} cm^{-3}
$n++$	$>10^{19}$ cm^{-3}
Classification	Acceptor concentration (N_a)
p	10^{13}–10^{17} cm^{-3}
$p+$	10^{18}–10^{19} cm^{-3}
$p++$	$>10^{19}$ cm^{-3}

1.7.4 Effect of Doping on the Fermi Level

As mentioned at the beginning of Sect. 1.7.2, doping alters the position of the Fermi energy level. The objective of this subsection is to establish a quantitative relationship between the Fermi energy level and dopant concentration.

From Eq. (1.69) and using the Boltzmann approximation, we have

$$n_0 = N_c \exp\left(-\frac{E_c - E_F}{k_B T}\right), \tag{1.104}$$

which can be rearranged as

$$E_c - E_F = k_B T \ln\left(\frac{N_c}{n_0}\right). \tag{1.105}$$

For an n-type semiconductor where the donor concentration is much greater than the intrinsic carrier concentration ($N_d \gg n_i$) and acceptor concentration is negligible ($N_a \approx 0$), the equilibrium electron concentration is approximately equal to the donor concentration: $n_0 \approx N_d$. Substituting into the equation gives

$$E_c - E_F \approx k_B T \ln\left(\frac{N_c}{N_d}\right). \tag{1.106}$$

Equation (1.106) shows that the energy difference between the conduction band edge and the Fermi level decreases logarithmically as donor concentration increases. **Thus, as donor concentration increases, the Fermi level moves closer to the conduction band.** For an $n++$ semiconductor, its Fermi level can enter the conduction band if $N_d > N_c$.

Similarly, for a p-type semiconductor where the acceptor concentration dominates ($N_a \gg n_i$ and $N_d \approx 0$), the hole concentration is approximately $p_0 \approx N_a$. Applying the same approach yields

$$E_F - E_v \approx k_B T \ln\left(\frac{N_v}{N_a}\right). \tag{1.107}$$

Equation (1.107) indicates that the distance between the Fermi level and the valence band edge also decreases logarithmically with increasing acceptor concentration. **Therefore, as the acceptor concentration increases, the Fermi level shifts closer to the valence band (Fig. 1.21).** For a $p++$ semiconductor, its Fermi level can enter the valence band if $N_a > N_v$.

Finally, let us introduce a key property of the Fermi level: **In thermal equilibrium, the Fermi level is the same at every point in a connected system.**

To qualitatively understand why this is true, consider two materials, A and B, each with different intrinsic Fermi levels such that $E_{F_i}^A > E_{F_i}^B$. Now imagine bringing these two materials into physical contact to form a single system.

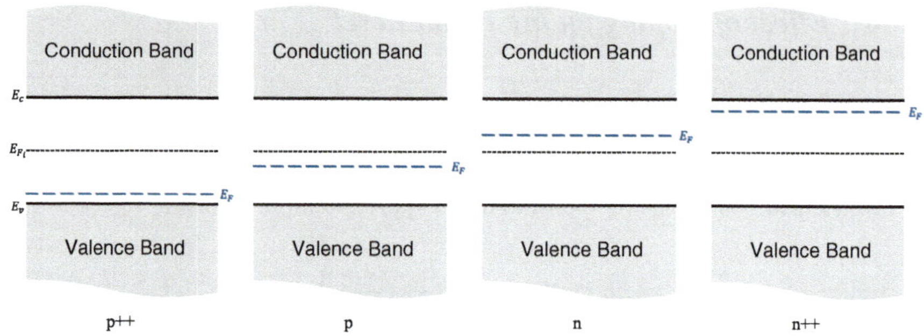

Fig. 1.21 Variation of the Fermi energy level (E_F) with doping

Initially, because material A has a higher Fermi level, its electrons are at a higher total energy (including both kinetic and potential contributions). As a result, electrons from material A will tend to move into material B, occupying available energy states just above $E_{F_i}^{B}$. This electron transfer raises the Fermi level in material B and lowers it in material A, since B gains electrons and A loses them, filling and emptying available energy states, respectively.

This redistribution continues until the Fermi levels in both regions become equal. At this point, there is no net movement of electrons between the two materials—the system has reached thermal equilibrium.

We will use this important principle later when analyzing the behavior of a PN junction in Chap. 2, Sect. 2.2.

1.8 Charge Carrier Transport in Semiconductors

For the purpose of radiation detection, our primary interest lies in the electrical properties of semiconductors, particularly the current density, which depends on both the density and velocity of charge carriers. In Sects. 1.6–1.7, we examined the densities of the two types of charge carriers in semiconductors: electrons in the conduction band and holes in the valence band. We also explored how these densities vary with system parameters such as temperature and dopant concentration.

The main objective of this section is to derive the expressions for the velocities of electrons and holes, or, equivalently, to understand how charge carriers transport (i.e., move) within a semiconductor.

1.8.1 Drift

When an external electric field is applied to a semiconductor, it exerts an electric force on the charge carriers. For electrons in the conduction band, there are many available (empty) quantum states that allow them to gain energy and accelerate. Similarly, holes in the valence band can also be accelerated by the electric field. We refer to **the net movement of charge carriers due to an *external* electric field as drift.**

This drift motion of charge carriers gives rise to what is known as the drift current. For example, if the charge carriers are positively charged holes with a concentration of p_0, based on its definition, the current density induced by the drift of holes is given by

$$J_{p|\mathrm{drift}} = e\, p_0\, \bar{v}_p, \tag{1.108}$$

where $\bar{v}_p$ is the average drift velocity of holes.

Similarly, if the charge carriers are negatively charged electrons with a density of n, the current density induced by the drift of electrons is given by

$$J_{n|\mathrm{drift}} = (-e)\, n_0\, \bar{v}_n, \tag{1.109}$$

where $\bar{v}_n$ is the average drift velocity of electrons.

As we saw earlier in Eqs. (1.24) and (1.36), the acceleration of electrons and holes in a semiconductor under an external electric field is related to their effective masses as follows:

$$F_p = e\,\mathcal{E} = m_p^*\, a_p \tag{1.110}$$

$$F_n = (-e)\,\mathcal{E} = m_n^*\, a_n \tag{1.111}$$

where $\mathcal{E}$ denotes the external electric field. According to these relations, the velocity of an electron or hole would continuously increase as long as the electric field is applied.

However, in reality, charge carriers frequently undergo collisions, scattering, and experience Coulomb interactions with the lattice atoms. These interactions act to slow down or temporarily stop the carriers' motion. After each scattering event, the carriers accelerate again until the next interaction occurs, creating a cycle of acceleration and deceleration. As a result, charge carriers attain an average drift velocity that, under relatively low electric field conditions, is proportional to the applied electric field. For holes, this relationship is expressed as

$$\boxed{\bar{v}_p = \mu_p\,\mathcal{E}}\,, \tag{1.112}$$

Table 1.5 Physical properties of some semiconductor materials, including electron and hole mobilities at $T = 300\,\mathrm{K}$

Material	μ_n [cm²/(V·s)]	μ_p [cm²/(V·s)]	Bandgap (eV)	Z_{eff}	Density (g/cm³)
Si	1350	480	1.12	14	2.33
a-Si	1	0.005	1.8	14	2.33
GaAs	8500	400	1.44	32	5.32
Ge	3900	1900	0.67	32	5.33
CdTe	1100	100	1.52	50	6.06
$Cd_{0.9}Zn_{0.1}Te$	1000	90	1.57	49	5.90
HgI_2	100	4	2.32	62	6.36
TlBr	30	9	2.68	73	7.56
a-Se	0.005	0.14	2.2	34	4.26

a-Si amorphous silicon, *a-Se* amorphous selenium

where μ_p denotes the **hole mobility** of a given semiconductor material at a given temperature. The unit of mobility is often expressed in [cm²/(V·s)]. Its value is generally determined through experimental measurement rather than theoretical calculation. The experimental measurement of mobility relies on the following relation:

$$\mu_p = \frac{\bar{v}_p}{\mathcal{E}}. \tag{1.113}$$

For electrons, their average drift velocity is related to the **electron mobility**, μ_n, by the following expression:

$$\boxed{\bar{v}_n = -\mu_n\,\mathcal{E}}\,. \tag{1.114}$$

The negative sign indicates that the net motion of negatively charged electrons is opposite to the direction of the applied electric field $\mathcal{E}$.

Both electron and hole mobilities are functions of temperature and doping concentration. Table 1.5 provides typical mobility values at room temperature for various intrinsic semiconductor materials.

Note that the average drift velocities of electrons and holes are linearly proportional to the electric field $\mathcal{E}$ only under relatively low field conditions. At higher electric fields (e.g., $\mathcal{E} > 10^4$ V/cm), this linear relationship no longer holds—the drift velocity begins to saturate and, in some cases, may even decrease due to high-field effects such as ionization and avalanche breakdown.

Substituting the expression for $\bar{v}_p$ from Eq. (1.112) into Eq. (1.108), we obtain

$$J_{p|\mathrm{drift}} = e\,p_0\,\mu_p\,\mathcal{E}. \tag{1.115}$$

Similarly, substituting the expression for $\bar{v}_n$ from Eq. (1.114) into Eq. (1.109) gives

$$
\begin{aligned}
J_{n|\text{drift}} &= (-e) \cdot n_0 \cdot (-\mu_n)\, \mathcal{E} \\
&= e\, n_0\, \mu_n\, \mathcal{E}.
\end{aligned} \tag{1.116}
$$

Note that both $J_{p|\text{drift}}$ and $J_{n|\text{drift}}$ point in the same direction (that of the electric field), despite electrons moving opposite to the field due to their negative charge.

In a semiconductor, both electrons and holes contribute to the total drift current. Therefore, the total drift current density is given by

$$
\begin{aligned}
J_{\text{drift}} &= J_{p|\text{drift}} + J_{n|\text{drift}} \\
&= \boxed{e\,(p_0\, \mu_p + n_0\, \mu_n)\, \mathcal{E}}\,.
\end{aligned} \tag{1.117}
$$

In general, the electrical resistivity (ρ) or electrical conductivity (σ) of a material characterizes its response to an external electric field. These properties can be derived from the drift current density. According to Ohm's law, the electrical current I through a system is related to the applied voltage V and the system's resistance R by

$$
I = \frac{V}{R}. \tag{1.118}
$$

Dividing both sides by the cross-sectional area A of the material gives the drift current density:

$$
J_{\text{drift}} = \frac{I}{A} \tag{1.119}
$$

$$
= \frac{V}{RA}. \tag{1.120}
$$

Next, divide both the numerator and denominator of (1.120) by the system's length L:

$$
\begin{aligned}
J_{\text{drift}} &= \frac{V/L}{RA/L} \\
&= \frac{\mathcal{E}}{\rho},
\end{aligned} \tag{1.121}
$$

where $\mathcal{E} = V/L$, and $\rho = RA/L$ is the resistivity of the material that builds the system. Recall the formula of J_{drift} shown earlier in Eq. (1.117); we can now equate this with Eq. (1.121):

$$
e\,(p_0\, \mu_p + n_0\, \mu_n)\, \mathcal{E} = \frac{\mathcal{E}}{\rho}. \tag{1.122}
$$

Solving for ρ, we obtain

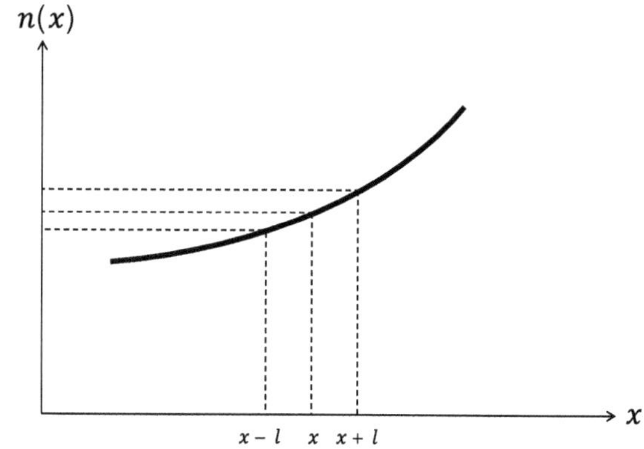

Fig. 1.22 A simplified one-dimensional semiconductor model where the electron concentration varies with spatial position x

$$\boxed{\rho = \frac{1}{e\,(p_0\,\mu_p + n_0\,\mu_n)}}\,. \tag{1.123}$$

The conductivity is the reciprocal of the resistivity:

$$\boxed{\sigma = e\,(p_0\,\mu_p + n_0\,\mu_n)}\,. \tag{1.124}$$

This expression shows that the conductivity of a semiconductor depends linearly on the carrier concentrations (p_0, n_0) and their respective mobilities (μ_p, μ_n). Consequently, in an extrinsic semiconductor, the conductivity increases with increasing impurity concentration, as impurities contribute additional charge carriers.

Is higher conductivity desirable for a radiation detector? The answer is not straightforward and depends on the specific application requirements. This question will be revisited in Sect. 1.9 of this chapter and in Chap. 3.

1.8.2 Diffusion

When two semiconductor materials with different electron concentrations are brought into contact, electrons tend to diffuse from the region of higher concentration to the region of lower concentration. This redistribution continues until the Fermi level becomes uniform across the entire system. The resulting movement of charge due to this concentration gradient is known as diffusion, and the corresponding flow of electrons generates a diffusion current.

To better understand the diffusion current, consider a one-dimensional system where the electron concentration n varies with position x, as illustrated in Fig. 1.22. In the absence of any external electric field, electrons possess thermal energy that allows them to move randomly in all directions. Let v_{th} denote the average thermal velocity of an electron and l the mean free path of an electron due to thermal motion.

At any given instant in time, half of the electrons located at position $x - l$ will be moving toward the right (positive x-direction), while half of the electrons at $x + l$ will be moving toward the left. The net electron flux through the plane at x, denoted by F_n, is given by

$$
\begin{aligned}
F_n &= \frac{1}{2} n(x - l)\, v_{th} - \frac{1}{2} n(x + l)\, v_{th} \\
&= \frac{1}{2} v_{th}\, [n(x - l) - n(x + l)].
\end{aligned}
\tag{1.125}
$$

Using a Taylor series expansion around x, we approximate

$$n(x - l) \approx n(x) - l \frac{\mathrm{d}n(x)}{\mathrm{d}x}, \tag{1.126}$$

$$n(x + l) \approx n(x) + l \frac{\mathrm{d}n(x)}{\mathrm{d}x}. \tag{1.127}$$

Substituting these into the expression for F_n:

$$
\begin{aligned}
F_n &\approx \frac{1}{2} v_{th} \left\{ \left[n(x) - l \frac{\mathrm{d}n(x)}{\mathrm{d}x} \right] - \left[n(x) + l \frac{\mathrm{d}n(x)}{\mathrm{d}x} \right] \right\} \\
&= -v_{th}\, l\, \frac{\mathrm{d}n(x)}{\mathrm{d}x}.
\end{aligned}
\tag{1.128}
$$

Since each electron carries a charge of $-e$, the net electron flux across the plane at x results in a **diffusion current density** of

$$J_{n|\text{diffuse}} = (-e) F_n = +e\, v_{th}\, l\, \frac{\mathrm{d}n(x)}{\mathrm{d}x} \tag{1.129}$$

or

$$J_{n|\text{diffuse}} = \begin{cases} e D_n \frac{\mathrm{d}n(x)}{\mathrm{d}x} & \text{(1D)} \\ e D_n \nabla n(\mathbf{x}) & \text{(3D)} \end{cases} \tag{1.130}$$

where $D_n = v_{th} l$ is the **electron diffusion coefficient** with units of cm^2/s.

Equation (1.130) is known as Fick's law. It shows that **the diffusion current density is directly proportional to the spatial gradient of electron concentration**.

Using a similar analysis, one can show that the hole drift current density is given by

$$J_{p|\text{diffuse}} = \begin{cases} -e D_p \frac{\mathrm{d}p(x)}{\mathrm{d}x} & \text{(1D)} \\ -e D_p \nabla p(\mathbf{x}) & \text{(3D)} \end{cases} \tag{1.131}$$

where D_p is the **hole diffusion coefficient** with units of cm^2/s.

Unlike drift current, which is caused by an electric field, diffusion is a passive process driven by the spatial nonuniformity of charge carrier concentration alone. Electrons and holes have their respective concentration gradients and diffusion coefficients; therefore, their diffusion current densities are often different.

The total current density due to electron motion comprises both drift and diffusion components. For a one-dimensional material, this is expressed as

$$J_n = J_{n|\text{drift}} + J_{n|\text{diffuse}} = e\,n(x)\,\mu_n\,\mathcal{E} + e\,D_n\,\frac{\mathrm{d}n(x)}{\mathrm{d}x}. \tag{1.132}$$

Similarly, the total current density due to hole motion is given by

$$J_p = J_{p|\text{drift}} + J_{p|\text{diffuse}} = e\,p(x)\,\mu_p\,\mathcal{E} - e\,D_p\,\frac{\mathrm{d}p(x)}{\mathrm{d}x}. \tag{1.133}$$

The total current density is the sum of the electron and hole contributions:

$$J = J_n + J_p. \tag{1.134}$$

In the context of radiation detection, it is often mistakenly assumed that drift dominates over diffusion in importance. Diffusion is sometimes viewed negatively, as it can lead to the charge sharing effect, which may degrade the spatial and energy resolution of a semiconductor detector. However, as discussed in Chap. 2, diffusion plays a crucial role—particularly in the formation and operation of rectifying junctions, such as PN junctions. Rather than being merely a source of performance limitations, diffusion is fundamental to the underlying physics of semiconductor radiation detectors.

1.8.3 Relationship Between Diffusion Coefficient and Mobility of Charge Carriers

At thermal equilibrium, there is no net flow of electrons in a semiconductor. According to Eq. (1.132),

$$en(x)\mu_n\mathcal{E} + eD_n\frac{\mathrm{d}n(x)}{\mathrm{d}x} = 0. \tag{1.135}$$

This is equivalent to

$$n(x)\mu_n\mathcal{E} + D_n\frac{\mathrm{d}n(x)}{\mathrm{d}x} = 0. \tag{1.136}$$

When the electron concentration $n(x)$ is uniform across the semiconductor, it is described by Eq. (1.69). However, when $\frac{\mathrm{d}n(x)}{\mathrm{d}x} \neq 0$, a spatial variation in charge density exists, which—according to the Poisson equation—gives rise to an internal electric field, $\mathcal{E}(x)$. Let $-eV(x)$ represent the additional potential energy experienced by electrons at position x due to this internal E-field. Taking this potential energy into account, Eq. (1.69) must be modified as follows:

$$n(x) = N_c \exp\left[-\frac{E_c - E_F - (-eV(x))}{k_B T}\right] = n_0 \exp\left[-\frac{eV(x)}{k_B T}\right]. \tag{1.137}$$

Differentiating $n(x)$ with respect to x yields

$$\frac{\mathrm{d}n(x)}{\mathrm{d}x} = \frac{\mathrm{d}}{\mathrm{d}x}\left\{n_0 \exp\left[-\frac{eV(x)}{k_B T}\right]\right\} = -\frac{e}{k_B T} n(x) \frac{\mathrm{d}V(x)}{\mathrm{d}x}. \tag{1.138}$$

Since $-\frac{\mathrm{d}V(x)}{\mathrm{d}x} = \mathcal{E}(x)$, Eq. (1.138) becomes

$$\frac{\mathrm{d}n(x)}{\mathrm{d}x} = \frac{e}{k_B T} n(x)\mathcal{E}(x). \tag{1.139}$$

Substituting this expression for $\frac{\mathrm{d}n(x)}{\mathrm{d}x}$ into Eq. (1.136), we get

$$n(x)\mu_n \mathcal{E} + D_n \left(\frac{e}{k_B T} n(x)\mathcal{E}(x)\right) = 0. \tag{1.140}$$

Factoring out $n(x)\mathcal{E}(x)$,

$$n(x)\mathcal{E}(x)\left(\mu_n + D_n \frac{e}{k_B T}\right) = 0. \tag{1.141}$$

Since $n(x) \neq 0$ and $\mathcal{E}(x) \neq 0$, it must follow that

$$\mu_n + D_n \frac{e}{k_B T} = 0, \tag{1.142}$$

which gives

$$\boxed{D_n = \mu_n \frac{k_B T}{e}}. \tag{1.143}$$

The above formula is known as the **Einstein relation**. It provides a fundamental link between electron mobility and electron diffusion coefficient at thermal equilibrium.

Table 1.6 Diffusion coefficients calculated using the Einstein relation at $T = 300$ K, where $\frac{k_B T}{e} = 0.026$ V

Material	μ_n [cm^2/(V·s)]	μ_p [cm^2/(V·s)]	D_n (cm^2/s)	D_p (cm^2/s)
Si	1350	480	35	12
GaAs	8500	400	221	10
Ge	3900	1900	101	49
CdTe	1100	100	29	2.6
$Cd_{0.9}Zn_{0.1}Te$	1000	90	26	2.3

For holes, the Einstein relation takes a similar form:

$$\boxed{D_p = \mu_p \frac{k_B T}{e}}. \quad (1.144)$$

Clearly, the diffusion coefficient is temperature-dependent, with higher temperatures leading to increased diffusion currents. Table 1.6 lists charge carrier diffusion coefficients of four semiconductor materials calculated using the Einstein relation.

1.9 An Elementary Semiconductor Radiation Detector

The basic principle behind using a semiconductor to detect external radiation such as X-rays is straightforward: The semiconductor absorbs the X-ray energy, which excites excessive electrons from the valence band to the conduction band. By measuring the resulting excess of electrons or holes, one can infer the energy fluence of the X-rays—this is the working mechanism of an energy-integrating detector (EID). If the detector is fast enough to measure the excess electrons or holes produced by individual X-ray photons, it operates as a photon-counting detector (PCD).

While the underlying concept is simple, practical implementation is significantly more complex. This section aims to highlight the challenges of using a uniform semiconductor material as a radiation detector, thereby motivating the exploration of *heterogeneous* semiconductor structures and devices in Chap. 2.

1.9.1 Interaction of X-Ray with Semiconductor

Let us consider a toy-model semiconductor detector, as illustrated in Fig. 1.23. When an X-ray photon with an initial energy E_0 is incident on the CdTe semiconductor, it can interact with the material through several mechanisms:

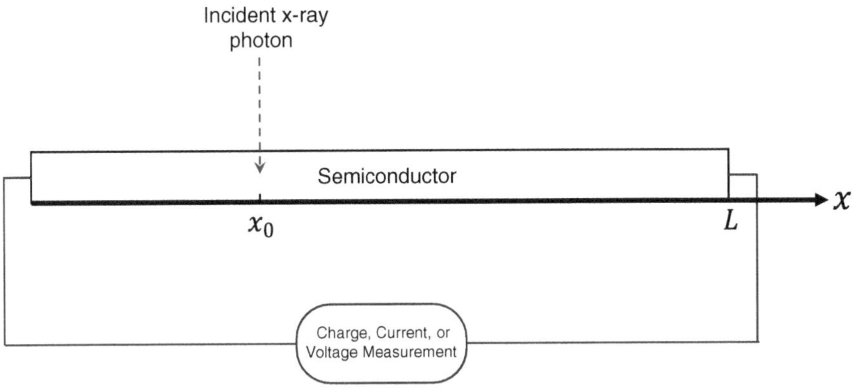

Fig. 1.23 A simplified quasi-one-dimensional (1D) semiconductor detector model without an external bias voltage

- Photoelectric effect (PE). Although the entire photon energy is transferred to the semiconductor, it is not necessarily deposited locally. Due to K-fluorescence, a portion of E_0 may be carried away by a secondary characteristic X-ray photon. This secondary photon may either undergo further interactions (PE, Compton scattering, etc.) at a different location within the semiconductor or escape entirely without further interaction.
- Compton scattering (CS). In this inelastic scattering process, a fraction of the photon's energy is transferred to the semiconductor. The amount transferred depends on the scattering angle and is inherently stochastic. The scattered photon may undergo additional interactions at different locations, depositing more energy, or it may escape the semiconductor.
- Rayleigh scattering (RS). This is an elastic scattering event in which no energy is transferred to the material. However, the photon's direction is changed, potentially leading to subsequent interactions—such as PE or CS—at other locations, or the photon may escape the detector without depositing any energy.

Due to the stochastic nature of X-ray interactions with matter, the amount of energy deposited in the semiconductor by each incident photon varies randomly. This randomness contributes to what is known as **Swank noise** in the detector's output signal. For simplicity, we assume that the *average* energy deposited per incident photon is $\bar{E}$.

As illustrated earlier using the traffic-jam analogy in Fig. 1.12, the injection of energy $\bar{E}$ into the semiconductor excites valence-band electrons into the conduction band, producing an excess of charge carriers (electrons and holes) beyond those generated by thermal excitation. The creation of these electron–hole pairs occurs on an extremely short timescale, typically within **a few picoseconds** (10^{-12} **s**). Even for a fixed energy input $\bar{E}$, the exact number of radiation-induced excess carriers is **stochastic**. However, the mean value, $\overline{\Delta n}$, is proportional to the injected energy $\bar{E}$.

Table 1.7 Typical electron–hole pair creation energies (ϵ) in several semiconductor materials

Material	Si	GaAs	Ge	CdTe	$Cd_{0.9}Zn_{0.1}Te$	a-Se
E_g (eV)	1.12	1.44	0.67	1.52	1.57	$\approx 2.0 - 2.2$
ϵ (eV)	3.62	4.25	2.95	4.43	4.64	$\approx 15 - 18$

Some may intuitively assume that $\overline{\Delta n} = \frac{\bar{E}}{E_g}$, where E_g is the bandgap energy. This, however, oversimplifies the underlying physics. Electrons are not always excited from the very top of the valence band to the very bottom of the conduction band. Moreover, part of the deposited energy is dissipated through lattice excitations (i.e., phonon generation), which do not contribute to electron–hole pair production. As a result, the actual energy required to create a single electron–hole pair is greater than the bandgap energy. This leads to the definition of the **average electron–hole pair creation energy**:

$$\epsilon = \frac{\bar{E}}{\overline{\Delta n}}. \tag{1.145}$$

As shown in Table 1.7, ϵ is typically about three times the bandgap energy. Because semiconductors have a narrower bandgap than insulators, they can generate significantly more electron–hole pairs per unit of absorbed X-ray energy. Compared to gas-filled detectors, for example, the difference is even more pronounced: The average energy required to produce an electron–ion pair in an ionization chamber is approximately 30 eV.

Why is this an advantage? First, a larger mean number of excess charge carriers, $\overline{\Delta n}$, results in a stronger electrical signal. This stronger signal is more easily distinguished from the background electronic noise of the readout circuitry, making it more measurable. Second, the variance of Δn, denoted by $\sigma^2_{\Delta n}$, is typically proportional to $\overline{\Delta n}$:

$$\sigma^2_{\Delta n} = F\bar{n}, \qquad \text{where } F \text{ is known as the Fano factor.} \tag{1.146}$$

From this, the signal-to-noise ratio (SNR) of Δn is given by:

$$\begin{aligned} \mathrm{SNR}_{\Delta n} &= \frac{\overline{\Delta n}}{\sigma_{\Delta n}} = \frac{\overline{\Delta n}}{\sqrt{F\overline{\Delta n}}} \\ &= \frac{\sqrt{\overline{\Delta n}}}{\sqrt{F}}. \end{aligned} \tag{1.147}$$

The energy resolution of a radiation detector is fundamentally determined by the SNR of Δn. Therefore, increasing $\overline{\Delta n}$ improves the SNR and enhances the detector's energy resolution.

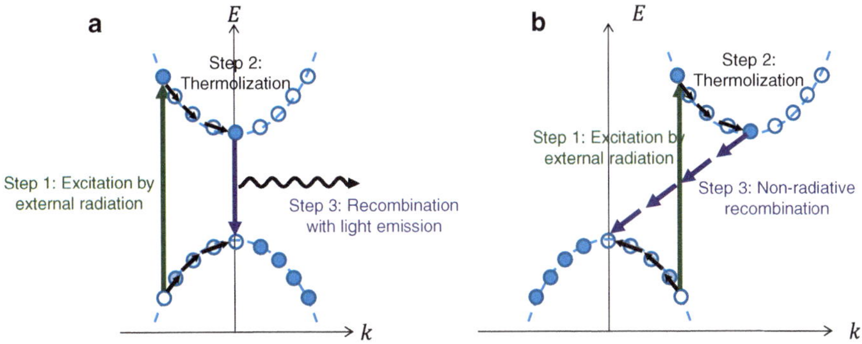

Fig. 1.24 (**a**) In a direct-bandgap semiconductor, electrons and holes excited by external radiation can recombine, releasing energy that may partially be emitted as light (photons). (**b**) In an indirect-bandgap semiconductor, electron–hole recombination typically does not produce secondary radiation. Instead, the energy is primarily released as heat through phonon interactions

Once the excess electrons and holes are created, how do they generate a detectable signal? The answer depends critically on whether an external electric field is present—a topic explored in the next three subsections.

1.9.2 The Concept of Charge Carrier Recombination and Lifetime

The creation of excess electrons and holes drives the semiconductor out of thermal equilibrium. Naturally, these carriers tend to recombine—either by electrons falling back to the valence band and pairing with holes or by becoming trapped in energy states within the bandgap introduced by impurity atoms (trapping centers). In the absence of an external force, such as an electric field, these recombination processes dominate.

As illustrated in Fig. 1.24, when an electron with higher potential energy recombines with a hole, energy is released in accordance with the principle of energy conservation. The form in which this energy is released depends on the material's bandgap structure. In direct bandgap materials, such as those used in scintillators, the recombination can result in the emission of visible light. In contrast, in indirect bandgap materials like Si, the released energy is typically converted into phonons, i.e., thermal excitation of the crystal lattice, leading to a slight increase in temperature. In principle, by measuring this temperature change (a calorimetric approach), one could estimate the number of recombination events. However, this method is impractical in most applications due to the extremely small temperature changes involved.

In this context, scintillator materials have an important advantage as X-ray detectors: They do not require an external electric field. Instead, they harness the natural recombination of electrons and holes to emit visible photons, which can then be detected by light sensors. In contrast, semiconductor detectors rely on an external electric field to separate electrons and holes before they recombine, thereby enabling charge collection. This requirement for an external field is a practical disadvantage of semiconductor detectors compared to scintillators.

Even in the presence of a strong external electric field, charge carrier recombination can still occur in a semiconductor. Let $\Delta n(t)$ denote the concentration of excess (radiation-induced) electrons at time t, and let κ represent the recombination rate per electron.

Assume that no external electric field is applied and that the incident radiation is spatially uniform, such that $\Delta n(t)$ is uniform throughout the semiconductor and diffusion can be neglected. Under these assumptions, the time evolution of the excess electron concentration is governed by

$$\frac{\mathrm{d}\Delta n(t)}{\mathrm{d}t} = -\kappa \Delta n(t), \tag{1.148}$$

The solution is

$$\Delta n(t) = \Delta n(0) \exp(-\kappa t). \tag{1.149}$$

The mean lifetime of the radiation-induced electrons, τ_n, can be defined as the average time an excess electron exists before recombination. It is given by

$$\begin{aligned} \tau_n &= \frac{\int_0^\infty t\,\mathrm{d}[\Delta n(t)]}{\int_0^\infty \mathrm{d}[\Delta n(t)]} \\ &= \frac{\int_0^\infty t\,\mathrm{d}[\Delta n(0)\exp(-\kappa t)]}{\int_0^\infty \mathrm{d}[\Delta n(0)\exp(-\kappa t)]} \\ &= \frac{1}{\kappa}. \end{aligned} \tag{1.150}$$

Therefore, lifetime τ_n is the reciprocal of the recombination rate. Equation (1.148) can therefore be rewritten as

$$\frac{\mathrm{d}\Delta n(t)}{\mathrm{d}t} = -\frac{\Delta n(t)}{\tau_n}. \tag{1.151}$$

Similarly, for excess holes,

$$\frac{\mathrm{d}\Delta p(t)}{\mathrm{d}t} = -\frac{\Delta p(t)}{\tau_p}, \tag{1.152}$$

where τ_p is the mean lifetime of radiation-induced holes.

Table 1.8 Typical electron and hole lifetimes in some high-purity, detector-grade single-crystal semiconductors

Material	τ_n	τ_p
Si	~100 μs–10 ms	~100 μs–10 ms
GaAs	~100 ns–1 μs	~50–500 ns
Ge	~50 –1000 μs	~50–1000 μs
CdTe	~5–10 μs	~1–5 μs
$Cd_{0.9}Zn_{0.1}Te$	~1–10 μs	~0.1–2 μs

As shown in Table 1.8, electrons and holes often exhibit different mean lifetimes. Due to their lower mobility and heavier effective mass, holes are generally more prone to being captured or trapped by defects, which shortens their lifetime. Carrier lifetime also depends on the energy levels of impurity states within the bandgap. For instance, a trap located closer to the valence band more readily captures holes, further reducing their lifetime. In compound semiconductors such as CdTe, the complex band structure can create conditions where holes are less stable or more susceptible to non-radiative recombination. Additionally, material imperfections such as dislocations, grain boundaries, and other crystal defects often trap one carrier type more efficiently than the other, contributing to the asymmetry in carrier lifetimes.

1.9.3 Elementary Semiconductor Detector Without External E-Field

The 1D detector model presented in Sect. 1.9.2 assumes uniform radiation exposure, resulting in spatially uniform distributions of excess electrons and holes. This simplified model is designed to decouple charge recombination from charge diffusion, thereby facilitating a clear definition of the mean carrier lifetime.

In this subsection, we consider a slightly more realistic 1D detector model (still without external E-field), in which excess electrons and holes are generated locally at position $x = 0$ and $t = 0$ by an incident X-ray photon. The initial spatial distributions of the excess carriers are described by the following Dirac delta functions:

$$\Delta n(x, t = 0) = N_0\, \delta(x); \tag{1.153}$$

$$\Delta p(x, t = 0) = P_0\, \delta(x). \tag{1.154}$$

Due to the nonuniformity of these initial conditions, diffusion must now be taken into account. Considering both diffusion and recombination, the time evolution of the excess electron concentration $\Delta n(x, t)$ is governed by the following partial differential equation:

$$\frac{\partial \Delta n(x,t)}{\partial t} = D_n \frac{\partial^2 \Delta n(x,t)}{\partial x^2} - \frac{\Delta n(x,t)}{\tau_n}. \tag{1.155}$$

This a partial differential equation (PDE). To solve it, we can multiply both sides by $e^{\frac{t}{\tau_n}}$:

$$\frac{\partial \Delta n(x,t)}{\partial t} e^{\frac{t}{\tau_n}} = D_n e^{\frac{t}{\tau_n}} \frac{\partial^2 \Delta n(x,t)}{\partial x^2} - \frac{\Delta n(x,t)}{\tau_n} e^{\frac{t}{\tau_n}}. \tag{1.156}$$

This is equivalent to

$$\frac{\partial \Delta n(x,t)}{\partial t} e^{\frac{t}{\tau_n}} + \frac{\Delta n(x,t)}{\tau_n} e^{\frac{t}{\tau_n}} = D_n e^{\frac{t}{\tau_n}} \frac{\partial^2 \Delta n(x,t)}{\partial x^2}, \tag{1.157}$$

or

$$\frac{\partial}{\partial t}\left[\Delta n(x,t) e^{\frac{t}{\tau_n}}\right] = D_n \frac{\partial^2}{\partial x^2}\left[\Delta n(x,t) e^{\frac{t}{\tau_n}}\right]. \tag{1.158}$$

Define $f(x,t) = \Delta n(x,t) e^{\frac{t}{\tau_n}}$; Eq. (1.156) can be written as

$$\frac{\partial f(x,t)}{\partial t} = D_n \frac{\partial^2 f(x,t)}{\partial x^2}, \tag{1.159}$$

with an initial condition $f(x,0) = A\delta(x)$.

To handle this PDE with delta-function initial condition, we perform the Fourier transform of $f(x,t)$ with respect to x:

$$\tilde{f}(k,t) = \int_{-\infty}^{\infty} f(x,t) e^{-\imath 2\pi k x} \, dx. \tag{1.160}$$

Then, taking the Fourier transform of both sides of Eq. 1.159,

$$\frac{\partial}{\partial t} \tilde{f}(k,t) = -D_n (2\pi k)^2 \tilde{f}(k,t). \tag{1.161}$$

This is now a first-order ordinary differential equation with the following general solution:

$$\tilde{f}(k,t) = \tilde{f}(k,0) \cdot e^{-D_n (2\pi k)^2 t}. \tag{1.162}$$

Based on the initial condition $f(x,0) = N_0 \delta(x)$, we know that $\tilde{f}(k,0) = N_0$. Therefore,

$$\tilde{f}(k,t) = N_0 e^{-D_n (2\pi k)^2 t}. \tag{1.163}$$

Take the inverse Fourier transform of $\tilde{f}(k, t)$:

$$f(x, t) = \int_{-\infty}^{\infty} N_0 \, \mathrm{e}^{-D_n (2\pi k)^2 t} \mathrm{e}^{\mathrm{i} 2\pi k x} \, \mathrm{d}k \tag{1.164}$$

$$= \frac{N_0}{\sqrt{4\pi D_n t}} \exp\left(-\frac{x^2}{4 D_n t}\right). \tag{1.165}$$

Therefore, the solution of $\Delta n(x, t)$ is

$$\Delta n(x, t) = f(x, t) \exp\left(-\frac{t}{\tau}\right)$$

$$= \frac{N_0}{\sqrt{4\pi D_n t}} \exp\left[-\left(\frac{x^2}{4 D_n t} + \frac{t}{\tau_n}\right)\right]. \tag{1.166}$$

In solution of $\Delta n(x, t)$ in Eq. (1.166), the $f(x, t)$ term reflects the contribution of diffusion; the exponent decay term, $\exp\left(-\frac{t}{\tau_n}\right)$, reflects the contribution of recombination.

The solution of $\Delta p(x, t)$ takes a similar form:

$$\Delta p(x, t) = \frac{P_0}{\sqrt{4\pi D_p t}} \exp\left[-\left(\frac{x^2}{4 D_p t} + \frac{t}{\tau_p}\right)\right]. \tag{1.167}$$

Using the D_n and τ_n values for CdTe listed in Table 1.8, we calculate the excess electron concentration $\Delta n(x, t)$ at various times following an X-ray interaction at $t = 0$. As illustrated in Fig. 1.25, the number of excess electrons decays over time due to recombination. Even 10 μs after the radiation event, the number of electrons that reach either end of the 1D detector remains negligible. As will be discussed in Chap. 3, the induced electric current peaks when the electrons reach the detector boundaries. In this case, it is clear that diffusion alone is insufficient to produce a detectable signal.

This example demonstrates that extracting an electrical signal from a semiconductor without an external electric field is highly ineffective. In contrast, as shown in the following subsection, applying a strong electric field induces electron drift, enabling a substantial portion of the electrons to reach the detector boundaries before recombination occurs.

As a side note, in scintillator materials—which are wide bandgap insulating solid materials—the electron–hole recombination generates the desired secondary signal in the form of light photons. These photons can then be detected using light sensors. In fact, a faster electron–hole recombination rate is preferable in scintillators, and no external bias voltage is required across the material. This is a key advantage of using scintillators as radiation sensors.

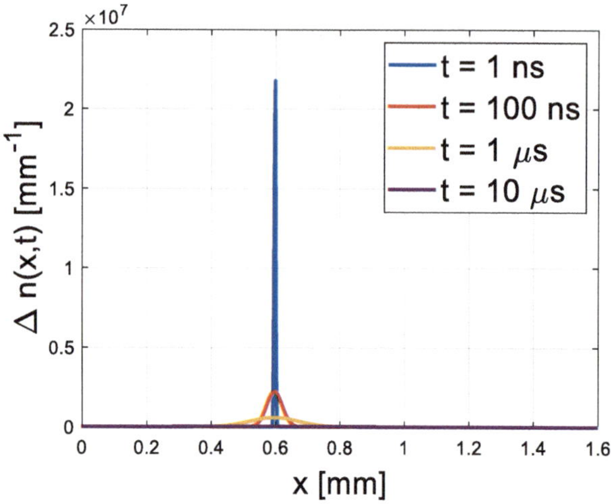

Fig. 1.25 Spatiotemporal distribution of excess electron density, calculated using Eq. (1.166), for a 1D CdTe detector with a thickness of 1.6 mm. The initial X-ray interaction occurs at $x_0 = 0.6$ mm and $t = 0$, with no external bias voltage applied

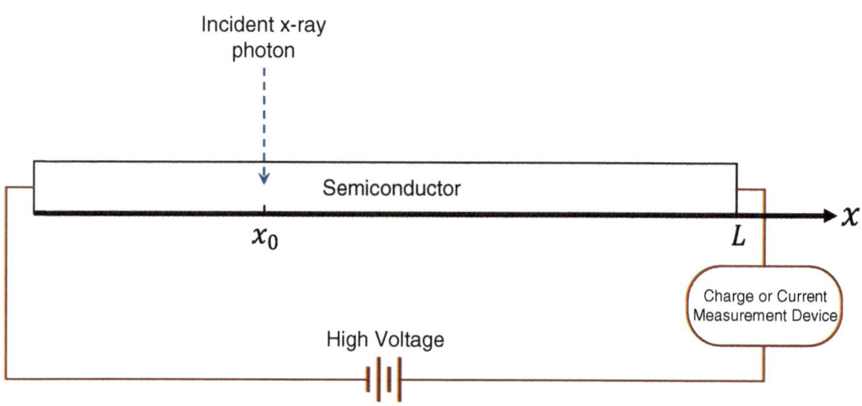

Fig. 1.26 A quasi-1D semiconductor detector model with an external bias voltage

1.9.4 Elementary Semiconductor Detector with External E-Field

As discussed in the previous subsection, diffusion alone is insufficient to generate a strong enough electrical signal from external radiation. As illustrated in Fig. 1.26, applying a high voltage across the semiconductor sensor creates a strong electric field $\mathcal{E}$, introducing an additional transport mechanism that affects the temporal and spatial evolution of $\Delta n(x, t)$: drift.

Assuming a uniform electric field directed along the $-x$ axis, electrons will drift toward the $+x$ direction. According to the continuity equation, the temporal variation of $\Delta n(x,t)$ is governed by the electron drift current density $J_{n|\text{drift}}$, as given by

$$\begin{cases} \nabla \cdot \mathbf{J}_{n|\text{drift}}(\mathbf{x},t) + \frac{\partial[(-e)\Delta n(\mathbf{x},t)]}{\partial t}\Big|_{\text{drift}} = 0 & \text{3D} \\ \frac{\partial J_{n|\text{drift}}(x,t)}{\partial x} + \frac{\partial[(-e)\Delta n(x,t)]}{\partial t}\Big|_{\text{drift}} = 0 & \text{1D} \end{cases} \tag{1.168}$$

Therefore, in the 1D case, the drift-induced temporal variation of $\Delta n(x,t)$ is

$$\begin{aligned} \frac{\partial \Delta n(x,t)}{\partial t}\bigg|_{\text{drift}} &= \frac{1}{-e}\frac{\partial J_{n|\text{drift}}(x,t)}{\partial x} \\ &= -\mu_n \mathcal{E}\frac{\partial \Delta n(x,t)}{\partial x}. \end{aligned} \tag{1.169}$$

With drift taken into account, the expression for $\frac{\partial \Delta n(x,t)}{\partial t}$ from Eq. (1.155) becomes

$$\frac{\partial \Delta n(x,t)}{\partial t} = D_n \frac{\partial^2 \Delta n(x,t)}{\partial x^2} - \mu_n \mathcal{E}\frac{\partial \Delta n(x,t)}{\partial x} - \frac{\Delta n(x,t)}{\tau_n}. \tag{1.170}$$

To simplify this PDE, perform a variable substitution, $x' = x - \mu_n \mathcal{E} t$, and assume the solution of $\Delta n(x,t)$ takes the form

$$\Delta n(x,t) = f(x',t)\exp\left(-\frac{t}{\tau_n}\right). \tag{1.171}$$

Substituting into Eq. (1.170), the left-hand side becomes

$$\begin{aligned} \frac{\partial \Delta n(x,t)}{\partial t} &= \frac{\partial}{\partial t}[f(x',t)\,\mathrm{e}^{-\frac{t}{\tau_n}}] \\ &= \mathrm{e}^{-\frac{t}{\tau_n}}\frac{\partial}{\partial t}f(x',t) + \mathrm{e}^{-\frac{t}{\tau_n}}\frac{\partial f(x',t)}{\partial x'}\frac{\partial x'}{\partial x} + f(x',t)\frac{\partial}{\partial t}\mathrm{e}^{-\frac{t}{\tau_n}} \\ &= \left[\frac{\partial}{\partial t}f(x',t) - \mu_n \mathcal{E}\frac{\partial f(x',t)}{\partial x'} - \frac{1}{\tau_n}f(x',t)\right]\mathrm{e}^{-\frac{t}{\tau_n}}. \end{aligned} \tag{1.172}$$

Similarly, the right-hand side of Eq. (1.170) becomes

$$\begin{aligned} &D_n \frac{\partial^2 \Delta n(x,t)}{\partial x^2} - \mu_n \mathcal{E}\frac{\partial \Delta n(x,t)}{\partial x} - \frac{\Delta n(x,t)}{\tau_n} \\ &= \left[D_n \frac{\partial^2 \Delta f(x',t)}{\partial x'^2} - \mu_n \mathcal{E}\frac{\partial f(x',t)}{\partial x'} - \frac{\Delta f(x',t)}{\tau_n}\right]\mathrm{e}^{-\frac{t}{\tau_n}}. \end{aligned} \tag{1.173}$$

Equating (1.172) and (1.173), we get

$$\frac{\partial}{\partial t} f(x', t) = D_n \frac{\partial^2 \Delta f(x', t)}{\partial x'^2}. \tag{1.174}$$

This equation has the same form as Eq. (1.159), and therefore, the solution $f(x', t)$ is of the same form as in Eq. (1.166):

$$\begin{aligned}\Delta n(x, t) &= \frac{N_0}{\sqrt{4\pi D_n t}} \exp\left[-\left(\frac{x'^2}{4D_n t} + \frac{t}{\tau_n}\right)\right] \\ &= \frac{N_0}{\sqrt{4\pi D_n t}} \exp\left[-\left(\frac{(x - \mu_n \mathcal{E})^2}{4D_n t} + \frac{t}{\tau_n}\right)\right].\end{aligned} \tag{1.175}$$

For holes, the temporal and spatial variation of their excess concentration takes a similar form, with the drift direction reversed (since holes drift opposite to electrons in the electric field):

$$\Delta p(x, t) = \frac{P_0}{\sqrt{4\pi D_p t}} \exp\left[-\left(\frac{(x + \mu_p \mathcal{E})^2}{4D_p t} + \frac{t}{\tau_p}\right)\right]. \tag{1.176}$$

Figure 1.27 shows the temporal evolution of $\Delta n(x)$ in a 1D CdTe detector under a 1000-V bias, which generates an electric field of $\mathcal{E} = 6.26 \times 10^3$ V/cm. This field drives electron drift at a velocity of $\mu_n \mathcal{E} = 6.9 \times 10^6$ cm/s. Consequently, within just 14 ns, the peak of $\Delta n(x)$ reaches the far end of the detector. At this time, recombination leads to only a 0.3% loss of electrons. This example demonstrates the advantage of applying a high bias voltage across the semiconductor sensor to enable rapid and efficient charge collection.

1.9.5 Issues with the Elementary Semiconductor Detector

As discussed in the previous subsection, one might think that a semiconductor detector can be constructed simply by applying a high voltage across a semiconductor and using a charge or current meter to measure the electrical signal induced by external radiation. In this picture, incoming X-ray photons generate electron–hole pairs that drift under the applied bias, producing an additional electric current and thereby changing the material's apparent conductivity due to the radiation.

However, this physical picture has an important limitation: A semiconductor, even in the absence of external radiation, inherently conducts electricity—hence the name semi*conductor*, rather than insulator. When a high voltage is applied, a steady current known as the dark current flows through the material, adding to the radiation-induced signal.

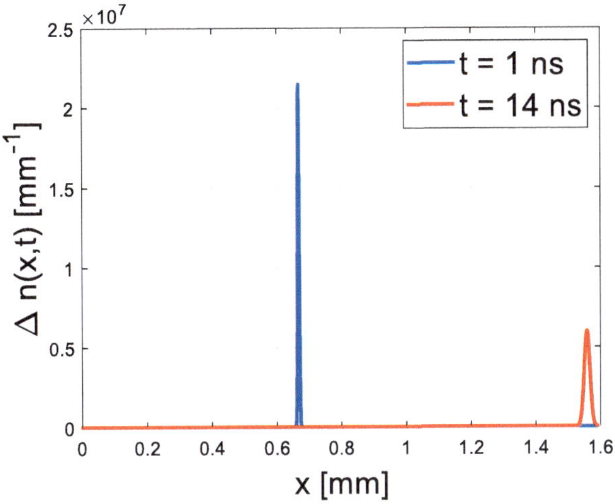

Fig. 1.27 Temporal evolution of the excess electron density, $\Delta n(x)$, calculated using Eq. (1.175), for a 1D CdTe detector with a thickness of 1.6 mm. The X-ray interaction is assumed to occur at $x_0 = 0.6$ mm at time $t = 0$. An external bias voltage of 1000 V is applied across the detector. At $t = 14$ ns, the area under the $\Delta n(x)$ curve remains at 99.7% of its initial value at $t = 0$, indicating minimal recombination loss within the 14-ns time window

To illustrate the dark current problem, consider intrinsic Si at room temperature. Using Eq. (1.123), the resistivity of intrinsic Si is calculated as

$$\begin{aligned}\rho &= \frac{1}{e\,(p_0\,\mu_p + n_0\,\mu_n)}\\ &= \frac{1}{1.6\times 10^{-19}\cdot(1.5\times 10^{10}\cdot 480 + 1.5\times 10^{10}\cdot 1350)}\\ &= 2.3\times 10^5[\Omega\cdot\mathrm{cm}].\end{aligned} \tag{1.177}$$

Applying a bias voltage of 100 V across a 0.05-cm-thick Si wafer produces a steady dark current density of

$$\begin{aligned}J_{\mathrm{dark}} &= \frac{\mathcal{E}}{\rho}\\ &= \frac{100}{0.05\cdot 2.3\times 10^5}\\ &= 0.0086[\mathrm{A/cm^2}]\\ &= 8.6[\mathrm{mA/cm^2}].\end{aligned} \tag{1.178}$$

In contrast, when a 100-keV X-ray photon is fully absorbed in silicon, the drift of the generated electrons and holes produces a current density given by Eq. (1.117):

$$\begin{aligned} J_{\text{drift}} &= e\,(\Delta p\,\mu_p + \Delta n\,\mu_n)\,\mathcal{E} \\ &= 1.6 \times 10^{-19} \left(\frac{100 \times 10^3}{3.62} \cdot 480 + \frac{100 \times 10^3}{3.62} \cdot 1350 \right) \cdot \frac{100}{0.05} \\ &= 1.6 \times 10^{-8} [\text{A/cm}^2] \\ &= 1.6 \times 10^{-5} [\text{mA/cm}^2]. \end{aligned} \tag{1.179}$$

Thus, J_{drift} is five orders of magnitude smaller than J_{dark}. When this small signal is superimposed on the much larger dark current, isolating the contribution of J_{drift} becomes challenging. Moreover, this background signal is not deterministic: It is subject to noise due to fluctuations in charge carrier concentration and temperature, making detection of photon-induced current even more challenging.

In addition, the steady dark current introduces several practical problems. Since current is continuously flowing, the high-voltage power supply must constantly deliver power, leading to nonnegligible energy consumption. Over time, this results in self-heating, which in turn increases the intrinsic carrier concentration and exacerbates the dark current—a feedback loop that further degrades detector performance.

As a result, the elementary homogeneous semiconductor detector model, as shown in Fig. 1.26, is generally unsuitable for high-sensitivity X-ray detection tasks, such as single-photon counting detector for CT application. This simple design only becomes viable if the input light is strong and steady (e.g., in optical flux measurements), or **if the homogeneous semiconductor material exhibits extremely high resistivity.**

One such material is CdTe grown using either the traveling heater method (THM) [8] or the high-pressure Bridgman (HPB) method [1]. These techniques can yield high-purity, low-defect, *p*-type CdTe crystals with resistivities on the order of 10^9 Ω·cm. Under a 1000-V bias applied across a 2-mm-thick CdTe layer, the resulting dark current is approximately

$$\begin{aligned} J_{\text{dark}} &= \frac{\mathcal{E}}{\rho} \\ &\approx \frac{1000}{0.2 \cdot 10^9} \\ &= 5 [\mu\text{A/cm}^2]. \end{aligned} \tag{1.180}$$

Based on the carrier mobilities in CdTe listed in Table 1.6, electrons and holes generated by a 100-keV X-ray photon can produce a drift current of

$$
\begin{aligned}
J_{\text{drift}} &= e\,(\Delta p\,\mu_p + \Delta n\,\mu_n)\,\mathcal{E} \\
&= 1.6 \times 10^{-19} \left(\frac{100 \times 10^3}{4.43} \cdot 100 + \frac{100 \times 10^3}{4.43} \cdot 1100 \right) \cdot \frac{1000}{0.2} \\
&= 2 \times 10^{-2} [\mu\text{A/cm}^2],
\end{aligned} \tag{1.181}
$$

which is two orders of magnitude lower than the dark current. If a capacitor is used to block the steady-state dark current while allowing the time-varying drift current to pass, it becomes feasible to build a radiation detector using a homogeneous CdTe crystal. However, the energy resolution of such detectors is generally poor due to the substantial dark current: While a capacitor can block the DC component of the dark current, it cannot suppress the high-frequency noise embedded within it. This noise contributes to signal degradation, thereby reducing the detector's energy resolution.

In summary, for medical X-ray detectors, both of the following criteria must be satisfied—though the second is often overlooked:

1. **In the presence of radiation**, the detector should generate a strong, measurable electric signal.
2. **In the absence of radiation**, the detector should produce minimal to no electric signal (i.e., low dark current).

A **heterogeneous semiconductor structure**, such as a PN junction, can fulfill both requirements simultaneously. This will be the focus of the following chapter.

1.10 Summary of Chap. 1

1. The presence of a forbidden bandgap in the energy–momentum (E-k) diagram of a semiconductor arises from the periodic arrangement of atoms and the resulting periodic potential. The curvature of the E-k diagram is material-specific and is inversely proportional to the effective mass of electrons or holes.
2. In an undoped semiconductor, there are $\sim 10^{23}$ electrons per cm^3. However, only about 10^6–10^{13} electrons can contribute to electrical signal formation at room temperature. The vast majority of electrons remain in the valence band and cannot be accelerated to form a current.
3. At any nonzero temperature T, some valence electrons may acquire enough thermal energy to cross the bandgap into the conduction band. At thermal equilibrium, the concentration of conduction electrons in an intrinsic semiconductor is given by

$$
n_0 = 2 \left(\frac{m_n^* k}{2\pi \hbar^2} \right)^{\frac{3}{2}} T^{\frac{3}{2}} \exp\left(-\frac{E_c - E_F}{kT} \right).
$$

The concentration of holes (vacancies in the valence band) is given by

$$p_0 = 2\left(\frac{m_p^* k}{2\pi\hbar^2}\right)^{\frac{3}{2}} T^{\frac{3}{2}} \exp\left(-\frac{E_F - E_v}{kT}\right).$$

Both n_0 and p_0 depend strongly on the temperature and the material's band structure, particularly the effective masses.

4. One of the most attractive features of semiconductors is that, by adding only a small amount of impurity atoms (electron donors or acceptors), the material's conductivity can be greatly adjusted. The concentration of the majority charge carrier in a doped (extrinsic) semiconductor is approximately $|N_d - N_a|$, where N_d and N_a are the concentrations of donor and acceptor atoms, respectively.
5. In an intrinsic semiconductor, the Fermi level lies near the middle of the bandgap. In contrast, for an extrinsic semiconductor, the Fermi level shifts with doping: It moves closer to the conduction band with more donor atoms and closer to the valence band with more acceptor atoms.
6. Regardless of doping level, the following fundamental semiconductor equation holds at thermal equilibrium:

$$n_0 \, p_0 = n_i^2.$$

7. One mechanism of charge transport in semiconductors is drift. Under an electric field, electrons and holes acquire an average drift velocity, and the resulting drift current density is proportional to the electric field and the respective carrier mobility.
8. Another important transport mechanism is diffusion. The diffusion current density is proportional to the spatial gradient of the carrier concentration. The diffusion coefficient is related to mobility and temperature via the Einstein relation.
9. When X-rays interact with a semiconductor, their energy creates a stochastic number of electron–hole pairs. The mean number of pairs generated is proportional to the X-ray energy deposited and inversely proportional to ϵ (average energy required to create one electron–hole pair, typically about three times the bandgap energy.
10. Over time, electrons and holes recombine, with a recombination rate determined by their mean lifetimes. To collect the excess carriers before recombination, a strong electric field is needed to induce drift toward the electrodes.
11. Applying a high voltage across a homogeneous semiconductor usually generates a large dark current, which can obscure the small signal from radiation-induced carriers. Therefore, for high-sensitivity detection, a heterogeneous semiconductor structure discussed in the next chapter is often preferred.

Appendix 1: Derivation of the Fermi Function and the Boltzmann Function

Derivation of the Fermi Function

To derive the Fermi–Dirac distribution from first principles, consider a system with a set of discrete energy levels indexed by l, where each level has energy E_l and contains n_l particles. Let the lth energy level have a degeneracy g_l, meaning it consists of g_l distinct sublevels. According to the Pauli exclusion principle, each sublevel can be occupied by at most one fermion. Given that the fermions are indistinguishable, the number of distinct ways to distribute n_l fermions among the g_l sublevels at energy E_l is

$$\omega_l = \frac{g_l!}{n_l!(g_l - n_l)!}. \tag{1.182}$$

Considering all energy levels, the total number of microstates Ω_F corresponding to a set of $\{n_l\}$ fermions is the product of the individual contributions from each energy level:

$$\Omega_F = \prod_l \omega_l = \prod_l \frac{g_l!}{n_l!(g_l - n_l)!}. \tag{1.183}$$

We wish to find a set of $\{n_l^*\}$ with which Ω_F is maximized, subject to the constraint that the total number of particles is fixed at N, namely, $\sum_l n_l = N$, and the total energy of the particles is fixed at E, namely, $\sum_l n_l E_l = E$. This constrained optimization problem can be formulated using the Lagrange multiplier method:

$$\{n_l^*\} = \underset{\{n_l\}}{\arg\max} \left\{ \Omega_F(n_l) - \alpha\left(\sum_l n_l - N\right) - \beta\left(\sum_l n_l E_l - E\right) \right\}, \tag{1.184}$$

where α and β are two numerical multipliers. By observing the formula of Ω_F in Eq. (1.183), it is easier to work on the natural logarithm of Ω_F that converts the product operation to summation operation. Since logarithm is a monotonic function, a set of $\{n_l^*\}$ that maximizes $\ln \Omega_F$ should also maximize Ω_F. Therefore, the optimization problem in (1.184) is equivalent to [9]

$$\begin{aligned} \{n_l^*\} &= \underset{\{n_l\}}{\arg\max}\, f(n_l) \\ &= \underset{\{n_l\}}{\arg\max} \left\{ \ln\Omega_F(n_l) - \alpha(\sum_l n_l - N) - \beta(\sum_l n_l E_l - E) \right\} \end{aligned}$$

$$= \arg\max_{\{n_l\}} \left\{ \sum_l [\ln g_l! - \ln n_l! - \ln(g_l - n_l)!] \right.$$

$$\left. -\alpha(\sum_l n_l - N) - \beta(\sum_l n_l E_l - E) \right\}. \tag{1.185}$$

Using Stirling's approximation for the factorials,

$$\begin{aligned} \ln g_l! &\approx g_l(\ln g_l - 1) \\ \ln n_l! &\approx n_l(\ln n_l - 1) \\ \ln(g_l - n_l)! &\approx (g_l - n_l)(\ln(g_l - n_l) - 1), \end{aligned} \tag{1.186}$$

$f(n_l)$ in Eq. (1.185) becomes

$$f(n_l) \approx \sum_l [g_l(\ln g_l - 1) - n_l(\ln n_l - 1) - (g_l - n_l)(\ln(g_l - n_l) - 1)]$$

$$-\alpha\left(\sum_l n_l - N\right) - \beta\left(\sum_l n_l E_l - E\right). \tag{1.187}$$

The value n_l^* that maximizes f is obtained by setting the derivative of f with respect to n_l equal to zero:

$$\begin{aligned} 0 &= \left.\frac{\partial f(n_l)}{\partial n_l}\right|_{n_l = n_l^*} \\ &= -(\ln n_l^* - 1) - 1 + (\ln(g_l - n_l^*) - 1) + 1 - \alpha - \beta E_l \\ &= \ln(g_l - n_l^*) - \ln n_l^* - \alpha - \beta E_l. \end{aligned} \tag{1.188}$$

This yields

$$\ln\left(\frac{g_l}{n_l^*} - 1\right) = \alpha + \beta E_l, \tag{1.189}$$

which can be rearranged as

$$\frac{g_l}{n_l^*} = 1 + \exp(\alpha + \beta E_l). \tag{1.190}$$

Therefore, the average number of fermions occupying each sub-quantum energy level, namely, the Fermi–Dirac function, is given by

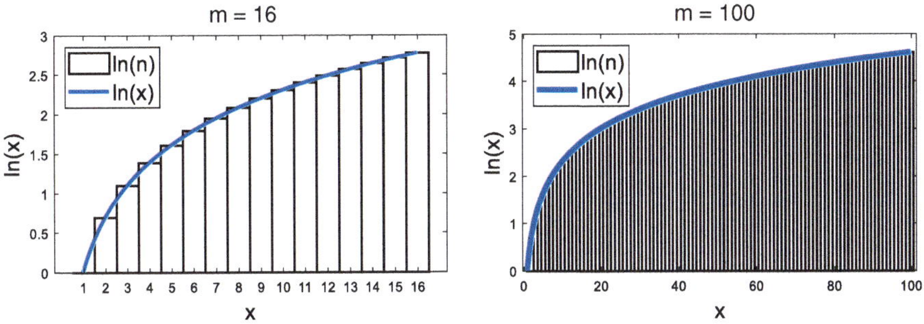

Fig. 1.28 Approximation of $\ln m!$ by area under the $\ln x$ curve. n is an integer $\in [1, m]$

$$\boxed{f_F(E_l) = \frac{n_l^*}{g_l} = \frac{1}{1 + \exp(\alpha + \beta E_l)}}. \tag{1.191}$$

Here, $f_F(E_l)$ denotes the Fermi–Dirac probability distribution function.

Proof of Stirling's approximation

The natural logarithm of the factorial of an integer m can be decomposed into the following summation series:

$$\ln m! = \ln 1 + \ln 2 + \cdots + \ln m. \tag{1.192}$$

As shown in Fig. 1.28, the result of this summation series is given by adding the areas of individual rectangles. When $m \gg 1$, the summation of these areas can be approximated by the area under the $\ln x$ curve:

$$\begin{aligned} \ln m! &\approx \int_1^m \ln x \mathrm{d}x \\ &= (x \ln x - x)|_1^m \\ &\approx (m \ln m - m). \end{aligned} \tag{1.193}$$

Derivation of the Boltzmann Distribution Function

To derive the Boltzmann distribution function, we begin by considering a system in which the number of particles occupying the energy level E_l is denoted by n_l, and the degeneracy (i.e., number of available sublevels) at that energy level is g_l. The number of distinct ways to distribute these n_l distinguishable particles among the g_l available sublevels is $g_l^{n_l}$.

When all energy levels are considered, the total number of arrangements for distributing the particles across all sublevels becomes

$$\prod_l g_l^{n_l}. \tag{1.194}$$

However, since the particles in a Maxwell–Boltzmann (classical) system are mutually distinguishable, exchanging two particles occupying different energy levels results in a distinct configuration. Therefore, the total number of permutations of all N particles contributes a factor of $N!$.

Nevertheless, we must correct for overcounting due to the indistinguishable arrangements of particles within the same energy level. For each level E_l, there are $n_l!$ ways to permute particles among themselves without leading to new physical configurations—these are already accounted for in $g_l^{n_l}$. To correct for this overcounting, we divide by $\prod_l n_l!$.

Thus, the total number of distinct ways, Ω_B, to distribute N distinguishable particles among energy levels with degeneracy g_l is given by

$$\Omega_B = N! \prod_l \frac{g_l^{n_l}}{n_l!}. \tag{1.195}$$

To find the most probable distribution n_l^*, we maximize Ω_B subject to the constraints of $\sum_l n_l = N$, and $\sum_l n_l E_l = E$; we use the Lagrange multiplier method [9]:

$$\begin{aligned} n_l^* &= \arg\max_{n_l} f(n_l) \\ &= \arg\max_{n_l} \left[\ln\Omega_B - \alpha\left(\sum_l n_l - N\right) - \beta\left(\sum_l n_l E_l - E\right)\right]. \end{aligned} \tag{1.196}$$

This problem is solved by setting the derivative of $f(n_l)$ with respect to n_l to zero. Skipping the intermediate steps, we present the resulting expression for the most probable occupation number n_l^*:

$$n_l^* = g_l \exp(-\alpha - \beta E_l). \tag{1.197}$$

This leads to the expression for the Boltzmann distribution function:

$$\boxed{f_B(E_l) = \frac{n_l^*}{g_l} = \exp(-\alpha - \beta E_l)}. \tag{1.198}$$

To find expressions for the constants α and β, we begin by evaluating the natural logarithm of Ω_B in Eq. (1.195):

$$\begin{aligned}\ln \Omega &= \ln N! + \sum_l n_l \ln g_l - \sum_l n_l! \\ &\approx N(\ln N - 1) + \sum_l n_l \ln g_l - \sum_l n_l(\ln n_l - 1).\end{aligned} \tag{1.199}$$

Substituting the expression for n_l^* in Eq. (1.197) into the above and assuming $N \gg 1$, we obtain

$$\ln \Omega = (\alpha + 1)N + \beta E. \tag{1.200}$$

Solving for E, we get

$$E = \frac{\ln \Omega}{\beta} - \frac{N}{\beta} - \frac{\alpha N}{\beta}. \tag{1.201}$$

On the other hand, based on the fundamental law of thermodynamics, we have

$$\begin{aligned}E &= TS - PV + E_F N \\ &= T k_B \ln \Omega - N k_B T + E_F N,\end{aligned} \tag{1.202}$$

where S denotes the entropy.

By comparing Eqs. (1.201) and (1.202), we find

$$\beta = \frac{1}{k_B T}, \qquad \alpha = -\frac{E_F}{k_B T}. \tag{1.203}$$

Appendix 2: Boltzmann Approximation of the Fermi Distribution Function—Physical Meaning

For a Fermi–Dirac system composed of indistinguishable fermions, if the number of particles occupying the lth energy level, n_l, is much smaller than the number of available quantum states at that level, g_l, such that

$$n_l \ll g_l, \tag{1.204}$$

then the number of possible configurations (microstates) for arranging the fermions can be approximated as follows:

$$\begin{aligned}\Omega_F &= \prod_l \frac{g_l!}{n_l!(g_l-n_l)!} \\ &= \prod_l \frac{g_l(g_l-1)\cdots(g_l-n_l+1)}{n_l!} \\ &\approx \prod_l \frac{g_l^{n_l}}{n_l!} \\ &= \frac{\Omega_B}{N!},\end{aligned} \tag{1.205}$$

where Ω_B, defined in Eq. (1.195), is the number of microstates in a Boltzmann (classical) system.

Equation (1.205) reveals that **when the occupation probability of each quantum state is very small, the particles can be treated as mutually independent. In this limit, whether the particles are distinguishable or not becomes irrelevant** [9]. As a result, the total number of microstates for a Fermi–Dirac system differs from the Boltzmann case only by a factor of $N!$, accounting for the indistinguishability of fermions (i.e., permutations of identical particles do not yield new microstates).

If we use the approximation $\Omega_F \approx \Omega_B/N!$ to solve for the most probable occupation number n_l^* using the method of Lagrange multipliers (e.g., from Eq. 1.185), the resulting distribution is the same as that of a Boltzmann system. This is because the factor $N!$ is independent of n_l and does not influence the optimization.

In other words, under the condition $n_l \ll g_l$, the Fermi–Dirac distribution approaches the Boltzmann distribution:

$$\boxed{f_F(E_l) \approx f_B(E_l) \;\; \text{when} \;\; n_l \ll g_l}\,. \tag{1.206}$$

From Eq. (1.53), when $\frac{n_l}{g_l} \ll 1$ for all l, we have

$$1 + \exp\left(\frac{E - E_F}{k_B T}\right) \gg 1 \tag{1.207}$$

which means

$$E - E_F \gg k_B T. \tag{1.208}$$

Equation (1.208) provides the macroscopic condition for the validity of the Boltzmann approximation. The corresponding microscopic condition is $n_l \ll g_l$.

References

1. Al-Hamdi TK, McPherson SW, Swain SK, Jennings J, Duenow JN, Zheng X, Albin DS, Ablekim T, Colegrove E, Amarasinghe M, Ferguson A, Metzger WK, Szeles C, Lynn KG. CdTe synthesis and crystal growth using the high-pressure Bridgman technique. J Cryst Growth. 2020;534:125466. ISSN 0022-0248. https://doi.org/10.1016/j.jcrysgro.2019.125466
2. Ashcroft NW, Mermin ND. Solid state physics. Boston: Cengage Learning; 1976.
3. Bloch F. Über die Quantenmechanik der Elektronen in Kristallgittern. Zeitschrift für Physik. 1928;52:555–600.
4. de Broglie L. Research on the theory of quanta. PhD thesis, University of Paris (Sorbonne); 1924.
5. Griffiths DJ. Introduction to quantum mechanics. Englewood Cliffs: Prentice Hall; 1994.
6. Liu E-K, Zhu B-S, Luo J-S, et al. The physics of semiconductors (7th ed., in Chinese). Beijing: Publishing House of Electronics Industry; 2017.
7. Neamen D. An introduction of semiconductor devices. New York: McGraw-Hill Education; 2005.
8. Vijayakumar P, Amaladass EP, Ganesan K, Sarguna RM, Roy V, Ganesamoorthy S. Development of travelling heater method for growth of detector grade CdZnTe single crystals. Mater Sci Semicond Process. 2024;169:107897. ISSN 1369-8001. https://doi.org/10.1016/j.mssp.2023.107897.
9. Wang Z. Thermodynamics and statistical physics (6th ed., in Chinese). Beijing: Higher Education Press, 2020.

Chapter 2
Basic Semiconductor Detector Structure and Elements

Contents

2.1 Introduction

As discussed in the previous chapter, introducing a small concentration of donor or acceptor atoms into a semiconductor can significantly alter its energy band structure and electrical conductivity. By exploiting this property, heterogeneous semiconductorstructures such as diodes and transistors can exhibit highly useful optoelectronic

K. Li, *Essential Physics of Medical X-ray Sources and Detectors*,
https://doi.org/10.1007/978-3-032-27800-5_2

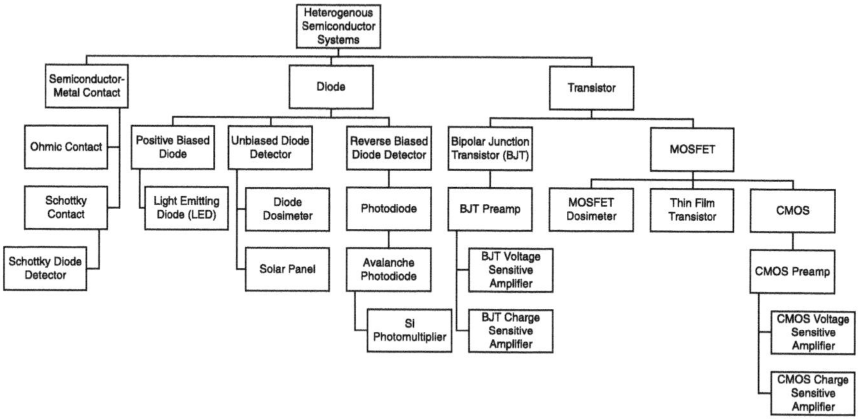

Fig. 2.1 A hierarchical overview of semiconductor structures and devices utilized in X-ray detectors

characteristics, making them ideal for radiation detection applications.

In this chapter, we explore how these heterogeneous semiconductor structures fulfill two essential requirements for radiation detection:

1. Generate a strong, measurable electrical signal in the presence of radiation.
2. Produce minimal or no signal in the absence of radiation (i.e., maintain a low dark current).

We begin by reviewing the physics of the fundamental building block of these devices—the PN junction—and then explain the operating principles of the semiconductor diode detector. This is followed by a discussion of several common types of semiconductor diode detectors, including photodiodes, avalanche photodiodes, and silicon photomultipliers.

In addition to serving as radiation sensors, heterogeneous semiconductor structures play a crucial role in the readout circuits paired with non-semiconductor radiation detectors. Examples include thin-film transistor (TFT) arrays in flat-panel detectors that use scintillators. Therefore, this chapter also covers some of the fundamental semiconductor circuit elements, summarized in Fig. 2.1. Understanding their working principle will help clarify the signal processing mechanisms in photon counting detectors, which will be discussed in detail in the next chapter.

2.2 PN Junction

In this section, we will explore the structure and electrical properties of a PN junction, which is the fundamental building block of nearly all semiconductor optoelectronic devices—including photodiodes and direct-conversion X-ray detectors.

As the name suggests, a PN junction is formed by bringing a p-type semiconductor into contact with an n-type semiconductor. The invention of the PN junction is commonly credited to researchers at Bell Labs and Purdue University, who developed germanium diodes used in radar systems during World War II [2]. While it is widely known that a PN junction can function as a rectifier, our goal is to go beyond this basic understanding. We aim to uncover the underlying physics of the PN junction and learn how to apply it specifically in the context of radiation detection.

2.2.1 *Basic Structure*

When a p-type semiconductor doped with acceptor atoms is brought into contact with an n-type semiconductor doped with donor atoms, a steep gradient in the concentration of electrons and holes forms near the p-n interface. As discussed in Sect. 1.8.2, a gradient in charge carrier density leads to diffusion: Electrons from the n-region diffuse into the p-region, leaving behind an equal number of positively charged ionized donor atoms. Similarly, holes from the p-region diffuse into the n-region, leaving behind negatively charged ionized acceptor atoms.

This movement of carriers results in a region near the junction where mobile charge carriers are depleted and immobile ionized dopants remain. The resulting net positive charge on the n-side and net negative charge on the p-side form what is known as the **space charge region**, which establishes an electric field directed from the n-side to the p-side (Fig. 2.2a).

As the PN junction approaches thermal equilibrium, all conduction-band electrons in the n-region and valence-band holes in the p-region are swept out of the space charge region. If any remain, the electric field continues to act until they are removed. For this reason, the space charge region is also referred to as the **depletion region**.

Although a charge carrier density gradient still exists at the two edges of the depletion region, the electric field generated by the space charges produces a force that goes against the "diffusion force." In the end, the two forces balance each other and the pn junction reaches an equilibrium state in which charge carriers no longer diffuse.

To simplify our analysis of the electric field in the depletion region at equilibrium, we assume that the space charge region abruptly ends at $x = +x_n$ on the n-side and at $x = -x_p$ on the p-side. Admittedly, this is a crude approximation, but such simplifications are commonly used in physics to help build a clear understanding of the underlying physical picture.

Within the region $-x_p \leq x \leq x_n$, the space charge density, $\rho(x)$, is assumed to be (Fig. 2.2b)

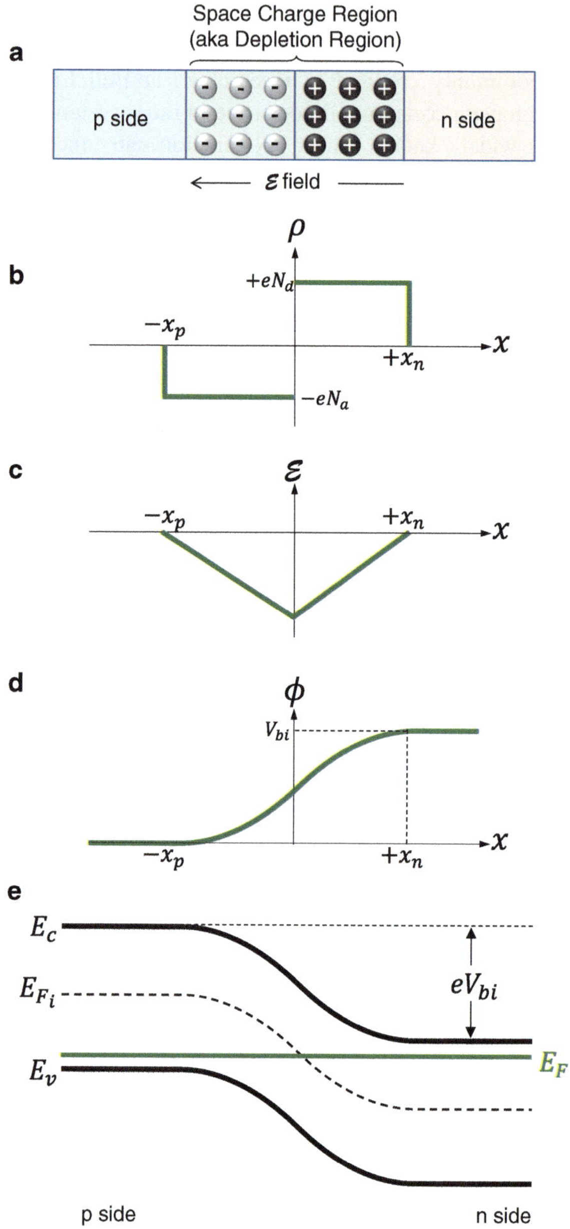

Fig. 2.2 A PN junction in thermal equilibrium, with no external voltage applied, is considered. The p-side is uniformly doped with acceptor impurities at a concentration N_a, and the n-side is uniformly doped with donor impurities at a concentration N_d. (**a**) Illustration of the space charge region (also known as the depletion region). An electric field generated by the immobile ionized donors and acceptors in this region prevents further diffusion of holes from the p-side and electrons from the n-side into the depletion region. (**b**) Distribution of the space charge density within the depletion region. (**c**) Electric field across the depletion region. (**d**) Electric potential across the PN junction. (**e**) Energy band diagram for **electrons** in the PN junction under thermal equilibrium with no applied voltage

$$\rho(x) = \begin{cases} -eN_a & -x_p \le x \le 0 \\ +eN_p & 0 < x \le x_n. \end{cases} \tag{2.1}$$

Note that $N_a\,x_p = N_d\,x_n$ due to the charge neutrality of the whole system.

By charge neutrality, the total positive and negative charge in the depletion region must balance:

$$N_a\,x_p = N_d\,x_n. \tag{2.2}$$

According to Poisson's equation:

$$\begin{aligned} \frac{\mathrm{d}^2\phi(x)}{\mathrm{d}x^2} &= -\frac{\mathrm{d}\mathcal{E}(x)}{\mathrm{d}x} \\ &= -\frac{\rho(x)}{\epsilon}, \end{aligned} \tag{2.3}$$

where ϕ denotes the electric potential, and ϵ is the permittivity of the semiconductor. The electric field $\mathcal{E}$ in the p region is given by integrating $\rho(x)$:

$$\begin{aligned} \mathcal{E}(x) &= \int_{-\infty}^{x} \frac{\rho(x')}{\epsilon}\,\mathrm{d}x' \\ &= \int_{-x_p}^{x} \frac{-eN_a}{\epsilon}\,\mathrm{d}x' \\ &= -\frac{eN_a}{\epsilon}(x + x_p), \end{aligned} \tag{2.4}$$

where $-x_p \le x \le 0$.

Similarly, one can show that the electric field in the n-region is given by

$$\begin{aligned} \mathcal{E}(x) &= \mathcal{E}(0) + \int_{0}^{x} \frac{eN_d}{\epsilon}\,\mathrm{d}x' \\ &= -\frac{eN_a}{\epsilon}x_p + \frac{eN_d}{\epsilon}x. \end{aligned} \tag{2.5}$$

Equations (2.4)–(2.5) show that there exists an electric field in the depletion region even when no voltage is applied to the semiconductor. Figure 2.2c plots the electric field in the depletion region. The $\mathcal{E}$ field is a linear function in each region and its magnitude reaches its maximum at the p-n interface.

Based on the electric field $\mathcal{E}(x)$, we can now derive the electric potential $\phi(x)$ in the PN junction. In the p-region ($-x_p \leq x \leq 0$), the potential is given by

$$\begin{aligned}\phi(x) &= -\int_{-\infty}^{x} \mathcal{E}(x')\,\mathrm{d}x' \\ &= \int_{-x_p}^{x} \frac{eN_a}{\epsilon}(x + x_p)\,\mathrm{d}x' \\ &= \frac{eN_a}{\epsilon}\left(\frac{x^2}{2} + x_p x\right) + C',\end{aligned} \tag{2.6}$$

where C' is an integration constant. Since we are interested in the potential difference across the junction rather than its absolute value, the specific value of C' is not critical. For convenience, we set the reference potential such that $\phi(-x_p) = 0$. Substituting $x = -x_p$ into Eq. (2.6) gives

$$C' = \frac{eN_a}{2\epsilon}x_p^2. \tag{2.7}$$

Therefore, the potential in the p-region becomes

$$\phi(x) = \frac{eN_a}{2\epsilon}\left(x + x_p\right)^2 \qquad -x_p \leq x \leq 0. \tag{2.8}$$

In the n-region ($0 < x \leq x_n$), a similar integration yields

$$\phi(x) = \frac{eN_d}{\epsilon}\left(x_n x - \frac{x^2}{2}\right) + \frac{eN_a}{2\epsilon}x_p^2 \qquad 0 < x \leq x_n. \tag{2.9}$$

Figure 2.2d shows the resulting quadratic dependence of the electric potential across the depletion region. The potential difference between the two ends of the depletion region, $\phi(x_n) - \phi(-x_p)$, is known as the **built-in potential barrier**, denoted as V_{bi}. Using Eqs. (2.8) and (2.9), we find

$$V_{bi} = \phi(x_n) - \phi(-x_p) = \frac{e}{2\epsilon}\left(N_d x_n^2 + N_a x_p^2\right). \tag{2.10}$$

This built-in potential represents the internal electric potential that opposes further diffusion of carriers in thermal equilibrium.

An alternative formula of the build-in potential V_{bi}
For the p-side of the pn junction (to the left of the space charge region shown in Fig. 2.3), according to Eq. (1.91),

$$p_0 = n_i \exp\left(-\frac{E_F - E_{F_{i,p}}}{k_B T}\right). \tag{2.11}$$

where $p_0 \approx N_a$ (acceptor concentration). For the n-side of the pn junction (to the right of the space charge region shown in Fig. 2.3), according to Eq. (1.90),

$$n_0 = n_i \exp\left(\frac{E_F - E_{F_{i,n}}}{k_B T}\right), \tag{2.12}$$

where $p_0 \approx N_d$ (donor concentration).

As equilibrium, E_F is a constant across the pn junction, and $eV_{bi} = E_{F_{i,p}} - E_{F_{i,n}}$. Therefore,

$$\begin{aligned} V_{bi} &= \frac{1}{e}(E_{F_{i,p}} - E_{F_{i,n}}) \\ &= \frac{1}{e}[-(E_F - E_{F_{i,p}}) + (E_F - E_{F_{i,n}})] \\ &= \frac{k_B T}{e}\left(\ln\frac{N_a}{n_i} - \ln\frac{N_d}{n_i}\right) \\ &= \frac{k_B T}{e}\ln\frac{N_a N_d}{n_i^2}. \end{aligned} \tag{2.13}$$

For a Si at room temperature, $n_i = 1.5 \times 10^{10}\,\text{cm}^{-3}$. At typical doping levels of $N_a = 2 \times 10^{16}\,\text{cm}^{-3}$ and $N_d = 5 \times 10^{15}\,\text{cm}^{-3}$,

$$\begin{aligned} V_{bi} &= \frac{1.38 \times 10^{-23} \times 300}{1.6 \times 10^{-19}} \ln\frac{2 \times 10^{16} \times 5 \times 10^{15}}{(1.5 \times 10^{10})^2} \\ &= 0.0259 \times 26.82 \\ &= 0.695\ [\text{V}]. \end{aligned} \tag{2.14}$$

Since the potential energy of an electron is given by $-e\phi(x)$, the electron's potential energy is higher on the p-side than on the n-side of the junction. This potential barrier prevents electrons in the n-region (where they are the majority carriers) from crossing into the p-region (Fig. 2.2e).

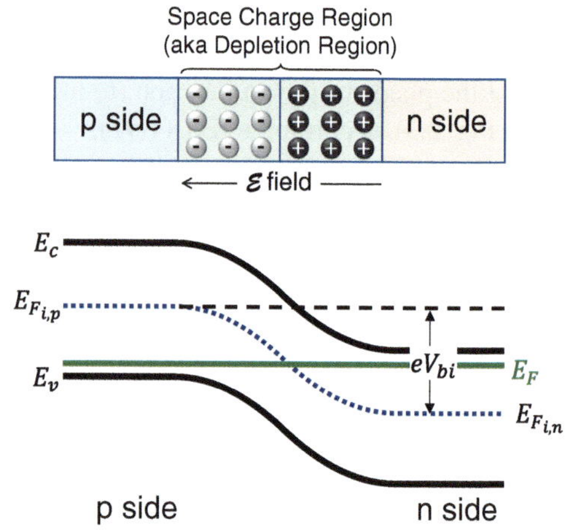

Fig. 2.3 Energy band diagram for electrons in a PN junction at thermal equilibrium. The Fermi energy (E_F) is a constant throughout the junction

Similarly, the potential energy of holes is higher on the n-side than on the p-side, which prevents holes in the p-region (the majority carriers there) from diffusing into the n-region.

2.2.2 *Reverse Biasing and Forward Biasing of PN Junction*

In Sect. 2.2.1, we discussed the space charge distribution and electric field in a PN junction **under zero external bias**. At thermal equilibrium, a built-in potential barrier V_{bi} is established, preventing further diffusion of charge carriers between the p- and n-regions.

When an external bias voltage is applied, the junction is no longer in equilibrium—**the Fermi level is no longer constant throughout the device.** If the voltage on the n-side is made positive relative to the p-side, the PN junction is said to be reverse biased (Fig. 2.4).

The electric field induced by the applied reverse bias points in the same direction as the built-in electric field of the depletion region. These fields **add together**, increasing the potential barrier between the p- and n-regions. As a result, it becomes even more difficult for holes in the p-region and electrons in the n-region (**the majority carriers**) to cross the junction. Therefore, very little current flows through the junction under reverse bias. **This characteristic is highly desirable for radiation detectors: It enables the creation of a strong internal electric field with minimal dark current. Achieving this combination is difficult without utilizing the heterojunction structure.**

When the voltage applied to the p-side is positive relative to the n-side, the PN junction is said to be forward biased (Fig. 2.4). In this case, the electric field generated by the applied voltage opposes the built-in electric field of the depletion region. As a result, the net electric field is reduced, and the potential barrier between the p- and n-regions is lowered (Fig. 2.4). When the barrier becomes zero, holes

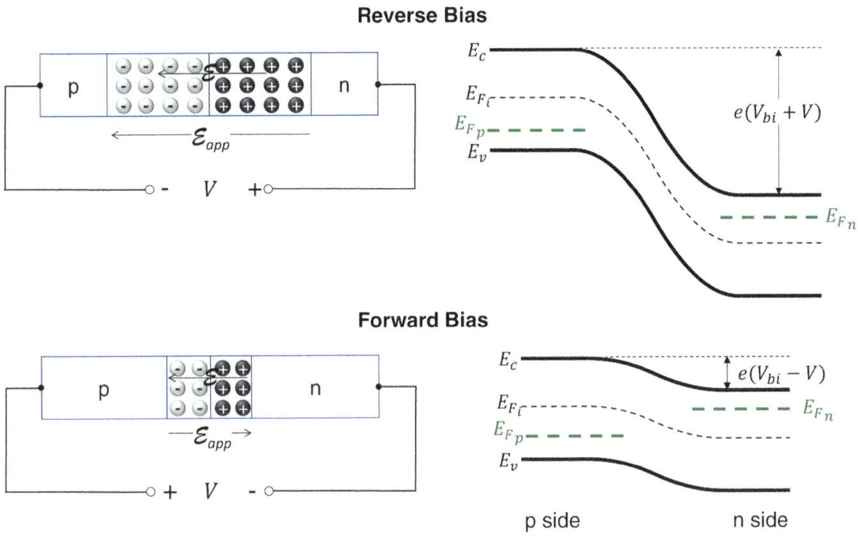

Fig. 2.4 Reversely biased and forwardly biased PN junction. $\mathcal{E}$ denotes the electric field generated by the space charge; $\mathcal{E}_{app}$ denotes the electric field from the applied voltage V. V_{bi} denotes the built-in potential. e denotes the elementary charge

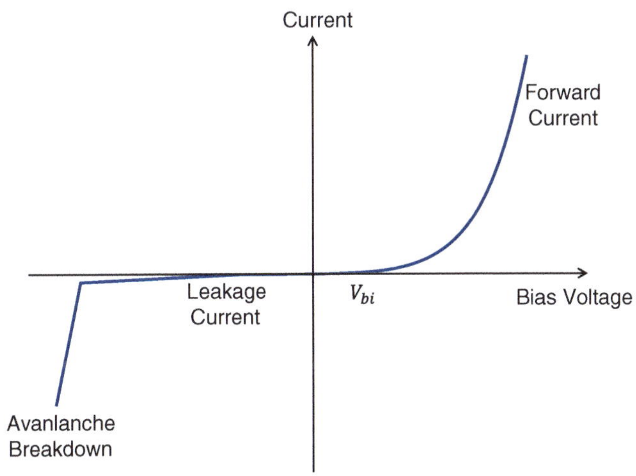

Fig. 2.5 A schematic illustration of the current–voltage characteristics of a PN junction diode

from the p-side can diffuse into the n-region, and electrons from the n-side can diffuse into the p-region. This leads to a steady current through the junction. **Only a small forward bias voltage is needed to generate a large current through the PN junction.**

Figure 2.5 shows a representative current–voltage (I–V) characteristic of a PN junction. Under forward bias (positive voltage), the current increases exponentially with the applied voltage; under reverse bias, leakage current is very small. Because of this distinctive I-V behavior, a PN junction is commonly referred to as a PN junction diode (di- meaning "two" and -ode meaning "path")—or simply, a diode.

One of its most common applications is as a rectifier in electronic circuits. When a PN diode is placed in series with an alternating current (AC) power source and a load resistor, during the positive half-cycle of the AC signal (forward bias), the diode conducts, allowing current to flow through the junction and the load resistor, resulting in a positive output voltage. During the negative half-cycle (reverse bias), the diode blocks the current, and the output voltage across the load drops to zero. By adding a smoothing capacitor, the pulsating output can be further converted into a steady direct current (DC) signal.

In this chapter, however, our primary interest lies in the use of the PN junction as a radiation detector, rather than as a rectifier. This application will be discussed in detail in Sect. 2.3.

2.2.3 Influence of Minority Carriers on PN Junction Characteristics

Under reverse bias (negative voltage), the leakage current through a PN junction is typically on the order of 10^{-14}–10^{-12} A. This leakage current is primarily caused by the movement of minority charge carriers, which were omitted in our analysis in the previous subsection.

As shown earlier in Fig. 2.4, under reverse bias, majority carriers (electrons in the n-region and holes in the p-region) encounter a large energy barrier that prevents them from crossing the junction and contributing to current flow. However, the situation is different for minority carriers, and their behavior must be understood using a reversed view of the energy band diagram.

As illustrated in Fig. 2.6, minority electrons in the p-region occupy higher energy levels compared to the conduction band in the n-region. There is no potential barrier

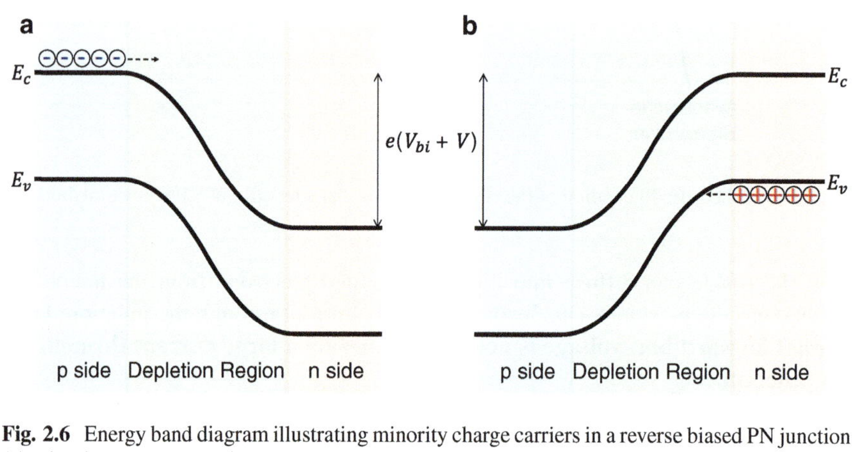

Fig. 2.6 Energy band diagram illustrating minority charge carriers in a reverse biased PN junction: (**a**) minority electrons on the p-side and (**b**) minority holes on the n-side

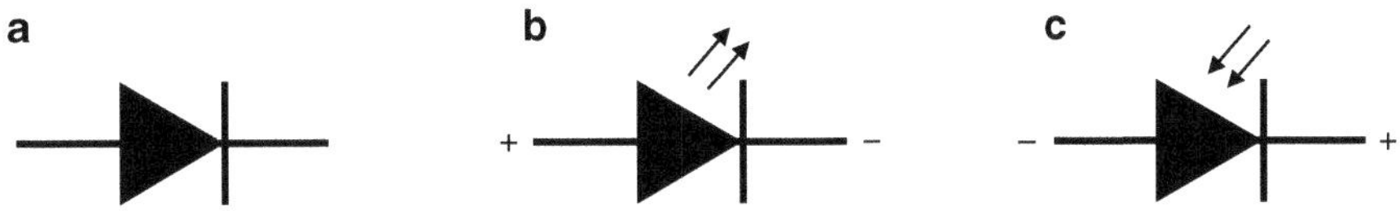

Fig. 2.7 Symbols for (**a**) standard diode, (**b**) light-emitting diode (LED), and (**c**) diode fetector. The + symbol indicates the anode and the − symbol denotes the cathode

preventing these electrons from diffusing into the depletion region, where they are swept across by the strong electric field, contributing to the reverse current. Similarly, minority holes in the n-region can diffuse into the depletion region and are likewise accelerated by the field, generating current.

From the fundamental semiconductor equation (see Eq. (1.92)), the minority carrier concentration is inversely proportional to the majority carrier concentration and is therefore several orders of magnitude smaller. This explains why the reverse leakage current is extremely small in a typical PN junction.

However, when the reverse bias voltage becomes sufficiently high, minority carriers diffused into the depletion region can gain enough kinetic energy from the electric field to ionize atoms upon collision. This process generates additional electron–hole pairs, leading to an avalanche multiplication effect. The result is a sudden and dramatic increase in current, known as avalanche breakdown, as shown in Fig. 2.5. This breakdown produces an extremely large dark current, which is highly undesirable for radiation detection. Consequently, the bias voltage applied to a semiconductor diode detector must be carefully limited and cannot be increased arbitrarily.

2.3 Using a PN Junction as a Light Emitter or Detector

The symbol for a standard PN junction diode is shown in Fig. 2.7a. This symbol reflects the rectifying behavior of the diode. It closely resembles the symbols for the light-emitting diode (LED) and the photodiode, which is not surprising—both are specialized subtypes of the diode.

Although this chapter focuses on radiation detection, presenting the diode detector alongside the LED helps reinforce key concepts—especially the advantages of operating a diode detector under reverse bias. Understanding how a diode emits light (as in an LED) provides useful contrast to how it detects light or ionizing radiation. The LED also shares certain similarities in working principle with scintillators (discussed in Chap. 4). For these reasons, this section begins with a very brief discussion on the basic principles of LEDs before introducing diode-based radiation detectors.

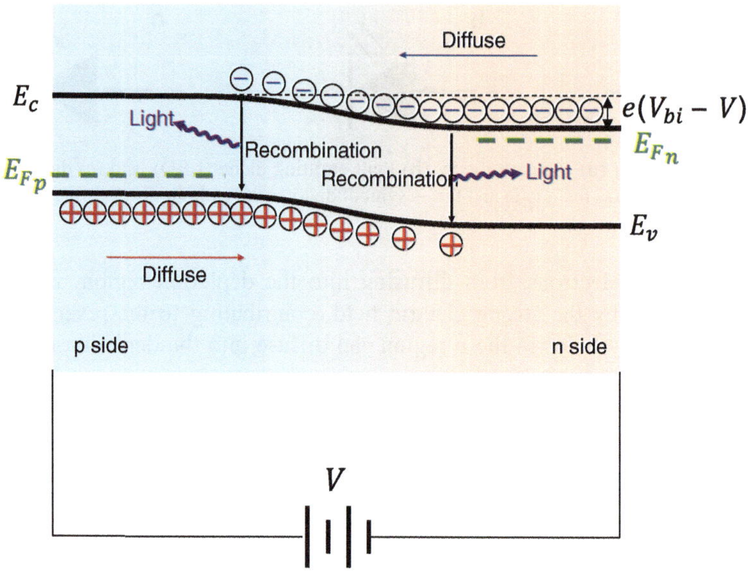

Fig. 2.8 Schematic illustration of the energy band diagram of an LED

2.3.1 LED: A Forward Biased Diode

As shown in the LED symbol in Fig. 2.7b, an LED must be forward biased to emit light: The p-side should be connected to the positive terminal of the voltage supply and the n-side to the negative terminal. Under forward bias, the system is no longer in equilibrium, and the Fermi level splits into two quasi-Fermi levels (E_{F_p} and E_{F_n}), as shown in Fig. 2.8.

The forward bias voltage V reduces potential barrier from V_{bi} to $V_{bi} - V$, allowing electrons on the n-side to diffuse into the p-side, where they are excess minority carriers. As discussed earlier in Sect. 1.9.2, these minority electrons can recombine with holes on the p-side, releasing energy. In a direct bandgap semiconductor, a significant portion of this recombination energy is emitted as electromagnetic radiation in the form of visible light. The energy—and thus the color—of the emitted light is determined by the material's bandgap energy. Therefore, different semiconductors can emit different colors. For instance, a GaAs-based LED emits infrared light due to its relatively small bandgap of 1.42 eV.

In principle, the forward voltage also allows holes from the p-side to diffuse into the n-side and combine with electrons there to emit light. In practice, however, the contribution of this process to light emission is generally less significant because the efficiency of hole injection is relatively low [6].

As long as the forward bias is maintained, electrons are continuously injected from the n-side to the p-side. This continuous injection sustains the recombination process and therefore supports steady light emission. This behavior contrasts with scintillators, where light is emitted in brief flashes following the absorption of X-ray photons and quickly decays (Chap. 4). In contrast, an LED produces continuous light through ongoing carrier injection from the external voltage source, maintaining a steady supply of mobile electrons and holes in the junction region.

2.3.2 Reverse Biased Diode Detectors

As discussed in Sect. 1.9 of Chap. 1, by measuring a physical quantity proportional to the number of electron–hole pairs (Δn) generated by incident radiation, one can estimate the energy fluence or other important information about the radiation.

In practice, however, accurately measuring Δn is nontrivial. This is because electrons and holes can recombine over time, diminishing the signal. To capture the charge carriers before recombination occurs, a strong electric field is required. This field rapidly separates the electrons and holes and causes them to drift quickly in opposite directions toward electrodes for collection.

Additionally, it is critical that in the absence of radiation, no significant current flows in the detector, even in the presence of a strong electric field. Otherwise, background currents would interfere with the measurement of the radiation-induced signal.

A reversely biased PN junction naturally satisfies both of these conditions. The reverse bias voltage establishes a strong electric field across the depletion region, which is free of charge carriers under equilibrium conditions. Without incident radiation, the leakage current, determined by the minority charge carrier concentration, remains negligible.

As illustrated in Fig. 2.9, when an X-ray photon interacts with the semiconductor and deposits energy within the depletion region, it produces excess charge carriers (electron–hole pairs). The strong electric field rapidly separates these carriers, causing electrons and holes to drift toward opposite ends of the diode, resulting in a transient drift current. As will be discussed in Chap. 3, this drift current can induce a measurable electrical signal in the detector system's readout circuit.

Therefore, a fundamental difference between a light-emitting diode (LED) and a light-detecting diode lies in the direction of the applied bias voltage. An LED emits light when forward biased, while it can function as a light detector when reverse biased, operating similarly to a photodiode. This dual functionality allows a simple LED to be repurposed as a light detector by reversing the bias polarity. An experimental demonstration of this principle is shown in Fig. 2.10.

It is important to note that external radiation can also generate excess charge carriers in the p- and n-sides outside the depletion region. However, these carriers do not contribute to the drift current, as there is virtually no electric field in these regions to drive their motion. This occurs because the depletion region has a very high resistance, causing nearly the entire reverse bias voltage to drop across it. As a result, the electric field in the surrounding neutral regions is effectively zero

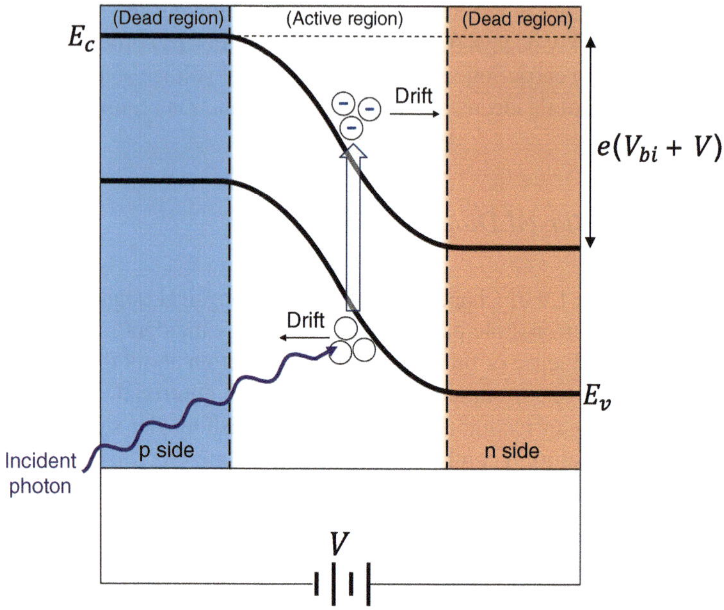

Fig. 2.9 Schematic illustration of the energy band diagram of a PN junction diode detector

(Fig. 2.2c). These areas outside the depletion region are therefore referred to as dead regions, and any energy deposited there by incident radiation is effectively lost.

So how should a diode detector be designed to avoid committing this "wasting-radiation crime?" The solution is to increase the % volume of the depletion region and reduce the % volume of the dead region. Let us assume the potential on the n-side of the reversely biased diode is V and the potential on the p-side is 0. Compared with the derivations of $\phi(x)$ in Eqs. (2.6)–(2.9), the boundary condition on the n-side changes from $\phi(x_n) = 0$ to $\phi(x_n) = V$, which results in a slight change of the formula of $\phi(x)$ on the n-side:

$$\phi(x) = \begin{cases} \frac{eN_a}{2\epsilon}\left(x + x_p\right)^2 & -x_p \leq x \leq 0 \\ -\frac{eN_d}{2\epsilon}\left(x - x_n\right)^2 + V & 0 < x \leq x_n. \end{cases} \tag{2.15}$$

Since the electrostatic potential $\phi(x)$ must be continuous across the junction, the solutions on either side must match at $x = 0$. This condition leads to the following equality:

$$\frac{eN_a}{2\epsilon}x_p^2 = -\frac{eN_d}{2\epsilon}x_n^2 + V. \tag{2.16}$$

Rearranging terms yields

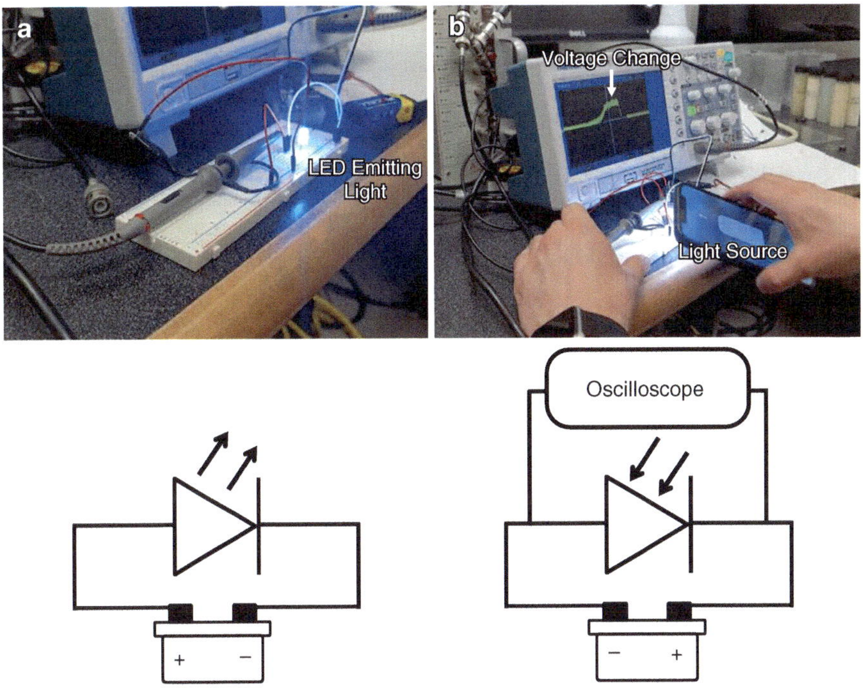

Fig. 2.10 (**a**) When an LED on a breadboard was forward biased using a battery, it emitted light, as expected. (**b**) Using the same circuit setup, the polarity of the bias voltage was reversed. When external light from a smartphone flashlight was directed onto the LED, a measurable voltage response was observed on the oscilloscope, clearly indicating the LED's ability to function as a light detector under reverse bias

$$N_a x_p^2 + N_d x_n^2 = \frac{2\epsilon V}{e}. \tag{2.17}$$

Applying the charge neutrality condition $N_a x_p = N_d x_n$, we substitute $x_p = \frac{N_d}{N_a} x_n$ into the equation above:

$$(N_d x_n) x_p + N_d x_n^2 = \frac{2\epsilon V}{e}. \tag{2.18}$$

Factoring out x_n, we obtain

$$(x_p + x_n) x_n = \frac{2\epsilon V}{e N_d}. \tag{2.19}$$

If the p-side of the PN junction is doped much more heavily than the n-side (i.e., $N_a \gg N_d$ (see Fig. 2.11), then from the charge neutrality condition $N_a x_p = N_d x_n$, it follows that $x_p \ll x_n$. In this case, the total depletion width becomes approximately

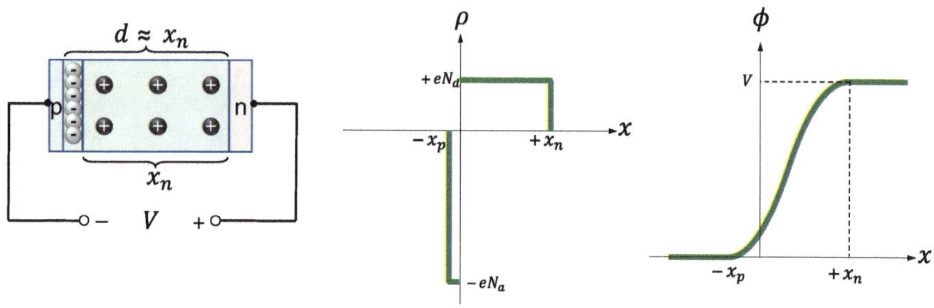

Fig. 2.11 If the p-side of the PN junction is heavily doped, the depletion region extends further into the n-side. The plot on the right shows the electric potential profile of this junction

$$(x_p + x_n)x_n \approx d^2 \approx \frac{2\epsilon V}{eN_d} \tag{2.20}$$

Substituting this into Eq. (2.19), we obtaini

$$d^2 \approx \frac{2\epsilon V}{eN_d}, \tag{2.21}$$

which leads to the widely used expression:

$$\boxed{d \approx \left(\frac{2\epsilon V}{eN_d}\right)^{1/2}}. \tag{2.22}$$

This result implies that the depletion width can be increased by:

1. Heavily doping one side of the junction (e.g., $N_a \gg N_d$)
2. Increasing the reverse bias voltage V

Both approaches reduce the extent of the dead region and expand the active (depleted) region where charge collection occurs.

As the reverse bias voltage increases, the depletion region eventually extends across nearly the entire thickness T of the high-purity side of the junction. This condition defines a fully depleted detector. From Eq. (2.22), the depletion voltage V_d—the minimum voltage required for full depletion—is given by

$$V_d = \frac{eN_d T^2}{2\epsilon} \tag{2.23}$$

In this configuration, the highly doped side is extremely thin, and the high-purity side is fully depleted, resulting in negligible dead region. This maximizes charge collection efficiency, which is critical for high-performance radiation detection.

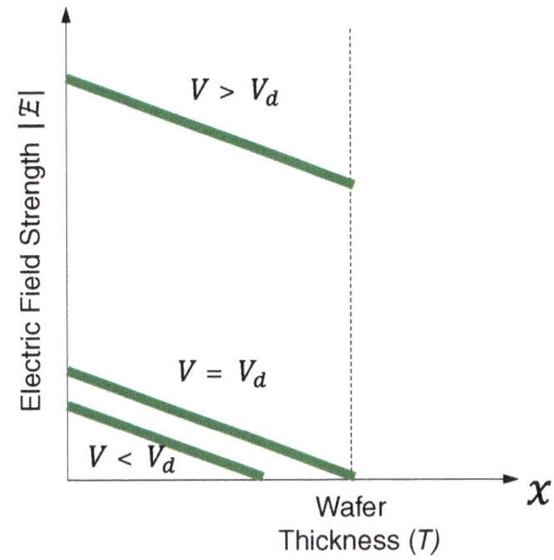

Fig. 2.12 The magnitude of the electric field, $|\mathcal{E}|$, in a reverse biased semiconductor diode detector is shown for three cases: when the applied bias voltage is below, equal to, and above the depletion voltage V_d

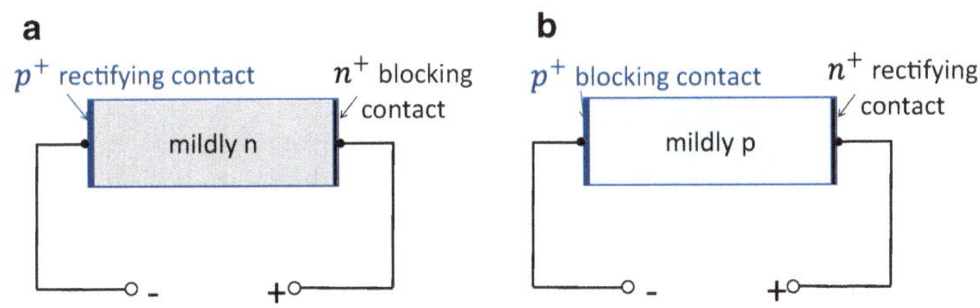

Fig. 2.13 Two configurations of planar semiconductor diode detectors

Once a semiconductor wafer is fully depleted, increasing the applied voltage further results in a uniform increase in the electric field throughout the wafer. At bias voltages significantly greater than the depletion voltage V_d, the electric field becomes nearly uniform across the entire wafer thickness (see Fig. 2.12). Under these conditions, the detector is said to be over-depleted [4].

Operating in the over-depleted regime offers several advantages, particularly the presence of a strong and uniform electric field throughout the active region, which enhances charge collection speed and efficiency. For this reason, most radiation detectors are intentionally over-depleted and operated at bias voltages well above V_d.

Figure 2.13 illustrates two configurations of semiconductor radiation detectors. To fully deplete a semiconductor wafer (e.g., silicon), fabrication typically begins with the highest-purity material available, either lightly n-type or lightly p-type. A *rectifying contact* is formed on one side of the wafer by creating a very thin, heavily doped layer of the opposite conductivity type, forming a PN junction. On the opposite side, a similarly thin and heavily doped layer of the same type is added to form a *blocking contact*, which serves to suppress leakage current by minimizing the injection of minority carriers across the junction.

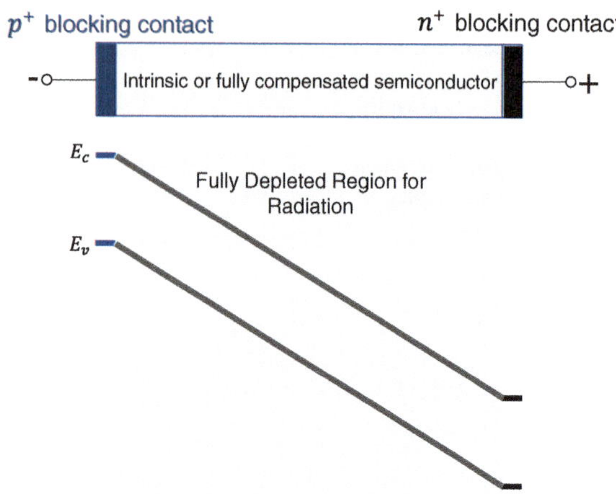

Fig. 2.14 Schematic of PIN diode detector configuration and its energy band diagram

2.3.2.1 PIN Diode Detector

Alternative to the two shown in Fig. 2.13, a semiconductor radiation detector can be built using an intrinsic or compensated semiconductor material. In this case, a heavily doped n-side is formed on one side of the wafer and connected to a positive bias voltage; a heavily doped p-side is formed on the opposite side of the wafer and connected to negative or zero bias (Fig. 2.14). Both of the two contacts serve as the blocking contacts. The electric field is uniform throughout the entire wafer, and the wafer is fully depleted even for very low values of applied voltage. This device is called a **p-i-n or PIN diode detector** (*i* stands for intrinsic).

An example of a compensated semiconductor material is chromium-doped GaAs (GaAs:Cr). Chromium acts as a deep acceptor in GaAs, introducing an energy level approximately 0.8 eV below the conduction band (i.e., near mid-gap). This deep level can trap free electrons, thereby compensating residual shallow donor impurities commonly present in GaAs. As a result, the Fermi level is pinned near the center of the bandgap, leading to a significant reduction in free carrier concentration. This yields extremely high resistivity (often exceeding $10^7\ \Omega \cdot \text{cm}$), which is highly desirable in radiation detector applications, as it minimizes leakage current and enhances detector performance.

2.3.3 Unbiased Diode Detector

To be clear, an *unbiased diode detector* does not refer to a detector that yields an *unbiased measurement* in a statistical sense. In fact, as we will show in this subsection, an unbiased diode detector can produce a biased estimate of the input radiation signal. Here, unbiased simply means that **no external bias voltage** is applied to the diode during operation.

Wait a moment—did we not argue in Sect. 2.3.2 that applying a reverse bias is beneficial because it increases the electric field strength and reduces the width of the dead region? So why would one want to use a diode detector without external bias?

To answer this, let us consider the example of a semiconductor solar panel. The primary goal of a solar panel is to harvest energy from incoming radiation (sunlight) and generate electricity, rather than to consume electrical power to produce a signal. In such cases, it is essential to *minimize power consumption*, ideally making the device **self-powered**.

As shown earlier in Fig. 2.3, even in the absence of an external bias voltage, a diode still possesses a built-in potential (approximately 0.7 V in a silicon diode), which creates an internal electric field. This field is sufficient to separate radiation-induced electron–hole pairs and drive their motion. While this internal field is weaker than that in a reverse biased diode—resulting in slower carrier drift and increased recombination losses—the benefit of being self-powered can outweigh these disadvantages in certain applications.

The solar panel is a prime example, where the trade-off favors low power consumption over high-speed carrier collection. Another example is radiation dosimetry in radiotherapy, where the incident X-ray energy fluence from medical linear accelerators is several orders of magnitude higher than in diagnostic imaging. In such high-radiation environments, the sheer number of charge carriers generated compensates for the slower carrier transport, allowing a usable signal to be obtained even with some recombination loss. Additionally, such recombination losses can be mathematically corrected when estimating the energy fluence or exposure level.

Due to the weak internal electric field in an unbiased diode, the drift velocity of excess charge carriers is low. This results in a longer duration of the output voltage pulse and a reduced signal amplitude (as discussed later in Chap. 3). As a consequence, unbiased diode detectors are generally not suitable for photon counting. Instead, they operate in charge (or energy) integrating mode, which is adequate—and often preferred—for radiation dosimetry applications [8].

In the absence of an external bias voltage, the depletion region remains relatively narrow, resulting in a larger dead region where photon interactions do not contribute to signal generation. However, for dosimetry purposes, photon loss in this dead region is not a significant concern. In fact, radiation dosimetry often requires low X-ray attenuation as it ensures minimal perturbation to the incident X-ray beam. This aligns with the requirements of cavity theory, which is used to convert measured exposure or charge to absorbed dose. Maintaining the integrity of the beam allows the dosimeter to more accurately represent the interaction conditions within human tissue or a reference medium.

With that said, the interaction of X-rays with the dead region introduces an issue: For a given total amount of input radiation, the accumulated output charge may depend on the input dose rate. Consider the n-side of the diode as an example (see Fig. 2.15). As the dose rate increases, more electron–hole pairs are generated per unit time, leading to increased occupancy of trap states (energy levels within the bandgap introduced by impurities or defects). As these traps become saturated, the probability of trap-assisted (indirect) recombination decreases.

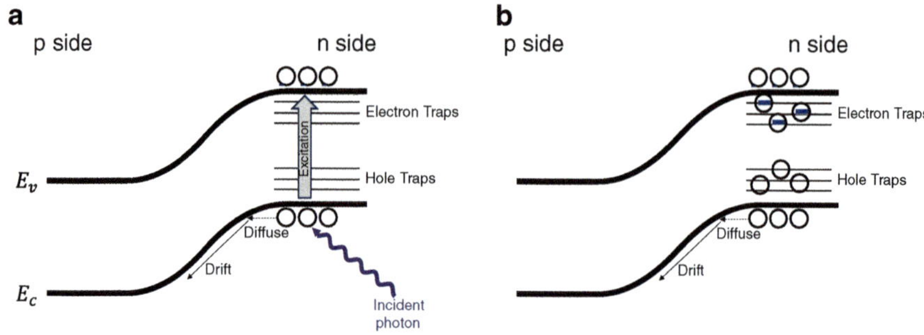

Fig. 2.15 (**a**) Energy band diagram showing the behavior of holes in an unbiased diode detector. Excess holes generated by incident photons on the n-side, within a distance shorter than the diffusion length from the depletion region, can diffuse into the depletion region, where they are further accelerated by the built-in electric field. (**b**) At high input photon flux, a significant number of trapping centers become occupied by excess charge carriers. This saturation reduces the probability of trap-assisted indirect recombination for newly generated carriers, thereby increasing their lifetime. As a result, more holes can reach the depletion region, leading to an enhanced diffusion current

Fig. 2.16 Photo of two Hamamatsu S1223 series silicon PIN photodiodes. Image courtesy of Hamamatsu Photonics K.K.

A reduction in recombination results in longer minority carrier lifetimes. As shown in Sect. 1.9.3, this enhances the diffusion current, thereby increasing the magnitude of the output signal.

2.4 Photodiode: A Subtype of Diode Detector

As described in the previous section, a reverse biased or zero biased diode can directly convert incident radiation into an electrical signal. This radiation includes both ionizing radiation (e.g., X-rays) and nonionizing radiation (e.g., visible light). When a diode is specifically designed to convert visible or ultraviolet into an electrical signal, it is referred to as a photodiode (Fig. 2.16).

In X-ray imaging, the goal is to completely absorb the incident X-ray photons in the sensor so that the resulting electrical signal accurately reflects the information carried by the incoming radiation. To achieve this using only a semiconductor diode, the sensor must be sufficiently thick along the direction of the incoming X-rays. For example, 2.2 mm of CdTe is required to stop 90% of 100-keV X-ray photons. However, increasing the thickness of the semiconductor increases material costs and requires a higher bias voltage to maintain the necessary electric field—both of which add to the overall system cost.

An alternative approach uses a scintillator material to first convert the X-rays into visible light, which is then detected by a photodiode. The scintillator does not require any bias voltage to scintillate. As long as the total energy of the emitted visible light is proportional to the energy of the incident X-ray photon, the essential information can still be recovered with good accuracy. The physics of scintillator materials will be discussed in Chap. 4.

What is the basic principle of photodiodes?

A photodiode is essentially a semiconductor diode used as a radiation detector.

Wait... Isn't visible light different from "radiation?"

Not really. Visible light *is* a form of electromagnetic radiation. Although it is **nonionizing**—meaning it does not carry enough energy to ionize atoms—it is still radiation. The energy of a typical scintillation photon is about 3–4 eV, which is higher than the bandgap of silicon (Si) and thus sufficient to generate an electron–hole pair.

Photodiodes used in medical imaging are typically constructed with a Si PIN diode structure and operated under reverse bias. Si, with a bandgap of 1.12 eV, is well-suited for detecting a broad range of scintillation light commonly encountered in these applications. To maximize light absorption in the intrinsic (i) region, the p-type or n-type layer facing the incident light is made as thin as possible (see Fig. 2.17). When photons are absorbed in the i-region, they generate electron–hole pairs. These charge carriers are rapidly separated and collected by the electric field created by the reverse bias, resulting in a current. This current is subsequently amplified by a connected preamplifier to produce a usable output signal.

Due to the significantly lower energy of visible light photons compared to X-ray photons, the active region of a photodiode used to detect light can be much thinner than that of a detector designed for direct ionizing radiation detection. Consequently, the voltage required to establish a given electric field strength across the photodiode is also much lower.

For a silicon photodiode, the spectral response generally spans from the near ultraviolet to the near infrared (Fig. 2.18). It reaches a broad maximum in the visible to near-infrared range (typically around 800–900 nm), where absorption in the depletion region is efficient and carrier collection is strong. Beyond this range, the response drops sharply at long wavelengths because photons with energies below the silicon bandgap ($\sim$1.12 eV) cannot generate electron–hole pairs. The exact shape of the spectral response curve depends on device structure, depletion depth, surface treatments, and anti-reflection coatings. The spectral response curve

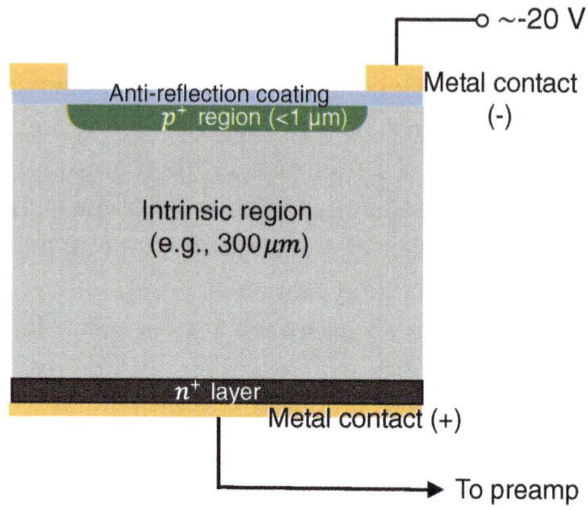

Fig. 2.17 Structure of a Si PIN photodiode

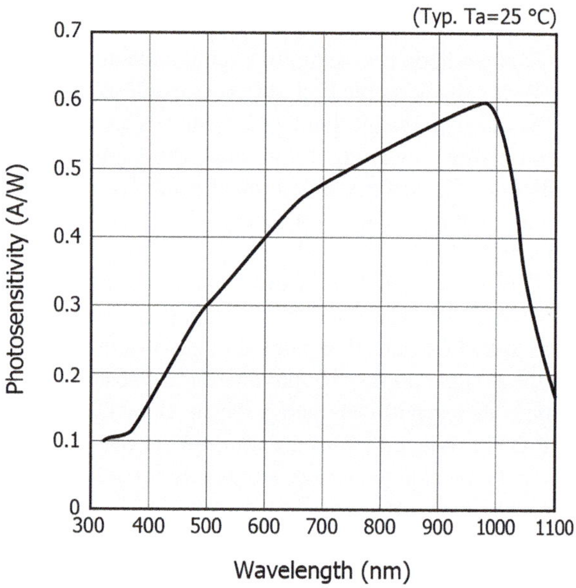

Fig. 2.18 Spectral response curve of the Hamamatsu S1223 series silicon PIN photodiode. Figure courtesy of Hamamatsu Photonics K.K.

is a key parameter for matching a photodiode to a given light source or scintillator, as it determines how efficiently photons at different wavelengths are detected.

When a 100-keV X-ray photon is absorbed by a scintillator, it typically produces only a few thousand visible light photons. These, in turn, can generate at most a few thousand electron–hole pairs in the photodiode. By contrast, in a direct-conversion semiconductor detector, the same 100-keV X-ray photon can produce tens of thousands of electron–hole pairs directly. The lower number of charge carriers in the scintillator–photodiode system results in a smaller induced current. As a result, electronic noise—from both the diode's leakage current and the readout amplifier—becomes a significant concern, particularly in single-photon counting mode. We will revisit this issue in more detail in Chap. 4.

2.5 Avalanche Photodiode: A Subtype of Photodiode

A green light photon emitted by a cesium iodide scintillator, with an energy of approximately 2.3 eV, can generate at most two electron–hole pairs in a silicon-based photodiode, given silicon's bandgap of 1.12 eV. In applications where a large number of photons are incident and the accumulated electrical signal is of interest—which is the case for most photodiode uses—the low number of charge carriers generated per photon is not a significant limitation. However, in applications requiring the detection of individual photons, this becomes a major challenge: The small number of photon-generated charge carriers can be comparable to or even smaller than those arising from thermally excited electron–hole pairs or background minority carriers, making single-photon sensing extremely difficult.

The small number of charge carriers (electron–hole pairs) generated by photon absorption in a conventional photodiode can be amplified through an avalanche process within a semiconductor subjected to a high electric field (typically on the order of 10^5 V/cm or higher). This process occurs in the avalanche breakdown region, as illustrated by the current–voltage characteristic of a PN junction in Fig. 2.5. In this region, charge carriers are accelerated by the strong electric field and gain sufficient kinetic energy to ionize other atoms through collision, thereby generating additional electron–hole pairs. This internal multiplication of carriers enhances the signal above the electronic noise floor, enabling improved energy resolution—an essential characteristic of photon-counting detectors. Typical avalanche gain factors range from ten to several hundred.

An avalanche photodiode (APD) is not necessarily a photodiode operated under a high reverse bias voltage; in fact, the required reverse bias can be as low as a few volts. Instead, the distinctive behavior of an APD arises from its internal structure, one common version of which is shown in Fig. 2.19. This structure consists of four thin semiconductor layers—n^+, p, π (a very lightly doped p-type layer), and p^+—typically made from materials such as silicon. The APD is reverse biased so that an electric field is established from the n^+ side to the p^+ side. A strong electric field develops at the interface between the p and n^+ layers, known as the *avalanche region*.

Incident light enters the APD through the thin n^+ and p layers and is primarily absorbed in the π region. Photon absorption in this region generates electron–hole pairs. The resulting electrons drift toward the n^+ side (the positive electrode or anode), while the holes drift toward the p^+ side (the negative electrode or cathode). As electrons enter the avalanche region, they are rapidly accelerated by the strong electric field, gaining enough kinetic energy to ionize additional atoms and produce secondary electron–hole pairs. Because the multiplication process is driven by the electric field strength, the internal gain of an APD is highly sensitive to the applied reverse bias voltage (Fig. 2.20).

The newly generated electrons continue drifting toward the n^+ contact, while the holes are swept toward the p^+ layer. At sufficiently high fields, these holes can

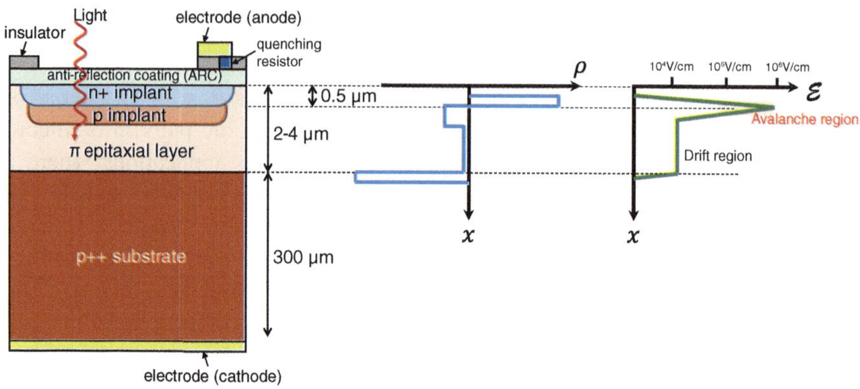

Fig. 2.19 Basic structure of a Si-based n-on-p avalanche photodiode (APD). A mild p-type (π-type) layer is epitaxially grown on a p++ substract, and then two neighboring regions within the expitaxial layer was doped to p-type and n+ type using the ion implantation technology. Due to the large gradient in change density at the interface between p and n+ layers, a strong electric field (e.g., 10^5 V/cm) is generated for charge multiplications. The π layer, known as the drift region, is where the majority of e-h was created by the incident light photons. Other APD structures such as p-on-n are reviewed in [7]

Fig. 2.20 Dependence of APD gain on reverse bias voltage for a Hamamatsu S17268 series device. Figure courtesy of Hamamatsu Photonics K.K.

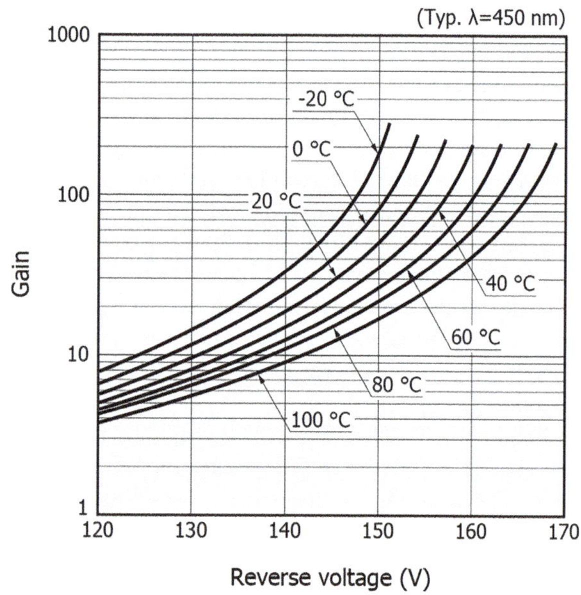

also initiate impact ionization, further contributing to the avalanche multiplication process.

Compared to conventional photomultiplier tubes (PMTs), which are not covered in this book but also provide high internal gain, APDs offer several advantages. These include insensitivity to magnetic fields, which is particularly important in PET detectors used in hybrid MR-PET systems, as well as favorable timing characteristics due to the fast rise time of the avalanche process. In optimized APD designs, rise time ranging from sub-nanosecond to a few nanoseconds can be achieved.

However, APDs also have certain disadvantages when compared to PMTs. Their gain is highly sensitive to both bias voltage and temperature, with a typical gain variation of approximately 2% per degree Celsius. Additionally, the stochastic nature of the avalanche multiplication process introduces extra noise. For example, consider a single photon that generates n initial electron–hole pairs. After avalanche multiplication, the total number of electrons collected at the output is N. Both n and N are random variables, and their means and variances are related as follows:

$$\bar{N} = \bar{n}\bar{g}; \tag{2.24}$$

$$\sigma_N^2 = \bar{g}^2\sigma_N^2 + \bar{n}\sigma_g^2, \tag{2.25}$$

where $\bar{g}$ and σ_g^2 denote the mean and variance, respectively, of the stochastic avalanche gain factor g.

Assuming that the number of input photons is Poisson-distributed with $\sigma_n^2 = \bar{n}$, according to Eqs. (2.24)–(2.25), the signal-to-noise ratio (SNR) of N is given by

$$\begin{aligned} \mathrm{SNR}_N &= \frac{\bar{N}}{\sigma_N} \\ &= \frac{\bar{n}\bar{g}}{\sqrt{(\bar{g}^2 + \sigma_g^2)\,\bar{n}}} \\ &= \frac{\sqrt{\bar{n}}}{\sqrt{1 + \frac{\sigma_g^2}{\bar{g}^2}}} < \sqrt{\bar{n}} = \mathrm{SNR}_n. \end{aligned} \tag{2.26}$$

Therefore, **the avalanche multiplication process actually degrades the SNR of the charge carrier number** due to the additional variance introduced by the stochastic gain factor g.

So why are APDs useful? The answer lies in the presence of electronic noise, σ_e^2, introduced by the readout circuit. Without any internal gain, the final SNR of the readout signal is

$$\begin{aligned} \mathrm{SNR}_{n'} &= \frac{\bar{n}}{\sqrt{\sigma_n^2 + \sigma_e^2}} = \frac{\bar{n}}{\sqrt{\bar{n} + \sigma_e^2}} \\ &= \frac{\sqrt{\bar{n}}}{1 + \frac{\sigma_e^2}{\bar{n}}}. \end{aligned} \tag{2.27}$$

With internal gain g, the final SNR of the readout signal becomes

$$\mathrm{SNR}_{N'} = \frac{\bar{N}}{\sqrt{\sigma_N^2 + \sigma_e^2}} = \frac{\bar{n}\bar{g}}{\sqrt{(\bar{g}^2 + \sigma_g^2)\bar{n} + \sigma_e^2}}$$
$$= \frac{\sqrt{\bar{n}}}{\sqrt{1 + \frac{\sigma_g^2}{\bar{g}^2} + \frac{\sigma_e^2}{\bar{n}\bar{g}^2}}}. \tag{2.28}$$

The condition under which $\mathrm{SNR}_{N'} > \mathrm{SNR}_{n'}$ is

$$\frac{\sigma_e^2}{\bar{n}} > \frac{\sigma_g^2}{\bar{g}^2} + \frac{\sigma_e^2}{\bar{n}\bar{g}^2}$$
$$\text{or equivalently,} \quad \frac{\bar{g}^2}{\sigma_g^2} > \frac{\bar{n}}{\sigma_e^2}. \tag{2.29}$$

Typically, σ_g^2 is proportional to $\bar{g}$, so the left-hand side of the inequality increases with increasing $\bar{g}$. When $\bar{g}$ becomes sufficiently large to satisfy this condition, the use of avalanche gain improves the overall SNR.

Proof of Eqs. (2.24)–(2.25)

Assume that each input quantum (in this case, a light photon) produces g output quanta (electrons), where g is a random variable. Given n input quanta, the total number of output quanta is

$$N = \sum_{i=1}^{n} g_i. \tag{2.30}$$

Note that the input events are mutually independent, so are the individual gain factors g_i. Let the probability mass function of the number of input quanta n be denoted by $P(n)$. The expected value of the total number of output quanta N can be calculated as

$$\bar{N} = \langle \sum_{i=1}^{n} g_i \rangle$$
$$= \sum_{n} P(n) \sum_{i=1}^{n} \langle g_i \rangle$$

(continued)

$$= \sum_{n} P(n) \sum_{i=1}^{n} \bar{g}$$
$$= \sum_{n} P(n) n \, \bar{g}$$
$$= \bar{n}\bar{g}.$$

By definition, the variance of N is given by

$$\begin{aligned}
\sigma_N^2 &= \langle (N - \bar{N})^2 \rangle \\
&= \langle N^2 \rangle - (\bar{N})^2 \\
&= \langle (\sum_{i=1}^{n} g_i)^2 \rangle - \bar{n}^2 \bar{g}^2 \\
&= \langle (n g_i^2 + n(n-1) g_i g_{j \neq i} \rangle - \bar{n}^2 \bar{g}^2 \\
&= \sum_{n} P(n)[n \langle g^2 \rangle + n^2 \bar{g}^2 - n \bar{g}^2] - \bar{n}^2 \bar{g}^2 \\
&= \bar{n} \langle g^2 \rangle + \langle n^2 \rangle \bar{g}^2 - \bar{n} \bar{g}^2 - \bar{n}^2 \bar{g}^2 \\
&= \bar{g}^2 (\langle n^2 \rangle - \bar{n}^2) + \bar{n} (\langle g^2 \rangle - \bar{g}^2) \\
&= \bar{g}^2 \sigma_n^2 + \bar{n} \sigma_g^2.
\end{aligned}$$

2.6 Silicon Photomultiplier: A Subtype of Avalanche Photodiode

The internal gain of an APD is highly dependent on the applied reverse bias voltage. When the voltage—and therefore the electric field in the avalanche region—is sufficiently high, the APD enters a Geiger-counting-like mode, in which the avalanche multiplication becomes self-sustaining. In this mode, the charge generated by the initial photon absorption is multiplied without bound, resulting in an effectively infinite internal gain. Regardless of how many electron–hole pairs are produced by the incoming photon, the APD produces a uniform, large output pulse. An APD operated in this regime is referred to as a **single-photon avalanche diode (SPAD)**. In contrast, an APD with a lower internal gain and produces an output signal proportional to the initial number of charge carriers is known as a linear APD.

Fig. 2.21 A photograph of SPAD cell arrays. Image courtesy of Hamamatsu Photonics K.K.

To detect ionizing radiation such as X-rays or gamma rays, a SPAD is typically coupled to a scintillator, which converts each high-energy photon into a burst of scintillation light photons. Because the number of scintillation photons is proportional to the energy of the incoming radiation, applications like positron emission tomography (PET) require not just photon detection but also accurate estimation of photon quantity. A single SPAD, however, is not suitable for this purpose, as it operates in binary mode and loses information about the initial scintillation light photon count.

To overcome this, many small SPAD cells—each typically sized around $50 \times 50\ \mu\text{m}^2$—are arranged in a 2D array (Fig. 2.21) to form a **silicon photomultiplier (SiPM)**, with a total active area typically around $10\,\text{mm}^2$ or larger [3, 4, 7]. The dimensions of each cell are small enough that, within a short measurement time window, at most one photon strikes any given SPAD cell. As a result, the number of SPAD cells that fire an avalanche is proportional to the number of incident scintillation photons. As illustrated in Fig. 2.22, by connecting all SPAD cells in parallel, the resulting analog output pulse has an amplitude proportional to the number of cells that registered an avalanche. This allows the SiPM to estimate the number of scintillation photons—and thus the energy of the incident gamma-ray photon. Consequently, SiPMs have become a popular photodetector in a variety of medical imaging applications, including time-of-flight PET (TOF-PET) systems.

Example

As we will learn in Chap. 4 (Basic Scintillator Physics), a typical scintillator material used in PET scanners is lutetium oxyorthosilicate (LSO), which on average produces 15,000 scintillation light photons from each absorbed

(continued)

511-keV gamma-ray photon. Among those scintillation light photons, approximately 2000 photons can be collected at the interface between the LSO crystal and the SiPM. The probability for each of the 39,984 SPAD cells in the SiPM shown in Fig. 2.22 to receive a light photon is only 0.05.

Compared with APDs, the timing properties of SiPM are even more preferable due to extremely fast and consistent pulse rise speed. As a single-cell SPAD, timing resolutions of tens of ps are achievable, while timing for arrays of SPADs (i.e., a SiPM) is often on the order of hundreds of ps (Fig. 2.23).

As mentioned at the beginning of this section, in an SPAD, the avalanche multiplication process is self-sustaining—once triggered, it can, in principle, continue indefinitely. This behavior is detrimental to radiation detection, where a continuous stream of input events must be detected independently. After each detection, the detector must quickly return to its initial state in order to register subsequent events without interference from previous ones.

To ensure that each avalanche is promptly terminated after the rising edge of the output pulse, a quenching circuit is required. This circuit halts the avalanche process and resets the SPAD for subsequent photon detection. As illustrated in Fig. 2.22, a common method of passive quenching involves connecting a resistor (typically typically a few hundred kΩ) in series with each SPAD cell. When an avalanche is triggered, the resulting surge in current causes a voltage drop across the quenching resistor. Since the total voltage between the power supply and ground is fixed, this increased voltage drop across the resistor leads to a corresponding decrease in the voltage across the SPAD itself. As the SPAD bias drops below the breakdown voltage, the internal electric field weakens, effectively quenching

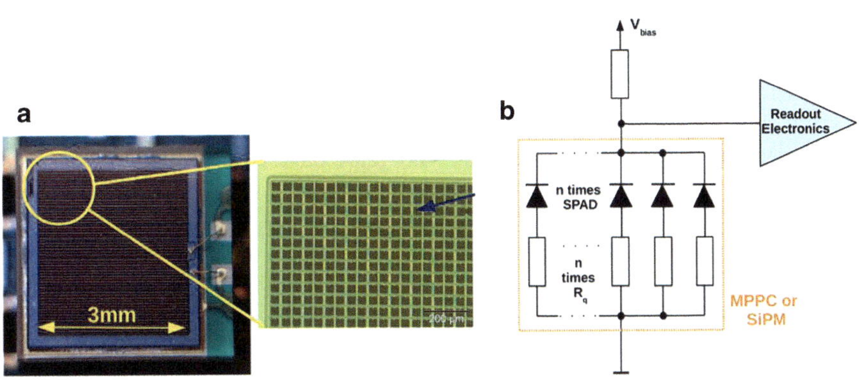

Fig. 2.22 (**a**) Picture of an analog SiPM with zoom on the individual SPADs (indicated by the arrow). (**b**) In an analog SiPM all SPADs with integrated quenching resistor are connected in parallel. This figure is taken from [3] with permission

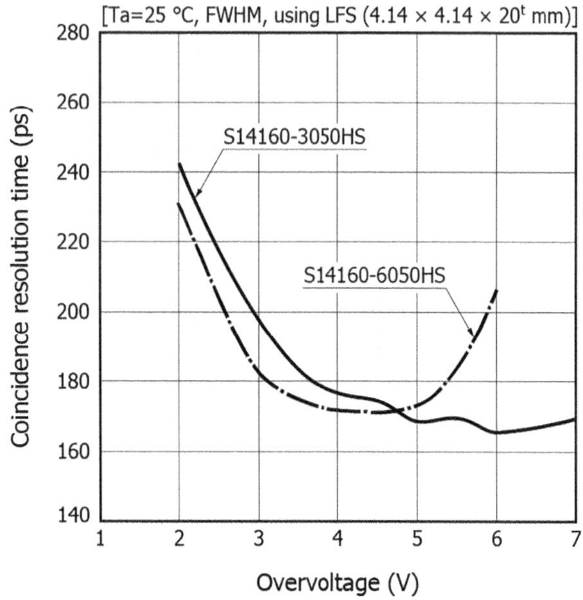

Fig. 2.23 Coincidence timing resolution of a Hamamatsu S14160 series SiPM. Figure courtesy of Hamamatsu Photonics K.K.

the avalanche. Once the avalanche current diminishes, the voltage across the SPAD gradually recovers back to a level above the breakdown threshold, thereby rearming the SPAD for the next detection event.

2.7 Semiconductor–Metal Contacts

In Sects. 2.3.1–2.6, we discussed how a voltage bias can be applied to a heterogeneous semiconductor device to form either a radiation sensor or a light source. The bias voltage is almost always delivered through a metallic conductor (e.g., Cu or Al wires). It is tempting to take for granted that when a metal is connected to a semiconductor, the interface does not impede current flow. This intuition comes from everyday experience: Since high school, we have learned that connecting a battery to a light bulb with metal wires is sufficient for current to flow.

However, when a metal meets a semiconductor, the situation becomes more subtle. Depending on the choice of metal and semiconductor, as well as the doping level and type of the semiconductor, the interface may present a barrier to current flow. Far from being a nuisance, this effect can be harnessed to design useful optoelectronic devices, including radiation detectors.

To understand the physical origin of such interface effects, we turn to the layman's version of the energy band diagram shown in Fig. 2.24a, where the horizontal axis represents spatial position instead of momentum. Two key physical quantities are needed:

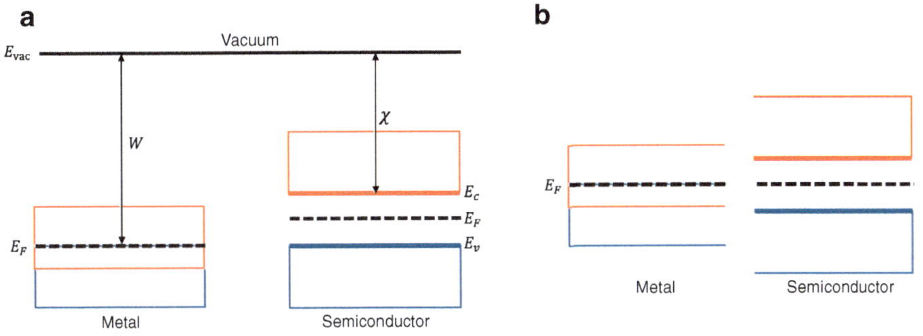

Fig. 2.24 (**a**) Definition of the work function (W) and electron affinity (χ) in solids. (**b**) When a metal and a semiconductor are brought into contact and reach thermal equilibrium, the Fermi level becomes uniform throughout the system outside the immediate interface region

Table 2.1 Work function (W) and electron affinity (χ) of some metals and semiconductors

Metal	W (eV)	Semiconductor	χ (eV)
Al	4.3	Si	4.0
Ag	4.3	Ge	4.1
Au	5.1	GaAs	4.0
Pt	5.7	CdTe	4.3
In	4.1	$Cd_{0.9}Zn_{0.1}Te$	4.3
Ti	4.3	a-Si	3.8
W	4.6	a-Se	4.3

- **Work function** (W): The minimum energy required to remove an electron from the Fermi level (E_F) of a material to the vacuum level just outside its surface. In other words, W is the energy difference between the vacuum level and the Fermi level:

$$W = E_{\text{vac}} - E_F. \tag{2.31}$$

- **Electron affinity** (χ): For a semiconductor, χ is the energy difference between the vacuum level and the conduction band minimum (E_c). Using the same convention,

$$\chi = E_{\text{vac}} - E_c. \tag{2.32}$$

Representative values of W and χ for several common metals and semiconductors are given in Table 2.1.

In analyzing metal–semiconductor or semiconductor–semiconductor junctions, a powerful general principle is that in thermal equilibrium, the Fermi level must be constant throughout the system. As shown schematically in Fig. 2.24b, even if the Fermi levels of two materials differ before contact, they must align after contact.

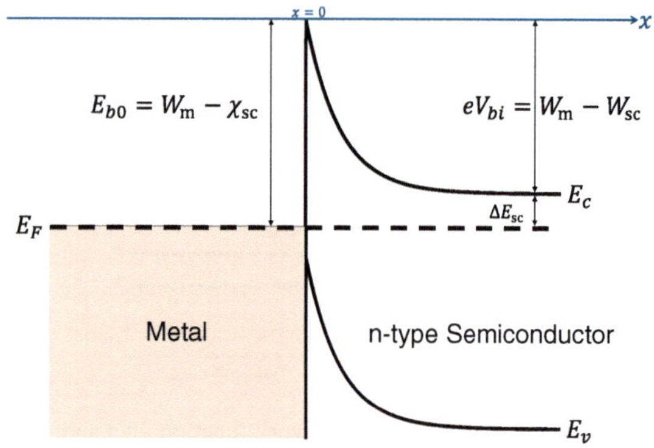

Fig. 2.25 Energy band diagram for an n-type semiconductor–metal Schottky junction with $W_m > \chi_{sc}$. No external bias voltage is applied

This principle forms the basis for analyzing Schottky barriers, ohmic contacts, and related interface phenomena.

2.7.1 Schottky Contact

As illustrated in Fig. 2.25, an n-type semiconductor is in contact with a metal. In n-type semiconductor, the Fermi level lies closer to the conduction band; here we assume it is located ΔE_{sc} below E_c. The effective work function of this doped semiconductor is therefore

$$W_{sc} = \Delta E_{sc} + \chi_{sc}. \tag{2.33}$$

Assume that this value is smaller than the work function of the metal, W_m. Consequently, electrons in the Si conduction band are at a higher energy than those in the metal. Consequentially, some electrons from the semiconductor flow into the metal, leaving behind immobile positive charges near the interface. This produces a space-charge region on the Si side and hence an electric field directed from the semiconductor toward the metal. Inside the metal, however, the redistributed electrons remain confined to the interface, so the electric field within the bulk of the metal is zero.

The redistribution of charge shifts the potential energies: On the semiconductor side, the potential energy of electrons decreases, while on the metal side it increases, until the Fermi levels align. In terms of the band diagram, the Si bands bend downward relative to the metal until thermal equilibrium is reached, i.e., the Fermi energy becomes continuous across the system. The potential difference is

$$\begin{aligned} eV_{bi} &= W_m - W_{sc} \\ &= W_m - \chi_{sc} - \Delta E_{sc}. \end{aligned} \tag{2.34}$$

At the interface, the metal must remain an equipotential, so the energy diagram on the metal side is flat. On the semiconductor side, by contrast, the positive space charge is distributed over a finite depletion width. Thus, at $x = 0$ (the metal–semiconductor boundary), there is an abrupt change in charge density that gives rise to the band bending. Moving further into the Si, the band bending gradually reduces the Fermi-level difference until equilibrium is reached.

The final band diagram is shown schematically in Fig. 2.25: A potential energy barrier, known as the **Schottky barrier**, is formed. If one measures from the conduction band edge on the Si side, the barrier height is the built-in potential eV_{bi} given in Eq. (2.34).

Alternatively, if referenced to the Fermi level on the metal side, the barrier height is

$$E_{b0} = W_\mathrm{m} - \chi_\mathrm{sc}. \tag{2.35}$$

For n-type semiconductors, if $E_{b0} > 0$ or, equivalently,

$$W_\mathrm{m} > \chi_\mathrm{sc}, \tag{2.36}$$

then the metal–semiconductor junction is called a **Schottky junction**. It exhibits a rectifying effect similar to that of a p-n junction:

1. **Reverse bias:** When a positive voltage V_R is applied to the semiconductor relative to the metal, the barrier height increases to

   $$e(V_{bi} + V_R). \tag{2.37}$$

   This increased barrier prevents the majority carriers in the semiconductor (electrons) from flowing into the metal. Only a very small leakage current flows, primarily due to minority carriers (holes) in the semiconductor.
2. **Forward bias:** When a positive voltage V_F is applied to the metal relative to the semiconductor, the barrier height is reduced to

   $$e(V_{bi} - V_F). \tag{2.38}$$

   This reduction allows electrons in the semiconductor conduction band to surmount the barrier by thermionic emission, or to tunnel through it, producing a forward current into the metal.

For p-type semiconductors, the condition to form a Schottky junction is different. Because the majority carriers are holes, we must consider the energy diagram in terms of the positive charges. A potential barrier for hole flow is formed when

$$W_\mathrm{m} < \chi_\mathrm{sc}. \tag{2.39}$$

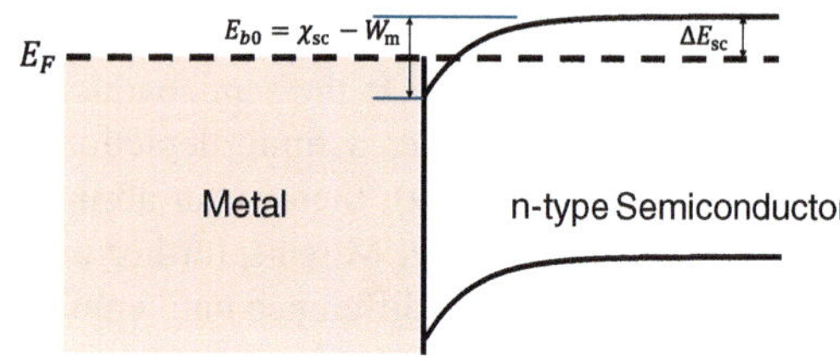

Fig. 2.26 Energy band diagram for an n-type semiconductor–metal junction with $W_m < \chi_{sc}$. This forms an ohmic contact

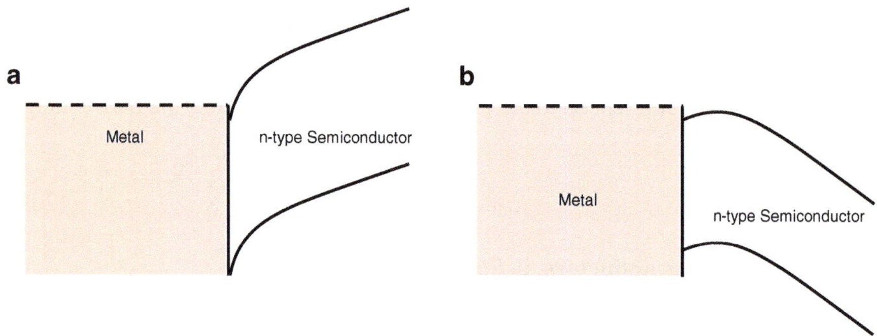

Fig. 2.27 Energy band diagram for an ohmic contact. (**a**) A positive voltage is applied to the metal side. (**b**) A positive voltage is applied to the semiconductor side

As discussed in Sect. 1.9.5, CdTe used for radiation detection is typically p-type. From the electron affinity and work function values in Table 2.1, an **In–CdTe contact** satisfies this condition and therefore can form a Schottky junction. When reverse biased, such a configuration can serve as a **Schottky radiation detector**.

2.7.2 *Ohmic Contact*

For n-type semiconductors, if

$$W_m \leq \chi_{sc}, \tag{2.40}$$

when the semiconductor makes contact with the metal, electrons from the metal flow into the semiconductor to establish thermal equilibrium and a uniform Fermi level (Fig. 2.26). This effectively makes the semiconductor more n-type. When a positive voltage is applied to the metal (Fig. 2.27a), there is no barrier to electrons flowing from the semiconductor into the metal. When a positive voltage is applied to the semiconductor (Fig. 2.27b), the barrier for electrons from the metal is ΔE_{sc}, which is very small, especially for moderately to heavily doped semiconductors. Electrons from the metal can easily overcome this small barrier to flow into the semiconductor. As a result, the contact exhibits negligible rectification and behaves as a conventional **ohmic contact**.

For p-type semiconductors, if

$$W_{\mathrm{m}} \geq \chi_{\mathrm{sc}} + E_g, \tag{2.41}$$

when the semiconductor makes contact with the metal, electrons flow from the semiconductor into the metal to achieve thermal equilibrium, leaving behind holes that increase the hole concentration near the interface. When a positive voltage is applied to the semiconductor, electrons from the metal can readily flow into the semiconductor, equivalent to holes moving from the semiconductor into the metal. Conversely, when a positive voltage is applied to the metal, the small barrier for holes to move from the semiconductor into the metal can be easily surmounted, especially in moderately to heavily doped p-type semiconductors. Therefore, the contact does not significantly rectify current and behaves as an **ohmic contact**.

Example: From Table 2.1, the work function of Pt is very close to $\chi_{\mathrm{sc}} + E_g$ for CdTe. Therefore, under typical doping conditions, a **Pt–CdTe contact behaves essentially as an ohmic contact**.

2.8 Transistors in Radiation Detectors

In addition to the PN junction, another fundamental semiconductor device used in digital radiation detectors is the transistor. Transistors are versatile components that, when integrated with other circuit elements, can perform a wide range of functions including signal amplification, charge storage, electrical switching, and digital logic operations. A transistor can even serve directly as a radiation sensor. These diverse capabilities make transistors indispensable in both analog front-end circuits (such as preamplifiers) and digital readout electronics.

We begin our discussion with the most classical form of transistor—the bipolar junction transistor—which was invented in 1947 by William Shockley, John Bardeen, and Walter Brattain at Bell Laboratories [1]. This invention marked a pivotal moment in the history of electronics, laying the foundation for modern semiconductor technology. In recognition of their groundbreaking work, all three scientists were awarded the Nobel Prize in Physics in 1956.

2.8.1 Bipolar Junction Transistor

A bipolar junction transistor (BJT) consists of three separately doped semiconductor regions and two closely spaced PN junctions, positioned such that their electrical behavior is interdependent. It is called a *bipolar* transistor because the operation involves the movement of **both electrons and holes** as charge carriers—unlike unipolar devices (such as field-effect transistors), which rely primarily on one type of carrier.

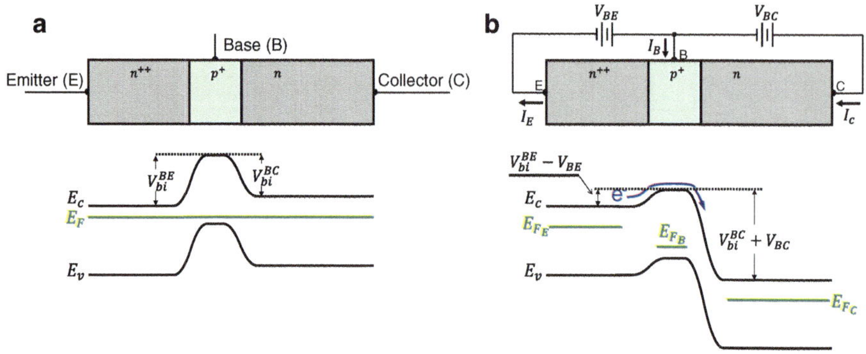

Fig. 2.28 An NPN bipolar transistor (**a**) under zero bias and (**b**) under forward-active mode bias. V_{bi}^{BE} denotes the built-in potential barrier between the base and emitter

The three regions can be arranged in two configurations: either two n-type regions separated by a thin p-type region, forming an NPN transistor, or two p-type regions separated by a thin n-type region, forming a PNP transistor. In this subsection, we will use the NPN configuration as the primary teaching example.

As illustrated in Fig. 2.28, an NPN transistor consists of [6]:

- A very heavily doped n-type region (n^{++}) called the **emitter (E)**
- A moderately doped p-type region (p^{+}) called the **base (B)**
- A lightly doped n-type region called the **collector (C)**

Typical doping concentrations for the emitter, base, and collector are approximately 10^{19}, 10^{17}, and $10^{15}\,\text{cm}^{-3}$, respectively. Due to this asymmetrical doping profile, an NPN or PNP transistor is not structurally symmetric, despite the apparent symmetry in its name.

The working principles of an NPN bipolar transistor are outlined as follows:

1. As discussed in Sect. 2.2 (PN Junction), a built-in potential barrier V_{bi} naturally forms at each PN junction due to the space charge within the depletion region. Accordingly, a potential barrier V_{bi}^{BE} exists between the base and emitter regions, and another barrier V_{bi}^{BC} exists between the base and collector regions (Fig. 2.28a).
2. If a power supply is connected such that its positive terminal is applied to the collector and its negative terminal to the base, the collector–base junction becomes reverse biased. Under this condition, no significant current flows, and the transistor remains in the OFF state, capable of storing charge (e.g., collected from a photodiode).
3. Introducing a second power supply with its positive terminal connected to the base and its negative terminal to the emitter causes the base–emitter junction to become forward biased. This forward bias allows a large number of electrons to be injected from the heavily doped emitter (n^{++}) into the left edge of the base (p^{+}). Meanwhile, because the base–collector junction remains reverse biased,

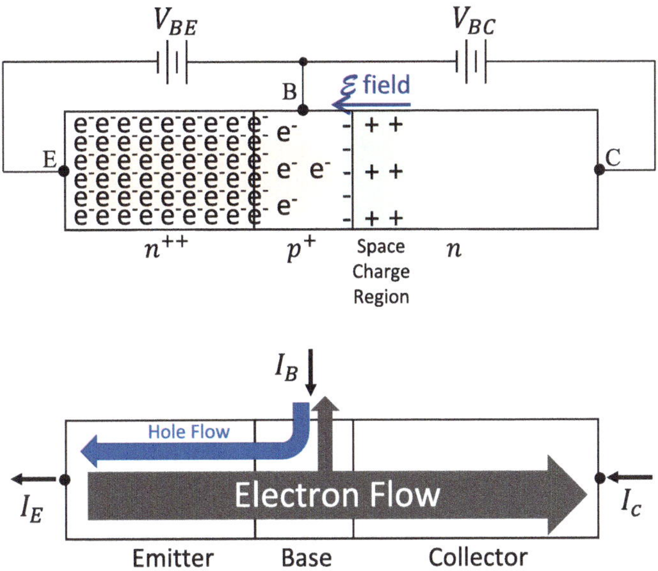

Fig. 2.29 When B-E junction is forwardly biased and B-C junction is reversely biased, electrons diffuse from emitter into base. 98% of them reach the collector, while 2% go to the base electrode. The holes in the p^+ region also form a small current by diffusing into the n^{++} region

electrons are effectively swept away from the right edge of the base as shown in Fig. 2.29, preventing accumulation there.

4. Due to the resulting electron concentration gradient across the base, electrons diffuse from the left to the right side of the base. Once they enter the depletion region of the base–collector junction, they are rapidly swept into the collector by the electric field. This transport of electrons from emitter to collector constitutes the collector current I_C, marking the transistor's transition to the ON state.
5. The effective potential barrier at the base–emitter junction is given by $V_{bi}^{BE} - V_{BE}$. Increasing V_{BE} reduces this barrier, allowing more electrons to be injected into the base. This enhances the electron gradient and increases the diffusion current. As a result,

$$I_C \propto \exp\left(\frac{eV_{BE}}{k_B T}\right). \tag{2.42}$$

Namely, **the current at the collector, I_C, is a strong function of V_{BE}.**
6. While the collector–base voltage V_{BC} influences the electric field strength in the base–collector depletion region (affecting how quickly electrons are swept into the collector), it does not control how many electrons are injected into the base from the emitter. Therefore, I_C **has negligible dependence on V_{BC}.**
7. Since the base–emitter junction is forwardly biased, holes from p^+ flow into n^{++} and form a current, I_B^{hole}. When electrons from n^{++} flow into p^+, the majority

of them ($\approx$ 98%) diffuse toward the collector in the n region; a small fraction (2%) of the electrons diffuse toward the base electrode, forming another current $I_B^{electron}$. The total current at the base is given by (Fig. 2.29)

$$I_B = I_B^{hole} + I_B^{electron}. \tag{2.43}$$

The emitter current is the sum of the collector current and the hole current:

$$I_E = I_B^{hole} + I_C. \tag{2.44}$$

Since the hole concentration in the p^+ base is much lower than the electron concentration in the n^{++} emitter, the resulting hole current into the emitter, I_B^{hole}, is much smaller than the collector current I_C, i.e.,

$$I_B^{\text{hole}} \ll I_C. \tag{2.45}$$

Additionally, the portion of electrons that recombine in the base and contribute to the base current, I_B^{electron}, is also negligible compared to I_C. Therefore, the total base current is much smaller than the collector current:

$$\boxed{I_B \ll I_C} \tag{2.46}$$

The ratio of the collector current to the base current defines the common-emitter current gain, denoted β:

$$\beta = \frac{I_C}{I_B}. \tag{2.47}$$

In practice, β typically ranges from 20 to 300, depending on the transistor design and operating conditions.

2.8.2 *Metal–Oxide–Semiconductor Field Effect Transistor (MOSFET)*

Similar to bipolar junction transistors (BJTs), a MOSFET (metal–oxide–semiconductor field-effect transistor) can also provide signal amplification and charge storage functionalities. Traditionally, the term "oxide" in MOSFET specifically refers to silicon dioxide (SiO_2), which serves as the insulating layer between the gate and the semiconductor body. More broadly, this type of device is also referred to as a metal–insulator–semiconductor field-effect transistor (MISFET), emphasizing the general structure rather than the specific silicon dioxide material.

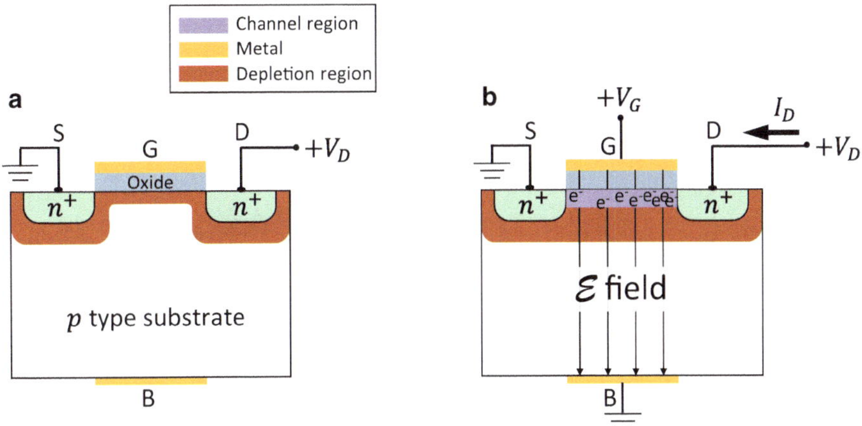

Fig. 2.30 Basic structure of an n-channel MOSFET. (**a**) Without any voltage applied to the gate. (**b**) With a gate voltage, V_G, a channel region is created between the two n^+ regions to enable the flow of electrons and the formation of a current I_D

As illustrated in Fig. 2.30, a MOSFET is typically fabricated on a p-type silicon substrate and consists of four terminals: The **gate (G)** and **body (B)** lie vertically aligned across the top and bottom of the device. The gate is formed by a conducting metal layer, which is electrically insulated from the semiconductor body by a thin layer of silicon dioxide. On either side of the gate oxide are two heavily doped n^+ regions, which form the **source (S)** and **drain (D)** terminals. The region of p-type silicon located underneath the gate oxide, between the source and drain, is referred to as the **channel region**.

The basic operating principles of the MOSFET are summarized as follows:

1. In Sect. 2.2 (PN junction), we discussed how a depletion region forms at each $n^+ - p$ junction, inhibiting charge flow across the junction. This depletion region extends more deeply into the p-type material, as described by Eq. (2.2). When a positive voltage is applied to the drain, the drain–body $(n^+ - p)$ junction becomes reverse biased, causing its depletion region to expand. Eventually, it connects with the depletion region at the source–body junction, as illustrated in Fig. 2.30a. In this state, the entire channel region between the source and drain is depleted of mobile charge carriers, and no current flows between the drain and source.
2. Now consider what happens when a positive voltage, $V_G > 0.5$ V, is applied to the gate. When a metal–oxide–metal structure is formed—such as the gate (metal), oxide (SiO_2), and body (semiconductor)—it behaves like a capacitor capable of storing electric charge. With the gate positively biased, free electrons are pulled away from the gate electrode and accumulate on the body-side electrode, creating an electric field directed from gate to body, as shown in Fig. 2.30b. This electric field repels the majority carriers (holes) in the p-type substrate deeper into the body and attracts the minority carriers (electrons)

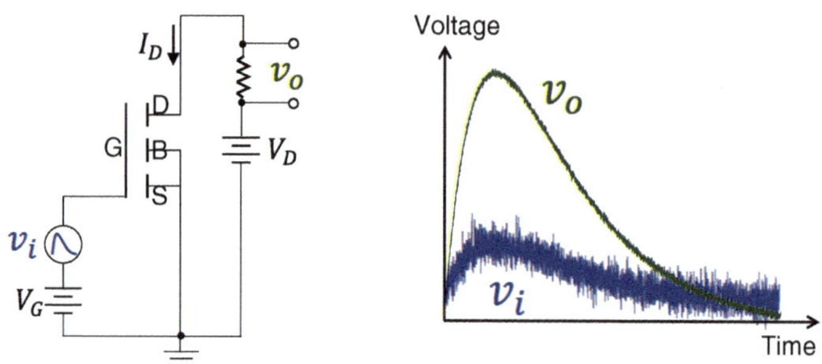

Fig. 2.31 Basic structure of an n-channel MOSFET. (**a**) Without any voltage applied to the gate. (**b**) With a gate voltage, V_G, a channel region is created between the two n^+ regions to enable the flow of electrons and the formation of a current I_D

toward the surface beneath the gate oxide. As a result, an inversion layer or **conductive channel of electrons forms in the region beneath the gate, bridging the source and drain.** This enables current flow between the source and drain. The number of electrons in the channel—and thus the magnitude of the drain current, I_D—depends on the strength of the electric field, which is controlled by the gate voltage V_G.

3. The term "field effect" in MOSFET stands for the control of current (I_D) flowing from the drain to the source via an electric field that is perpendicular to the direction of current flow. In the configuration shown in Fig. 2.30b, the drain current is carried by **electrons only**, so the device is called an **n-channel MOSFET**. Alternatively, if the device is fabricated on an n-type substrate with two p^+ terminals for the source and drain, the resulting device is a **p-channel MOSFET**, where **only holes** carry the current.
4. Because the drain current I_D responds strongly and nonlinearly to changes in the gate voltage V_G, a MOSFET can be used to **amplify small signals**: A small change in V_G results in a much larger change in I_D. This principle is widely used in analog circuits such as preamplifiers, which will be discussed in more detail in Sect. 2.9. An example of a MOSFET amplifier circuit is illustrated in Fig. 2.31.

MOSFETs offer several advantages over bipolar junction transistors (BJTs), especially in modern electronics and radiation detector systems:

- Lower power consumption: MOSFETs are voltage-controlled devices, requiring virtually no input current to the gate during steady-state operation (due to the insulating oxide layer). In contrast, BJTs are current-controlled devices, requiring a continuous base current, which leads to higher power consumption.
- High input impedance: The oxide layer between the gate and the channel in a MOSFET provides extremely high input impedance (in the range of MΩ to GΩ), minimizing loading on preceding circuit stages or minimizing dark current when used as a radiation dosimeter.

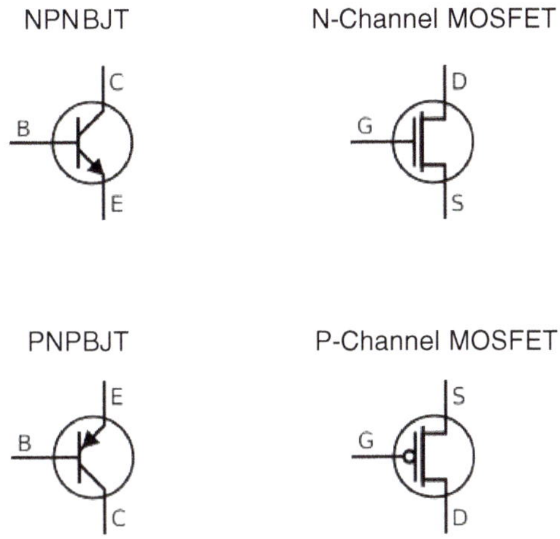

Fig. 2.32 Symbols of BJTs and MOSFETs

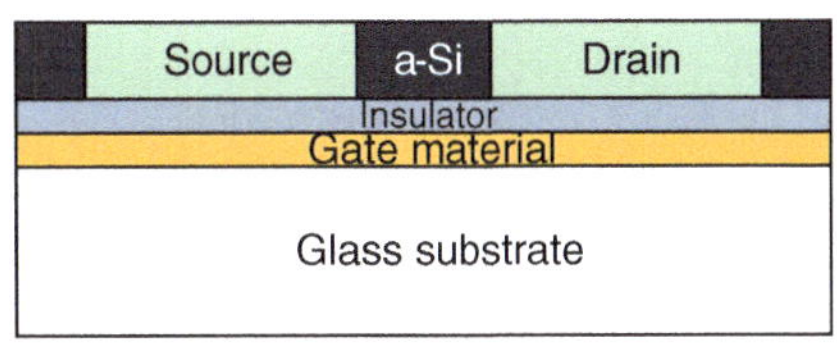

Fig. 2.33 Basic structure of a TFT

- Better scalability: MOSFETs are easier to scale down in size, making them more suitable for very-large-scale integration (VLSI) and modern microelectronics.
- Higher switching speeds: MOSFETs switch faster than BJTs due to the absence of minority carrier storage (which slows down BJTs during turn-off). This makes MOSFETs ideal for high-speed digital circuits and fast pulse applications.

Figure 2.32 illustrates representative symbols for BJTs and MOSFETs, with their input and output terminals labeled.

2.8.3 Thin-Film Transistor (TFT): A Subtype of MOSFET

At first glance, the term thin-film transistor (TFT) might suggest that it represents a distinct class of transistors, separate from BJTs and MOSFETs. However, this is not the case. A TFT is actually a specialized subtype of MOSFET, characterized by its method of fabrication. In essence, it is "a MOSFET fabricated by sequential deposition of thin films"—a description that is too lengthy for everyday use, hence the abbreviated term TFT.

TFT fabrication typically begins with a glass substrate, as shown in Fig. 2.33. Using techniques such as chemical vapor deposition (CVD) or physical vapor deposition (PVD), thin layers of materials, such as amorphous silicon (a-Si), insulating oxides, and conductive films like gold or indium tin oxide (ITO), are

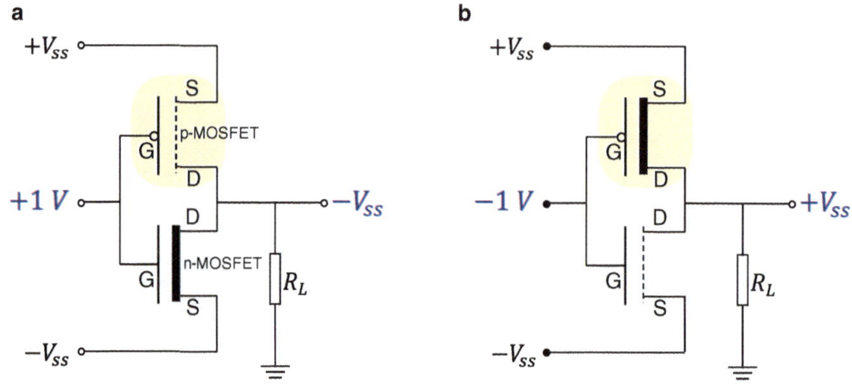

Fig. 2.34 Two logical states of a CMOS transistor. (**a**) When a small positive voltage is applied to the gate, the CMOS produces a negative output voltage. (**b**) When a small negative voltage is applied to the gate, the CMOS produces a positive output voltage

deposited in sequence. Combined with standard microfabrication processes such as photolithography and reactive ion etching, this approach enables the construction of large arrays of TFTs. The minimum feature size of each TFT is limited by the resolution of micro- and nanofabrication technology. As of 2025, feature sizes below 5 nm are now achievable in research settings. This enables the integration of billions to trillions of TFTs on a $1 \times 1\,\text{cm}^2$ chip.

Because TFTs are deposited layer-by-layer on glass rather than being grown from single-crystal Si wafers, they can be manufactured over large areas at low cost. This scalability and cost-effectiveness are major reasons why a-Si TFT-based flat-panel digital detectors have become widely adopted in diagnostic radiology and image-guided radiotherapy.

2.8.4 Complementary Metal–Oxide–Semiconductor (CMOS) Transistor

The key concept in CMOS (complementary metal–oxide–semiconductor) technology lies in the word "complementary." As shown in Fig. 2.34, each n-channel MOSFET is paired with a p-channel MOSFET, forming **a complementary pair**. According to the operating principles illustrated in Fig. 2.30, when a positive gate voltage (V_G) is applied, the n-MOSFET turns on (conducts) while the p-MOSFET turns off (nonconducting), resulting in an output voltage of $V_{\text{out}} = -V_{SS}$. Conversely, when a negative or low gate voltage is applied, the p-MOSFET conducts while the n-MOSFET is off, and the output voltage becomes $V_{\text{out}} = +V_{SS}$. Since V_{SS} can be much larger than the input signal V_G, this configuration allows for **voltage amplification**.

One of the most important advantages of CMOS technology is its **extremely low static power consumption**. Due to the complementary operation, for any given logic state or input, either the n-MOSFET or the p-MOSFET is always off, which prevents a direct conductive path between the supply voltages ($+V_{ss}$ and $-V_{ss}$). As a result, no static current flows, leading to **minimal power dissipation and heat generation.** Compared to TFTs and other transistor types that have inherent leakage or standing current, CMOS circuits also exhibit **significantly lower thermal noise**.

CMOS devices are typically fabricated on single-crystal Si wafers rather than on a-Si panels. This is due to two main reasons:

1. Ion implantation, a key step in CMOS fabrication, can easily form both n^+ and p^+ wells in crystalline Si, enabling high-precision doping profiles.
2. Hole mobility in a-Si is extremely low (see Fig. 1.5), making it impractical to fabricate both n- and p-channel MOSFETs in the same a-Si substrate.

Because single-crystal silicon wafers are smaller and more expensive than large-area glass substrates, the physical area of a single CMOS chip is typically limited. Nevertheless, the exceptional electrical performance, scalability, and energy efficiency of CMOS make it the dominant technology for both analog and digital integrated circuits.

In applications such as large-area medical imaging detectors, where extended coverage is required, multiple CMOS chips can be tiled or stitched together to form a larger active area. This modular approach enables the use of CMOS technology even in systems that demand high spatial resolution and low electronic noise across wide fields of view.

2.9 Semiconductor Preamplifier (Preamp)

As will be discussed in Chap. 3, except when an avalanche detector is used, the native electrical signal induced by external radiation in a radiation detector is extremely weak. It is too weak to be transmitted over long distances without experiencing significant degradation.

Signal transfer lines—whether traces in an integrated circuit or external cables—introduce electronic noise for several reasons:

- Electromagnetic interference: Signal lines can behave like antennas, unintentionally picking up electromagnetic noise from the surrounding environment. The weaker the original signal from the sensor, the more vulnerable it is to corruption by even small amounts of electromagnetic interference.
- Thermal noise: Materials generate random thermal voltage fluctuations due to the motion of charge carriers. As shown in Chap. 1, the number of charge carriers depends on temperature. Longer transfer lines have higher resistance, which contributes more thermal noise and may also result in increased heat generation, further exacerbating the effect.
- Capacitive and inductive coupling: Signal lines that run near other conductors or circuit elements can form unintended capacitive or inductive couplings. High-frequency noise can easily couple through these paths, especially in lines

with high parasitic capacitance or inductance. This makes it easier for nearby interference to affect the signal integrity.

In summary, the longer the transfer line, the more opportunity it has to accumulate noise. To mitigate this, a **preamplifier (preamp)** is placed in close physical proximity to the radiation sensor. The preamp boosts the weak signal by several orders of magnitude, raising it to a level that can be transmitted over longer distances with minimal degradation and noise interference.

To better understand the importance of preamplification, consider a microphone–speaker audio system. A microphone produces a very weak voltage signal. If this signal is sent directly through a cable to a power amplifier, the amplifier will amplify not only the voice signal but also any noise picked up along the cable. This results in unwanted hiss or hum from the speakers.

Therefore, it is essential to amplify the microphone signal locally—typically inside microphone housing—up to about 1 volt, commonly referred to as line level. This stronger signal is much less susceptible to noise during transmission. The power amplifier can then take the clean, line-level signal and drive the speakers efficiently.

2.9.1 Voltage-Sensitive Preamplifier

Figure 2.35 shows a simple toy-model voltage-sensitive preamplifier (VSA) based on a single NPN silicon BJT. The voltage source V_{cc} (e.g., 10 volts) provides DC

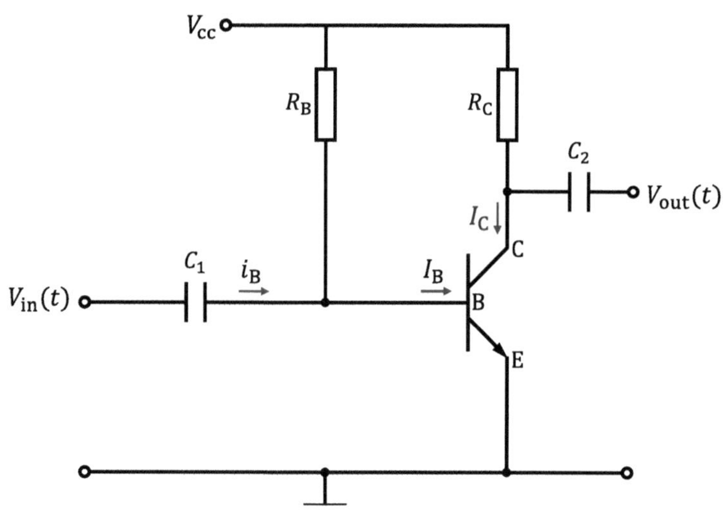

Fig. 2.35 Circuit diagram of a single NPN-based voltage-sensitive (pre)amplifier (VSA)

power to the collector (C) of the BJT through the resistor R_C (typically a few kΩ). The input signal V_{in} is a time-varying voltage pulse generated by a radiation detector (e.g., a photodiode). **Capacitors C_1 and C_2 are used to pass AC components of the signal (such as radiation-induced pulses) while blocking DC components (such as dark current).**

As discussed in Sect. 2.8.1, a BJT amplifies the base current I_B by a factor β, typically ranging from 20 to 300. However, for this amplification to occur, the base–emitter voltage V_{BE} must be positive and exceed the built-in potential of approximately 0.7 V. Since the input voltage V_{in} alone is too small to satisfy this condition (which is why amplification is needed), a DC bias is provided to the base. Specifically, the base is connected to V_{cc} via a resistor R_B (typically a few hundred kΩ), ensuring that the base–emitter junction is slightly forward biased.

According to Ohm's law, the base current is given by

$$I_B = \frac{V_{cc} - V_{BE}}{R_B}. \tag{2.48}$$

Since $V_{BE} \ll V_{cc}$, this can be approximated as

$$I_B \approx \frac{V_{cc}}{R_B}. \tag{2.49}$$

Thus, the base current is primarily determined by V_{cc} and the bias resistor R_B.

When the detector sends a small input signal $V_{in}(t)$, it induces a small, time-varying current $i_B(t)$ superimposed on the steady base current I_B. Assuming the effective input resistance of the base–emitter junction is R_{BE}, we can relate $i_B(t)$ to $V_{in}(t)$ as

$$i_B(t) = \frac{V_{in}(t)}{R_{BE}}, \tag{2.50}$$

where R_{BE} is the small-signal resistance of the forward biased base–emitter junction.

As shown in Sect. 2.8.1, the collector current is β-times the base current:

$$I_C(t) = \beta\,[I_B + i_B(t)]\,. \tag{2.51}$$

Hence, the voltage at the collector is

$$\begin{aligned} V_C(t) &= V_{cc} - I_C(t) R_C \\ &= V_{cc} - \beta\,[I_B + i_B(t)]\,R_C. \end{aligned} \tag{2.52}$$

Due to capacitor C_2, only the AC component of the collector voltage is passed to the output:

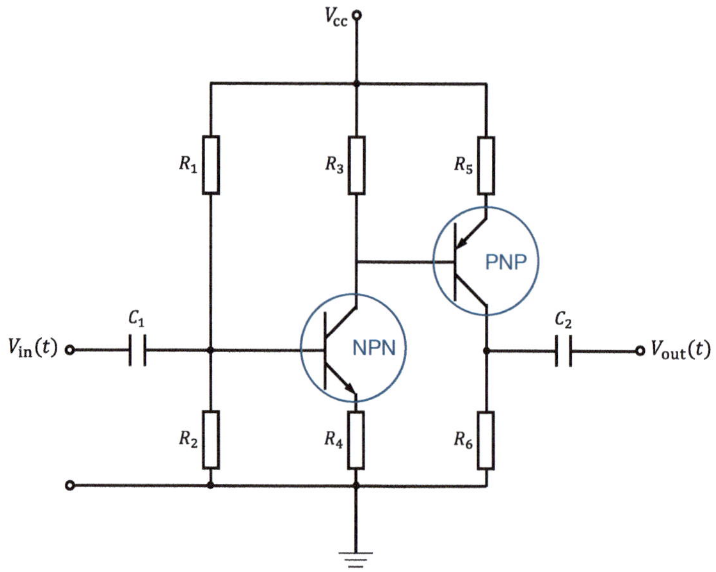

Fig. 2.36 Circuit diagram of a dual BJT (NPN and PNP)-based voltage-sensitive amplifier

$$\begin{aligned} V_{\text{out}}(t) &= -\beta\, i_{\text{B}}(t) R_{\text{C}} \\ &= -\beta \frac{R_{\text{C}}}{R_{\text{BE}}} V_{\text{in}}(t). \end{aligned} \tag{2.53}$$

Thus, the output voltage is inverted in phase relative to the input and amplified by a factor of $\beta \cdot \frac{R_C}{R_{BE}}$. This configuration is known as a **inverting preamplifier**.

Figure 2.36 shows an alternative VSA configuration, where two complementary BJTs (an NPN and a PNP) are used. The collector of the first BJT is connected to the based of the second BJT, such that the current amplified by the first transistor is amplified further by the second one. The amplification factors of the two BJTs effectively multiply, resulting in a much higher overall gain. This type of transistor assembly is known as the **Darlington transistor**. For the specific example shown in Fig. 2.36, the output voltage is in phase with the input, and this configuration is therefore referred to as a **non-inverting preamplifier**.

As an alternative to BJTs, preamplifiers can also be implemented using MOSFETs or CMOS transistors. An example of a CMOS-based amplifier is shown earlier in Fig. 2.34, which requires both positive and negative voltage supplies.

2.9.1.1 Operational Amplifier (Op Amp)

A VSA system may employ more than two BJTs or CMOS transistors to form an integrated device that provides the required gain and input/output characteristics. A

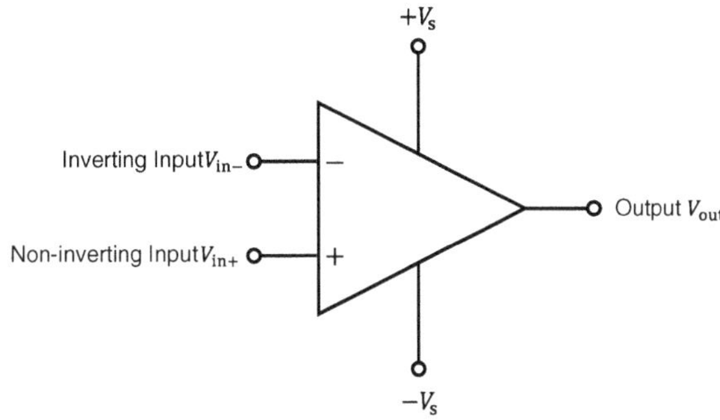

Fig. 2.37 Operational amplifier (op amp) symbol

well-known subtype of such an integrated device is the **operational amplifier (op amp)**, which has two voltage inputs (one inverting (−) and one non-inverting (+)), a single-ended voltage output, a power supply with a positive voltage (e.g., +15 V for BJT-based, +5 V for CMOS-based), and a power supplier with a negative voltage ((e.g., −15 V or −5 V). The operational amplifier was invented by Karl D. Swartzel Jr. in the 1940s, originally implemented with vacuum tubes (thermionic valves) rather than semiconductor transistors. The name "operational amplifier" reflects its early use in analog computing, where it performed mathematical operations.

In practice, unless one is designing an integrated op amp at the circuit level, most users are concerned primarily with its functional behavior. For this reason, preamplifiers and op amps are typically represented by a simple triangular symbol, as shown in Fig. 2.37, effectively treating the device as a black box.

An op amp has several basic properties:

1. High gain: The so-called **open-loop gain**, A, of an op amp is extremely large. The term "open loop" means that no feedback path is present from the output to the input. The op amp amplifies the voltage difference between its two inputs, $V_{\text{in}+}$ and $V_{\text{in}-}$. Even if this difference ($V_{\text{in}+} - V_{\text{in}-}$) is very small, the amplified output ideally saturates at one of the external voltage suppliers:

$$V_{\text{out}} = \begin{cases} V_{\text{s}+}, & \text{if } V_{\text{in}+} - V_{\text{in}-} > 0, \\ V_{\text{s}-}, & \text{if } V_{\text{in}+} - V_{\text{in}-} < 0. \end{cases} \tag{2.54}$$

   This makes the op amp well-suited for use as a **voltage comparator** between $V_{\text{in}+}$ and $V_{\text{in}-}$. We will revisit this application in Chap. 3. More importantly, when a feedback loop is applied, this property enables a wide range of useful circuit configurations, such as the one introduced in Sect. 2.9.2.
2. High input impedance: The **input impedance** of a device is the effective impedance seen by a signal source, defined as the ratio of the input voltage to the input current. For practical analysis, the input impedance of an op amp is treated as infinitely large compared to typical circuit resistances. As a result, essentially no current flows into or between the two input terminals.

3. Low output impedance: The **output impedance** of a device characterizes how much the output voltage drops when a load is connected. Because an op amp has very low output impedance, it can maintain nearly the same output voltage regardless of the connected load. In this sense, the op amp behaves as an excellent voltage source.

These basic properties, when combined with appropriate feedback networks, form the foundation for the many signal-processing and functional applications of op amps.

In practical circuit diagrams, the positive and negative power supplies depicted in Fig. 2.37 are often omitted for clarity. However, it is important to understand that **amplifiers increase electric power—not just voltage.** This additional power is sourced from the supply rails at V_{S+} and V_{S-}; without these supplies, amplification would not be possible.

2.9.2 *Charge-Sensitive Preamplifier*

Radiation sensors such as diode detectors inherently possess an input capacitance, denoted as C_{in}. Consequently, the relationship between the input voltage signal V_{in} and the total induced charge Q_{in} is given by

$$V_{\text{in}} = \frac{Q_{\text{in}}}{C_{\text{in}} + \text{other capacitance}}. \tag{2.55}$$

If C_{in} is constant, then V_{in} is directly proportional to the induced charge Q_{in}. However, in practice, the input capacitance of many radiation sensors may vary with time due to temperature changes or signal dynamics. In such cases, V_{in} no longer reliably reflects the input charge, making a voltage-sensitive amplifier an unsuitable choice.

To overcome this limitation, a **charge-sensitive amplifier** (CSA) is used. The key design feature of a CSA is the inclusion of a **feedback capacitor** C_f to op amp, as illustrated in Fig. 2.38. This capacitor enables the amplifier system to integrate the input charge, making the output voltage dependent solely on the total charge delivered, rather than the varying input capacitance.

Let the voltages at the inverting input and non-inverting input of the op amp be V_{in} and 0, respectively. Given the open-loop gain of the op amp, A, the relationship between V_{in} and V_{out} is

$$V_{\text{out}} = -A \cdot V_{\text{in}}. \tag{2.56}$$

The input charge Q_{in} delivered to the inverting input through the detector is stored across the feedback capacitor C_f, so by the capacitor charge–voltage relation,

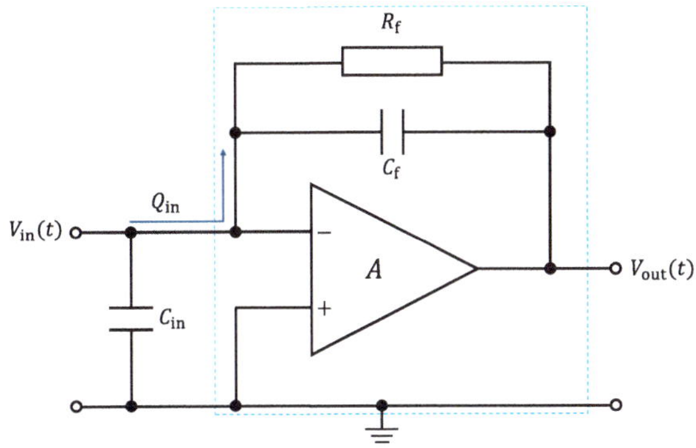

Fig. 2.38 Circuit diagram of a charge-sensitive amplifier (CSA). *A* denotes the open-loop gain of the op amp

$$\begin{aligned} Q_{\text{in}} &= C_f \cdot (V_{\text{in}} - V_{\text{out}}) \\ &= C_f \cdot (V_{\text{in}} + A V_{\text{in}}) \\ &= (A+1) C_f \cdot V_{\text{in}}. \end{aligned} \tag{2.57}$$

This shows that the feedback network behaves like an effective capacitor of value $(A+1)C_f$ at the input. This is known as the **Miller theorem** [5], which describes the apparent increase in capacitance due to the amplifier's gain.

Considering the sensor capacitance C_{in} connected in parallel with the amplifier input, the total effective capacitance seen at the input is

$$C_{\text{total}} = C_{\text{in}} + (A+1)C_f. \tag{2.58}$$

Using this total capacitance, the input voltage is related to Q_{in} by

$$V_{\text{in}} = \frac{Q_{\text{in}}}{C_{\text{total}}} \tag{2.59}$$

$$= \frac{Q_{\text{in}}}{C_{\text{in}} + (A+1)C_f}. \tag{2.60}$$

If we choose the feedback capacitor such that $C_f \gg C_{\text{in}}$ and assuming the amplifier gain $A \gg 1$, the expression simplifies to

$$V_{\text{in}} \approx \frac{Q_{\text{in}}}{A \cdot C_f}. \tag{2.61}$$

Substituting into the earlier expression of $V_{\text{out}} = -A \cdot V_{\text{in}}$, we obtain the final result:

$$\begin{aligned} V_{\text{out}} &= -A \cdot V_{\text{in}} \\ &\approx \frac{Q_{\text{in}}}{C_f}. \end{aligned} \tag{2.62}$$

Therefore, the output voltage of the CSA is linearly proportional to the input charge Q_{in} and is independent of the input capacitance C_{in}. This property eliminates a key limitation of voltage-sensitive amplifiers when used with sensors whose capacitance may vary.

Moreover, Eq. (2.62) shows that the output voltage is also independent of the time profile of the current pulse. This makes the CSA especially well-suited for radiation detection applications, where the total induced charge (rather than the shape of the pulse) carries the relevant physical information, such as the energy deposited by an incoming particle or photon.

To prevent the circuit from saturating after multiple pulses, a **high-value feedback resistor** R_f is typically placed in parallel with the feedback capacitor C_f. This resistor provides a discharge path for the accumulated charge, setting a decay time constant $\tau = R_f C_f$. This gradual discharge allows the CSA output to return to its baseline between events, ensuring proper operation during successive radiation pulses. We will revisit this point in Chap. 3.

2.10 Summary of Chap. 2

1. When a p-type semiconductor is brought into contact with an n-type semiconductor to form a PN junction diode, a depletion region forms at the interface. Although this region contains fixed space charges, it is depleted of *mobile* electrons and holes. These space charges generate an electric field that opposes the diffusion of electrons from the n-region to the p-region and of holes from the p-region to the n-region. The corresponding built-in potential is typically around 0.7 volts in Si PN junctions.
2. This built-in electric field allows a PN junction diode to function as a radiation detector: when external radiation liberates electron–hole pairs within the depletion region, the field drives these carriers apart, creating a measurable drift current.
3. The electric field strength and width of the depletion region can be increased significantly by applying a reverse bias voltage (making the n-region positive relative to the p-region). This enhances the drift velocity of charge carriers and increases the resulting current while also reducing the dark current.
4. A diode can also be formed by a metal–semiconductor junction. Depending on the metal's work function, the semiconductor's electron affinity, and its doping level, the contact may be either a rectifying Schottky contact or a non-rectifying ohmic contact—both of which have important applications in semiconductor radiation detectors.
5. A photodiode is a type of semiconductor detector specifically designed to detect optical signals. Because visible light photons carry significantly less energy than x-rays, the photodiode's active region can be thinner, and a lower external voltage is sufficient to establish the necessary electric field.

6. An avalanche photodiode (APD) is a specialized photodiode that employs a more abrupt doping profile to create a stronger internal electric field. This field provides enough energy to accelerate radiation-induced carriers to the point where they can trigger impact ionization, generating additional carriers and amplifying the signal.
7. A single-photon avalanche diode (SPAD) is a type of APD with an electric field strong enough that any single photon, regardless of its energy, can initiate a full avalanche, producing a uniform and strong output pulse. Arrays of Si-based SPADs form a SiPM, which provides extremely high gain and can resolve the number of photons generated by a scintillator in response to each external X-ray or gamma-ray photon.
8. Transistors—including BJTs, MOSFETs, TFTs, and CMOS devices—can be used either as direct radiation sensors or as circuit elements (e.g., in preamplifiers) within the readout electronics of a radiation detection system.

With this chapter, we have established a foundational understanding of the key components in semiconductor radiation detectors. In the next chapter, we will explore how these semiconductor elements can be integrated to implement X-ray photon counting detectors.

References

1. Nobel Prize in Physics. 1956 – Awarded jointly to William B. Shockley, John Bardeen, and Walter H. Brattain for their researches on semiconductors and their discovery of the transistor effect. 1956.
2. Bray R. The origin of semiconductor research at purdue, 2016. https://www.physics.purdue.edu/about/history/semi_conductor_research.html.
3. Gundacker S, Heering A. The silicon photomultiplier: fundamentals and applications of a modern solid-state photon detector. Phys Med Biol 2020; 65(17):17TR01. https://doi.org/10.1088/1361-6560/ab7b2d.
4. Knoll GF. Radiation detection and measurement. Hoboken: John Wiley & Sons; 2010.
5. Miller JM. Dependence of the input impedance of a three-electrode vacuumtube upon the load in the plate circuit. Sci Papers Bureau Standards 1920; 15(351):367–385.
6. Neamen D. An introduction of semiconductor devices. McGraw-Hill Education, New York; 2005.
7. Renker D, Lorenz E. Advances in solid state photon detectors. J Instrum 2009; 4(04):P04004; 2009. https://doi.org/10.1088/1748-0221/4/04/P04004.
8. Yorke E, Alecu R, Ding L, Fontenl D, Kalend A, Kaurin D, Masterson-McGary ME, Marinello G, Matzen T, Saini A, Shi J, Simon W, Zhu TC, Zhu XR, Rikner G, Nilsson G. AAPM Report No. 087 - Diode in vivo dosimetry for patients receiving external beam radiation therapy. Technical Report, AAPM, 2005.

Chapter 3
Semiconductor Photon-Counting Detector (PCD) Physics

Contents

K. Li, *Essential Physics of Medical X-ray Sources and Detectors*,
https://doi.org/10.1007/978-3-032-27800-5_3

3.1 Introduction

I am often asked by students and clinicians: *what fundamentally distinguishes a photon-counting detector (PCD) from an energy-integrating detector (EID), and what advantages does a PCD offer?* A common impression is that EIDs first convert x-rays into visible light and then into an electrical signal, whereas PCDs directly convert x-rays into electron–hole pairs. It is then assumed that this direct conversion alone gives PCDs better spatial resolution because there is no light spread. That impression, however, is not entirely accurate. The fundamental difference between PCDs and EIDs lies primarily in their operation mode and the design of the detector readout electronics, rather than in the sensor material itself. Still, this misconception is understandable: most x-ray PCDs, including those used in PCD-CT systems, employ semiconductor sensors, whereas many PCDs in γ-ray imaging applications, such as positron emission tomography (PET), are based on inorganic scintillator sensors.

3.1.1 Tomato Analogy

To understand the advantages of photon counting over energy integration, consider the following analogy.

Suppose Peter is a gardener who grew tomato plants from a seed package of mixed varieties. In early fall, he used a mechanical harvester to collect all the tomatoes from his garden and then measured the **total weight** of the harvest. This process represents integration—he obtained only the overall weight, without knowing the contribution from each type of tomato. The harvester may also pick up

Fig. 3.1 Left column: all tomatoes combined. Middle column: cherry tomatoes have higher market value than regular-sized tomatoes. Right column: tomatoes sorted by individual size

some leaves and stems, analogous to **electronic noise** contaminating the output of an EID.

One day, Peter visited a grocery store and noticed that small cherry tomatoes were sold at higher prices than medium-sized ones (Fig. 3.1). He realizes that he could make more profit by sorting the tomatoes one by one based on their **individual weight**, placing them into two baskets: cherry and non-cherry. During sorting, he also discarded the leaves and debris accidentally collected.

This sorting process is analogous to photon counting. Peter processes each tomato (quantum) individually, measures its property (weight), and classifies it accordingly. The outcome of counting is not only the totall number of items but also the number of items in each weight category.

3.1.2 The Essence of Photon Counting

As illustrated by the tomato analogy, the essence of photon counting is to observe or measure individual quanta to obtain richer information. In x-ray detection, this information may include the **number** of detected photons, the **energy** of each photon, the arrival **time** of photons, or even the **incident direction** or interaction depth of photons.

Therefore, the defining characteristic of a PCD is how it processes signals, not the material used for x-ray conversion.

That said, semiconductor sensors do offer several important advantages for photon counting: better spatial resolution due to the absence of light spread, a low mean electron–hole creation energy (ϵ in Table 1.7), and a faster electric pulse formation suitable for resolving individual photon events. For these reasons, this chapter uses semiconductor-based PCDs to illustrate the essential physics of PCDs.

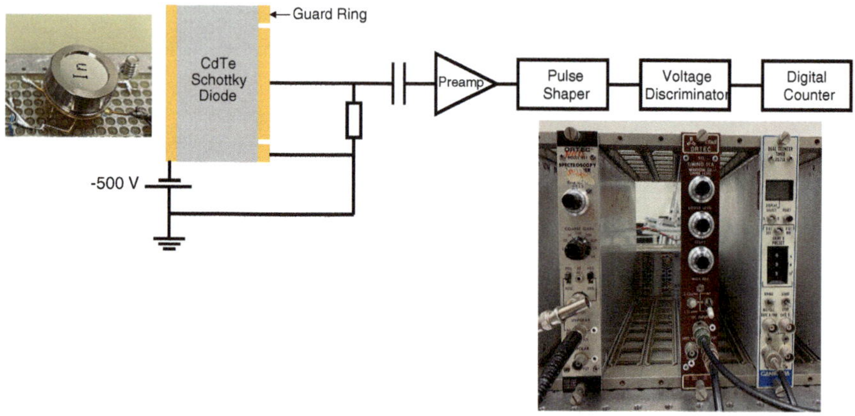

Fig. 3.2 Illustration of key components in a single-pixel, single-channel PCD system. The radiation sensor is a reverse-based CdTe Schottky diode (Acrorad Co. Ltd., Japan). The photograph in the lower left shows a pulse-shaping amplifier (Ortec Model 451 Spectroscopy Amplifier, AMETEK Inc.), a voltage discriminator (Ortec Model 551 Timing Single Channel Analyzer, AMETEK Inc.), and a counter (2017A Dual Counter Timer, Canberra Industries)

3.1.3 Basic Structure of a Semiconductor PCD

As illustrated in Fig. 3.2, a PCD typically consists of a semiconductor sensor, a preamplifier, a pulse processing or "pulse shaping" unit, a voltage discriminator (or trigger), and a digital counter.

As discussed in Chap. 2, a semiconductor diode can detect external radiation through its built-in electric field. However, such self-powered diode detectors are generally used for energy-integrating detection of high-dose radiation rather than for photon counting. The built-in electric field is usually too weak to drive radiation-induced electrons and holes into sufficiently fast drift motion to produce short, strong current pulses, which is required for photon-counting operation. Therefore, in a PCD, the semiconductor sensor is typically reverse biased using an external voltage supply.

A decoupling capacitor is placed between the output electrode of the sensor and the preamplifier to block unwanted background signals caused by the sensor's leakage current, particularly in sensors employing ohmic contacts.

The preamplifier and pulse shaper are analog signal processing units designed to produce strong, clean, and temporally narrow pulses that can reliably trigger the digital counting stage, which includes the voltage discriminator and counter. Each of these circuit components warrants detailed discussion on its own. In this chapter, we focus on the fundamental physical principles underlying these circuit elements rather than on their state-of-the-art engineering implementations. Readers interested in circuit-level details are referred to Glenn F. Knoll's book [10].

3.2 Formation of Initial Electric Pulses in Semiconductor PCDs

In Chap. 2, we learned that when an x-ray photon is absorbed in an unbiased or reverse-biased semiconductor diode, its energy is deposited in the semiconductor to generate electron–hole pairs. Under the influence of an electric field, the electrons and holes drift in opposite directions toward their respective electrodes.

When I first studied this process as a graduate student, I had a naive physics picture: electrons and holes were like two fleets of vessels—one carrying positive charges to the negative electrode and the other carrying negative charges to the positive electrode. According to this view, the electrodes would not register any charge until the carriers reached the "shore," meaning the electrode surface. In other words, I believed there was a time delay between the creation of electron–hole pairs and the generation of a measurable signal.

Later, when I revisited electrodynamics, I realized this picture was incorrect. There is no delay between the creation of charge carriers and the onset of an electric signal: the signal appears immediately once carriers begin moving. In essence, it is the *motion* of charge carriers, not their eventual arrival at the electrode, that induces the detector's initial output signal. The arrival of the carriers at the electrode instead signifies the termination of the initial electric pulse. This principle applies equally to photon-counting detectors and energy-integrating detectors with semiconductor light sensors.

Note that a similar process occurs in scintillator-based EIDs. When a scintillation light photon is absorbed in a light sensor, such as a photodiode, the resulting motion of the generated electrons and holes produces a current pulse. However, in the read-out circuit of an EID, the currents induced by many such light photons are integrated over time by a capacitor, thereby losing information about individual photons.

With this corrected understanding, the next scientific question is: how does the shape of the electric pulse quantitatively depend on the radiation input, detector material, and geometry? Remarkably, W. Shockley and S. Ramo provided a general solution to this problem in the 1930s. The Shockley–Ramo theorem predicts the induced signal at any detector electrode with arbitrary geometry, in response to the motion of a space charge along an arbitrary path.

Over the years of teaching this topic, I have found that, rather than immediately presenting the Shockley–Ramo theorem and asking students to memorize the formula, it is more effective to begin with **an experimental demonstration** and **a theoretical toy-model problem**.

The experimental demonstration is illustrated in Fig. 3.3. The device, designed by Jeff Radtke at the University of Wisconsin [13], consists of a vertical tube with a metal electrode at its bottom. The electrode is connected to an amplifier and then to Channel 1 of an oscilloscope. A Teflon puck is first rubbed with an animal pelt, acquiring a negative electrostatic charge, and then dropped into the tube toward the electrode. The puck is insulated from the electrode by a rubber ball, ensuring there is no direct electrical contact. Upon impact with the rubber ball, the puck bounces

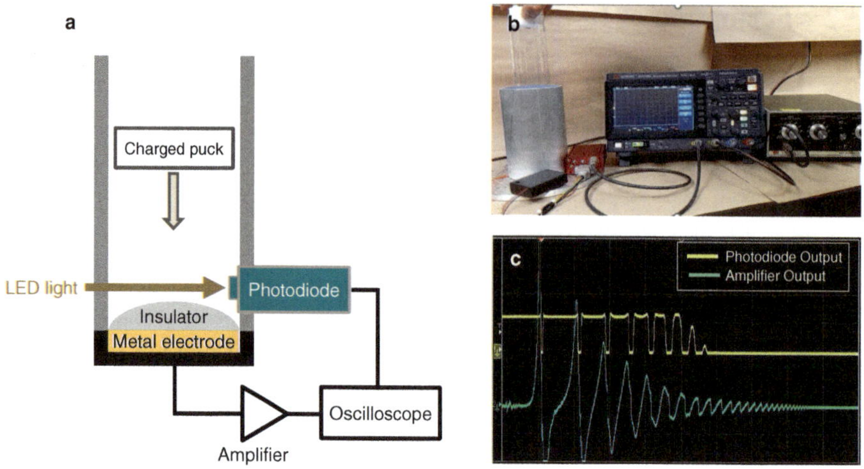

Fig. 3.3 An experiment, designed and implemented by Jeff Radtke at the University of Wisconsin, demonstrates how the motion of space charges induces an electric signal in an electrode. (**a**) Schematic diagram of the experimental setup. (**b**) Photograph of the experimental setup. (**c**) Screenshot of the oscilloscope display

upward and then falls back down due to gravity. This produces a damped oscillatory motion, but the puck never directly touches the electrode.

To precisely determine when the puck reaches the bottom of the tube, an LED–photodiode pair is installed near the rubber ball. The output of the photodiode is connected to Channel 2 of the oscilloscope. When the puck reaches the bottom, it blocks the LED light, causing the photodiode output to switch from high to low, thereby providing a precise time marker.

As shown in Fig. 3.3c, the oscilloscope waveforms clearly show that the electrode output peaks when the charged puck first reaches the bottom of the cylinder with maximum velocity. As the puck bounces upward, it produces a signal of opposite polarity. When it falls a second time, it generates another positive signal, but with reduced amplitude due to the damping caused by the nonelastic collision with the rubber ball.

Because the electrode and the puck are electrically insulated, this experiment demonstrates that:

1. The electrode output signal is generated by charge induction, not by direct charge transfer
2. The magnitude of the induced signal is positively correlated with the velocity of the moving charge
3. The polarity of the induced signal depends on the direction of the charge motion

To explain the physics underlying these observations, Sect. 3.2.1 analyzes a toy-model problem. This simplified electrostatic scenario is based on a parallel-plate detector, which provides an intuitive entry point to understanding the Shockley–Ramo theorem.

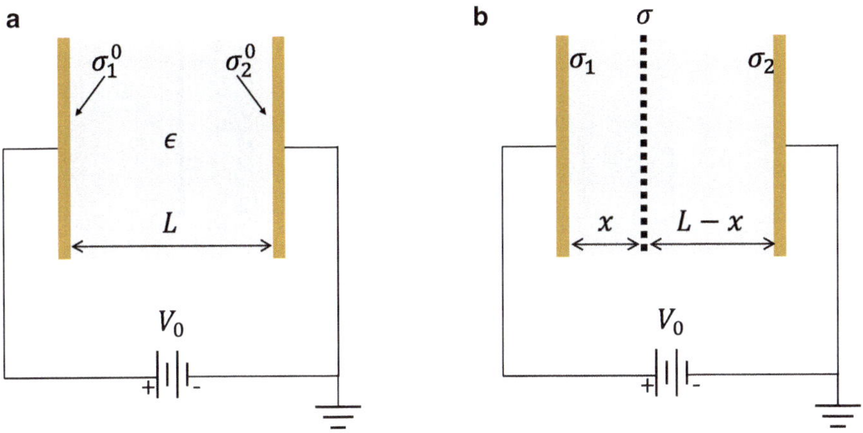

Fig. 3.4 (**a**) A planar detector with two parallel-plate electrodes. (**b**) A uniform sheet of charge having a surface charge density of σ is introduced to the medium between the two parallel-plate electrodes

3.2.1 Toy-Model Problem: Pulse Formation by Charge Motion in Parallel-Plate Detector

Figure 3.4 illustrates a toy-model detector consisting of two parallel planar electrodes. This system can be treated as a parallel-plate capacitor with capacitance:

$$C = \frac{\varepsilon A}{d}, \tag{3.1}$$

where ε is the permittivity of the semiconductor material, A is the electrode area, and d is the electrode separation.

For such a capacitor, the charge on each electrode depends not only on C and the applied bias voltage but also on any space charge that may exist between the electrodes. To build intuition, we first examine the simpler case with no space charge before moving to the more general scenario [Fig. 3.4a].

Case 1: No Space Charge

If a bias voltage V_0 is applied across the detector in the absence of any space charge, the electrodes acquire charges:

$$Q_1^0 = CV_0 = \frac{\varepsilon A V_0}{d},$$
$$Q_2^0 = -Q_1^0 = -\frac{\varepsilon A V_0}{d}. \tag{3.2}$$

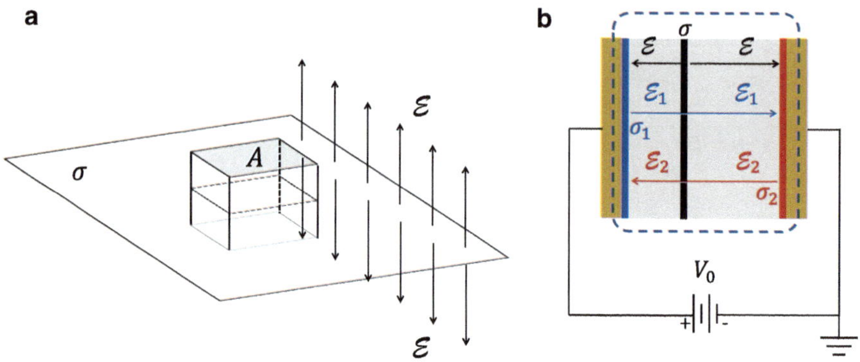

Fig. 3.5 (**a**) Suppose a sheet of charge has a uniform charge density of σ, and we draw a Gaussian pillbox enclosing an area of A on the sheet. Since the electric field generated by the uniform surface charges is always perpendicular to the sheet, according to Gauss's law, $2\mathcal{E} \cdot A = \frac{\sigma A}{\varepsilon}$, or $\mathcal{E} = \frac{\sigma}{2\varepsilon}$. (**b**) Individual components of electric fields generated by surface charges at different planes. The total electric field is given by the superposition of these individual electric fields

Assuming the electrodes are sufficiently large that edge effects can be neglected, the charges distribute uniformly over their surfaces. The corresponding surface charge densities are

$$\begin{cases} \sigma_1^0 = \frac{Q_1^0}{A} = \frac{\varepsilon V_0}{d}; \\ \sigma_2^0 = \frac{Q_2^0}{A} = -\frac{\varepsilon V_0}{d}. \end{cases} \tag{3.3}$$

Case 2: With Space Charge

Now consider a more complex situation in which a uniform sheet of charge, with surface charge density σ, is suddenly introduced into the detector (e.g., by an absorbed radiation event) at the position indicated in Fig. 3.4b. The total charge on this sheet is $q = \sigma A$.

Does the presence of this space charge alter the charges on the electrodes? The answer is *yes*. To see why, recall that the electric field produced by an infinite charged plane of density σ can be obtained from Gauss's law using the pillbox construction shown in Fig. 3.5a:

$$\mathcal{E} = \frac{\sigma}{2\varepsilon}. \tag{3.4}$$

Likewise, the electric fields generated by the electrode surface charge densities σ_1 and σ_2 are

$$\mathcal{E}_1 = \frac{\sigma_1}{2\varepsilon}, \tag{3.5}$$

$$\mathcal{E}_2 = \frac{\sigma_2}{2\varepsilon}. \tag{3.6}$$

By superposition, the net electric field between Electrode 1 and the sheet charge is [Fig. 3.5b]

$$\mathcal{E}_L = \mathcal{E}_1 - \mathcal{E}_2 - \mathcal{E} = \frac{1}{2\varepsilon}\left(\sigma_1 - \sigma_2 - \sigma\right), \tag{3.7}$$

while the net field between Electrode 2 and the sheet charge is

$$\mathcal{E}_R = \mathcal{E}_1 - \mathcal{E}_2 + \mathcal{E} = \frac{1}{2\varepsilon}\left(\sigma_1 - \sigma_2 + \sigma\right). \tag{3.8}$$

Relating field to potential, the applied bias V_0 must equal the total potential drop across the two regions:

$$\begin{aligned} V_0 &= \mathcal{E}_L x + \mathcal{E}_R (L - x) \\ &= \frac{1}{2\varepsilon}\left[(\sigma_1 - \sigma_2 - \sigma)x + (\sigma_1 - \sigma_2 + \sigma)(L - x)\right] \\ &= \frac{1}{2\varepsilon}\left[(\sigma_1 - \sigma_2)L + \sigma(L - 2x)\right]. \end{aligned} \tag{3.9}$$

This gives the first condition:

$$\boxed{\sigma_1 - \sigma_2 = \frac{2\varepsilon V_0}{L} - \sigma \frac{L - 2x}{L}}. \tag{3.10}$$

Next, consider the Gaussian surface drawn as a dashed boundary around the entire system in Fig. 3.5b. Because the electric field inside a conductor vanishes, no net flux crosses the boundary. Gauss's law then requires

$$\frac{Q_{\text{total}}}{\varepsilon} = \frac{A(\sigma_1 + \sigma_2 + \sigma)}{\varepsilon} = 0, \tag{3.11}$$

which yields the second condition:

$$\boxed{\sigma_1 + \sigma_2 = -\sigma}. \tag{3.12}$$

Solving Eqs. (3.10) and (3.12) simultaneously gives

$$\begin{cases} \sigma_1 = \frac{\varepsilon V_0}{L} - \sigma\left(1 - \frac{x}{L}\right); \\ \sigma_2 = -\frac{\varepsilon V_0}{L} - \sigma \frac{x}{L}. \end{cases} \tag{3.13}$$

By comparing Eqs. (3.3) and (3.13), we see that the presence of the sheet charge σ **induces** an additional surface charge density of $-\sigma\left(1-\frac{x}{L}\right)$ on Electrode 1 and $-\sigma\frac{x}{L}$ on Electrode 2. **The induced charge on each electrode therefore depends explicitly on the position x of the sheet charge.** However, their sum,

$$-\sigma\left(1-\tfrac{x}{L}\right)+\left(-\sigma\tfrac{x}{L}\right)=-\sigma, \tag{3.14}$$

is independent of x.

Now suppose the sheet charge moves from x to $x+\Delta x$. According to Eq. (3.13), this displacement modifies the electrode charge densities to

$$\begin{aligned}\sigma_1' &= \frac{\varepsilon V_0}{L}-\sigma\left(1-\frac{x+\Delta x}{L}\right)\\ &= \sigma_1+\sigma\frac{\Delta x}{L},\end{aligned} \tag{3.15}$$

$$\begin{aligned}\sigma_2' &= -\frac{\varepsilon V_0}{L}-\sigma\frac{x+\Delta x}{L}\\ &= \sigma_2-\sigma\frac{\Delta x}{L}.\end{aligned} \tag{3.16}$$

In other words, Electrode 1 gains an *additional* charge:

$$\begin{aligned}\Delta Q &= A(\sigma_1'-\sigma_1)\\ &= A\sigma\frac{\Delta x}{L}\\ &= q\frac{\Delta x}{L},\end{aligned} \tag{3.17}$$

while Electrode 2 loses the same amount. **Thus, the total induced charge on both electrodes remains constant, $(\sigma_1'+\sigma_2')A=-\sigma A=-q$, but the distribution shifts between them through the connecting circuit as the sheet charge moves.**

From the circuit's perspective, this redistribution appears as a transient flow of charge ΔQ through the cable (Fig. 3.6), i.e., an electrical pulse that an electrometer would detect.

The corresponding current I in the cable follows directly from the definition:

$$\begin{aligned}I &= \frac{\Delta Q}{\Delta t}\\ &= \frac{q}{L}\frac{\Delta x}{\Delta t}\\ &= \frac{q}{L}v,\end{aligned} \tag{3.18}$$

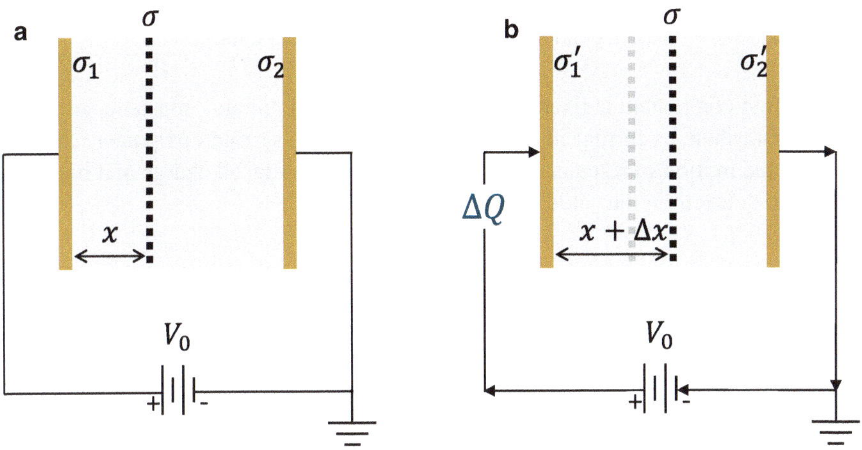

Fig. 3.6 When the uniform sheet of charge shown in (**a**) moves from x to $x + \Delta x$, an additional charge of ΔQ flows from Electrode 1 to Electrode 2 (**b**)

where v is the velocity of the moving space charge.

Summary of the Toy-Model Case

When a sheet of space charge q moves in a parallel-plate detector from x to $x + \Delta x$:

- The induced charges on the two electrodes are instantaneously redistributed; the magnitude of the change is

$$\Delta Q = q\frac{\Delta x}{L}. \tag{3.19}$$

- A current

$$I = q\frac{v}{L} \tag{3.20}$$

  flows through the cable connecting the electrodes, where v is the velocity of the moving charge.
- Since both ΔQ and I are linearly proportional to q, their measured values can be used to estimate the magnitude of q.
- A measurable signal (ΔQ or I) appears as soon as q is in motion ($v \neq 0$). The charge does not need to reach an electrode to generate a detectable signal.
- Neither ΔQ nor I depends on the bias voltage V_0. This may seem surprising, but Eq. (3.13) shows that the electrode charge can be decomposed into two components:

$$\pm\frac{\varepsilon V_0}{L} \quad \text{(bias-dependent)} \qquad \text{and} \qquad \text{(terms dependent on } q\text{)}. \tag{3.21}$$

The first component is fixed by the bias voltage and remains constant, so it does not contribute to current in the external circuit. The second component changes with the motion of q, causing the redistribution of induced charge and hence an electric current in the cable.

3.2.2 Pulse Formation in Arbitrary Detectors: The Shockley–Ramo Theorem

In Sect. 3.2.1, we analyzed a toy-model detector with planar geometry to demonstrate how the motion of charge in the inter-electrode region redistributes the induced charge on the electrodes and generates an electrical pulse in the connecting circuit. The planar geometry made it straightforward to derive closed-form expressions for the induced charge as a function of the charge location.

Real detectors, however, often do not have such simple geometries. Electrodes may differ in area or shape, and moving charges are not confined to uniform sheets parallel to the electrodes. In principle, the induced charge on an electrode can always be calculated from the instantaneous electric field $\boldsymbol{\mathcal{E}}$ at its surface when the charge occupies position $\mathbf{x}$:

$$Q = \varepsilon \oint_S \boldsymbol{\mathcal{E}} \cdot d\mathbf{S}. \tag{3.22}$$

where S is the electrode surface. For detectors of arbitrary geometry, however, evaluating $\boldsymbol{\mathcal{E}}$ directly is cumbersome, since one must account for the superposition of fields from both space charges and electrode charges. This motivates the search for a more convenient method.

Around 1938, William Shockley (yes, the same Shockley who co-invented transistors) and Simon Ramo (an engineer at General Electric) independently developed an elegant approach to calculate the induced charge and current in vacuum tubes. Their formulation, later generalized to arbitrary media (e.g., gases in ion chambers or semiconductors with space charges), is now known as the **Shockley–Ramo theorem** [15, 16].

The theorem states that when a charge q moves an infinitesimal distance $\boldsymbol{\Delta x}$, the change in induced charge on a given electrode is

$$\boxed{\Delta Q = -q\,[\varphi_w(\mathbf{x} + \boldsymbol{\Delta x}) - \varphi_w(\mathbf{x})]}\,, \tag{3.23}$$

where φ_w is the so-called weighting potential of that electrode. The weighting potential is defined as the potential generated by setting the potential of the electrode-of-interest to unity and setting the potentials of all other electrodes to zero and removing all space charges.

Based on Eq. (3.23), the induced current on the electrode is obtained as

$$
\begin{aligned}
I &= \lim_{\Delta t \to 0} \frac{\Delta Q}{\Delta t} \\
&= -q \lim_{\Delta t \to 0} \frac{\varphi_w(\mathbf{x} + \mathbf{\Delta x}) - \varphi_w(\mathbf{x})}{\Delta t} \\
&= -q \lim_{\Delta t \to 0} \frac{\varphi_w(\mathbf{x} + \mathbf{\Delta x}) - \varphi_w(\mathbf{x})}{\mathbf{\Delta x}} \cdot \frac{\mathbf{\Delta x}}{\Delta t} \\
&= -q[-\mathcal{E}_w(\mathbf{x})] \cdot \mathbf{v}(\mathbf{x}),
\end{aligned}
\tag{3.24}
$$

which simplifies to

$$
\boxed{I = q\,\mathcal{E}_w(\mathbf{x}) \cdot \mathbf{v}(\mathbf{x})}\,, \tag{3.25}
$$

where

$$
\mathcal{E}_w = -\frac{\mathrm{d}\varphi_w}{\mathrm{d}\mathbf{x}} \tag{3.26}
$$

is known as the weighting electric field or **weighting field** and $\mathbf{v}(\mathbf{x})$ is the instantaneous velocity of the moving charge. Note that this velocity is the actual velocity of the moving charge and is determined by the actual bias voltage instead of the hypothetical weighting potential.

A proof of the Shockley–Ramo theorem is provided in Appendix 1 at the end of this chapter for interested readers.

3.2.3 *Example Applications of the Shockley–Ramo Theorem*

Example 1

Let us revisit the parallel-plate detector problem analyzed in Sect. 3.2.1 and check whether the Shockley–Ramo theorem yields consistent results.

To compute the induced charge on Electrode 1 shown in Fig. 3.7, we first determine its weighting potential. By definition, this is obtained by setting the potential of Electrode 1 to unity, Electrode 2 to zero, and removing all space charges. The resulting electric field between the electrodes is uniform with magnitude $1/L$, giving the potential at position x:

$$
\varphi_{w,1}(x) = 1 - \frac{x}{L}. \tag{3.27}
$$

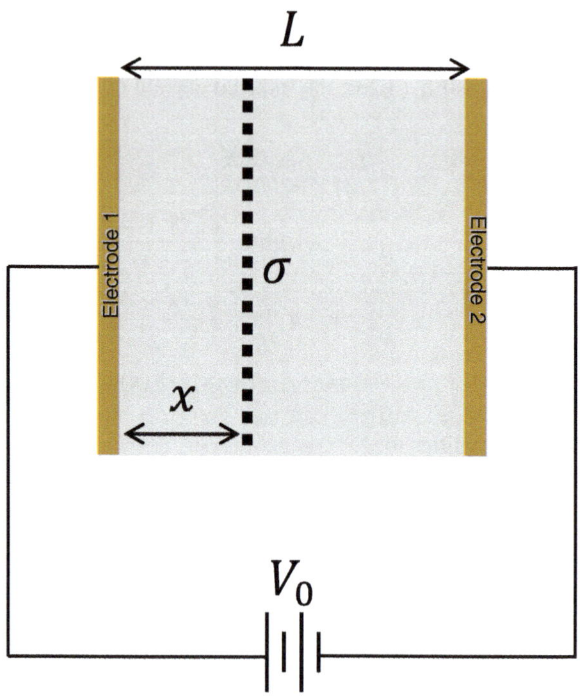

Fig. 3.7 A uniform sheet of charge having an area charge density of σ is placed in the medium between the two planar electrodes

Applying the Shockley–Ramo theorem (Eq. (3.23)), the change in charge on Electrode 1 as a charge q moves from x to $x + \Delta x$ is

$$\begin{aligned} \Delta Q_1 &= -q\left[\varphi_{w,1}(x+\Delta x) - \varphi_{w,1}(x)\right] \\ &= -q\left[\left(1 - \frac{x+\Delta x}{L}\right) - \left(1 - \frac{x}{L}\right)\right] \\ &= q\frac{\Delta x}{L}. \end{aligned} \tag{3.28}$$

This matches the earlier result in Eq. (3.17).

For Electrode 2, the weighting potential is obtained by setting Electrode 2 to unity and Electrode 1 to zero. Neglecting edge effects, we have

$$\varphi_{w,2}(x) = \frac{x}{L}. \tag{3.29}$$

Thus, the induced charge on Electrode 2 is

$$\Delta Q_2 = -q\left[\varphi_{w,2}(x+\Delta x) - \varphi_{w,2}(x)\right]$$

$$= -q\left(\frac{x+\Delta x}{L} - \frac{x}{L}\right)$$
$$= -q\frac{\Delta x}{L}$$
$$= -\Delta Q_1. \tag{3.30}$$

This is exactly what we expect: the induced charges on the two electrodes are equal in magnitude and opposite in sign. Compared with the step-by-step derivation in Sect. 3.2.1, the Shockley–Ramo theorem provides a more direct and elegant route to the same result.

Example 2

This example is a slightly modified version of Example 1: an x-ray is absorbed in the semiconductor material between the two planar electrodes, exciting a total of n electrons from the valence band to the conduction band and leaving an equal number of holes in the valence band. For simplicity, assume that the electron–hole pairs are uniformly generated in a plane at a distance x from Electrode 1, and assume that electron mobility is typically much higher than hole mobility.

Under the electric field created by the external bias voltage V_0, the electrons drift toward Electrode 1 with a velocity

$$v_n = -\mu_n\frac{V_0}{L}, \tag{3.31}$$

where μ_e is the electron mobility in the semiconductor. Similarly, the holes drift toward Electrode 2 with a velocity

$$v_p = \mu_p\frac{V_0}{L}. \tag{3.32}$$

Using the weighting potential of Electrode 1 [Eq. (3.27)], the total induced charge on Electrode 1 at time t is

$$\begin{aligned} Q_1(t) &= -n(-e)[\varphi_{w,1}(x - \mu_n V_0/Lt) - \varphi_{w,1}(x)] \\ &\quad - ne[\varphi_{w,1}(x + \mu_p V_0/Lt) - \varphi_{w,1}(x)] \\ &= ne\frac{v_n t}{L} + ne\frac{v_p t}{L}. \end{aligned} \tag{3.33}$$

Is this the complete answer? Not quite. After a certain time, the electrons reach Electrode 1 first, since electron mobility is assumed to be much higher than hole mobility. The critical arrival time for the electrons is

$$t_1 = \frac{x}{|v_n|}. \tag{3.34}$$

Beyond t_1, electrons stop contributing to $Q_1(t)$ because they are no longer in motion. Similarly, holes reach Electrode 2 after

$$t_2 = \frac{L - x}{v_p}, \tag{3.35}$$

after which they also stop contributing. Thus, the induced charge evolves as

$$Q_1(t) = \begin{cases} ne\frac{v_n t}{L} + ne\frac{v_p t}{L} & t < t_1 \\ ne\frac{x}{L} + ne\frac{v_p t}{L} & t_1 \leq t < t_2 \\ ne\frac{x}{L} + ne\frac{L-x}{L} = ne & t \geq t_2. \end{cases} \tag{3.36}$$

Two important observations follow:

1. At the end, the total induced charge equals the total charge of either the electrons ($-ne$) or the holes ($+ne$), but not both. **There may exist some misconception that the n electrons deposit $-ne$ on Electrode 1 and the n holes deposit $+ne$ on Electrode 2, yielding $2ne$. In reality, the induced signal accounts for only one set of charges, not both.**
2. The total induced charge ne is independent of the initial interaction depth x, provided that all electrons and holes are fully collected.

By differentiating $Q_1(t)$ with respect to time, we obtain the induced current flowing to Electrode 1:

$$I_1(t) = \frac{\mathrm{d}Q(t)}{\mathrm{d}t} = \begin{cases} ne\frac{v_n + v_p}{L} & t < t_1 \\ ne\frac{v_p}{L} & t_1 \leq t < t_2 \\ 0 & t \geq t_2. \end{cases} \tag{3.37}$$

Figure 3.8 shows the corresponding current–time curve, representing the electric pulse generated by the planar detector system. **The slow-moving holes produce a long tail in the pulse, which increases the probability of pulse pileup and thus limits the ability to resolve individual x-ray photons.**

To reduce the tail length, one must reduce t_2, given by

$$t_2 = \frac{L - x}{v_p} = \frac{L - x}{\mu_p V_0 / L}. \tag{3.38}$$

Therefore, one way to reduce t_2 is to increase the bias voltage V_0. This highlights the practical role of the bias voltage: although the Shockley–Ramo theorem shows that the magnitude of ΔQ is independent of V_0, the bias voltage strongly influences the pulse shape. If V_0 is too low, the pulse tail becomes longer, reducing the detector's ability to resolve closely spaced photon events.

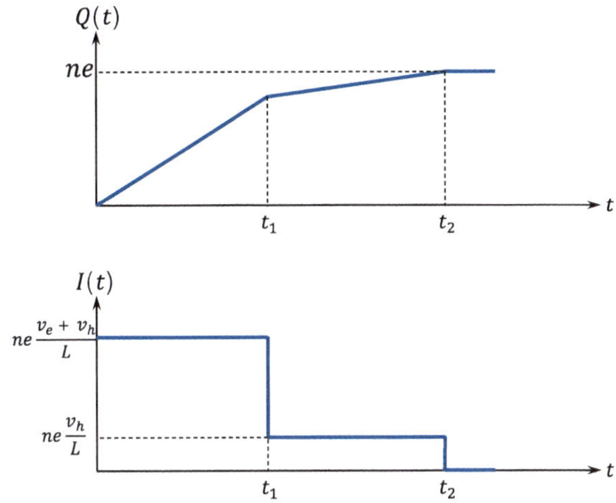

Fig. 3.8 Induced charge (top row) and induced current (bottom row) in a parallel-plate detector following the creation of n electron–hole pairs

Alternatively, t_2 can be shortened by reducing the hole transport distance $(L - x)$. This can be achieved by orienting Electrode 2 (the negatively biased electrode, or cathode) toward the incident x-rays. With a suitably chosen L, a larger fraction of x-ray interactions occurs near the cathode, thereby decreasing the distance holes must travel.

In materials (e.g., amorphous selenium) where holes have higher mobility and longer lifetimes, it is often preferable to orient Electrode 1 (the anode) toward the incident x-rays. This minimizes the transport distance for slow-moving electrons.

Example 3: Small-Pixel Effect

The previous example illustrates that slow-moving charge carriers (often holes) can produce an undesirable long tail in the electric pulse. Is there a way to suppress the contribution of holes and achieve so-called single-polarity sensing? This term refers to a detector system that is sensitive primarily to the motion of one type of charge carrier—typically electrons (or holes if electrons are slower).

A common approach to achieve single-polarity sensing is to exploit the **small-pixel effect**. As shown in Fig. 3.9, if the cathode is a continuous plate while the anode is divided into small pixels, the geometric asymmetry produces a highly nonlinear weighting potential for a given pixel. Here, the weighting potential is defined by setting the potential of the selected pixel to unity and all other pixels and the cathode to zero. The resulting potential gradient is steep near the pixelated electrode and almost flat at larger distances from it. This nonlinearity becomes more pronounced as the pixel size w decreases relative to the detector thickness L. Figure 3.9b shows that for $x < 0.4L$, the weighting potential $\varphi_w(x) \approx 0$.

Now consider an x-ray absorbed at depth x, generating n electron–hole pairs. Because of the exponential attenuation of x-ray intensity, most interactions occur

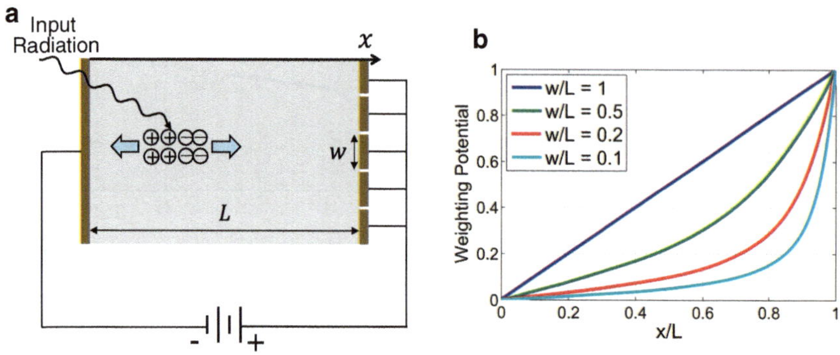

Fig. 3.9 The small-pixel effect. (**a**) A planar detector with a continuous cathode electrode on the left and pixelated anode electrodes on the right. Electrons will drift toward the anodes, and holes will drift toward the cathode. (**b**) Plot of the weighting potential for a pixelated anode for different pixel sizes

at shallow depths where $x < 0.4L$. Electrons drift toward the anode, from x to L, producing an induced charge:

$$\begin{aligned} Q_e &= -(-e)n, [\varphi_w(L) - \varphi_w(x)] \\ &\approx en[\varphi_w(L) - 0] \\ &= en. \end{aligned} \tag{3.39}$$

Holes drift toward the cathode, from x to 0, producing an induced charge:

$$\begin{aligned} Q_h &= -en, [\varphi_w(0) - \varphi_w(x)] \\ &\approx en[0 - 0] \\ &= 0. \end{aligned} \tag{3.40}$$

Therefore, the induced signal is dominated by the motion of electrons when the anode is pixelated with a size much smaller than the detector thickness. This design principle is widely adopted in direct-conversion semiconductor detectors, including both PCDs and energy-integrating detectors.

To apply the Shockley–Ramo theorem to pixelated detectors, the weighting potential or field of each pixel needs to be calculated. Unlike the first two examples, this calculation is more complex. Nevertheless, this third example is of the greatest practical relevance, especially for imaging detectors. Accordingly, the following section is dedicated to a detailed analysis of pulse formation in pixelated PCD systems.

3.3 Formation of Initial Pulses in Pixelated PCDs

Figure 3.10 provides a schematic illustration of a semiconductor photon-counting image detector. One side of the semiconductor sensor features a continuous electrode, while the opposite side contains an array of pixelated electrode pads. As discussed in Sect. 3.2.3, for materials such as CdTe and Si, where electron mobility is substantially higher than hole mobility, the pixelated electrodes are designed as anodes to minimize the influence of slow-moving holes on the induced signal.

Each pixelated anode is connected to its own preamplifier, shaping amplifier, and other circuit elements within an application-specific integrated circuit (ASIC). The semiconductor sensor and the ASIC are electrically connected via bump bonding, wire bonding, or through-silicon via (TSV) technology (Fig. 3.11). Since the semiconductor sensor and ASIC are typically fabricated separately and then assembled, such a detector is commonly referred to as a hybrid semiconductor detector [2].

Although the Shockley–Ramo theorem provides an elegant and simple mathematical framework, deriving closed-form formulas for the weighting field of pixelated electrodes from its definition is not straightforward. Consequently, implementing the Shockley–Ramo theorem often requires numerical solutions of partial differential equations (PDEs) with appropriate boundary conditions or other numerical approximations. An example of a numerical implementation of the Shockley–Ramo theorem is presented in Sect. 3.3.6.

In this section, we use the fundamental principles of electrostatics, including Coulomb's law and Gauss's law, to elucidate the electric pulse formation process

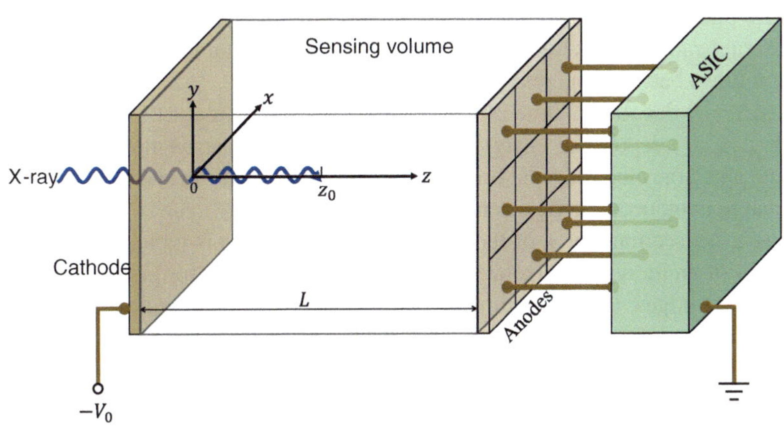

Fig. 3.10 Schematic illustration of a semiconductor-based PCD with a continuous cathode plane and pixelated anodes. ASIC: application-specific integrated circuit. The wire bonding between the pixelated anodes and ASIC is for illustration purpose only. In reality, the connection is usually achieved via bump bonding or through-silicon via

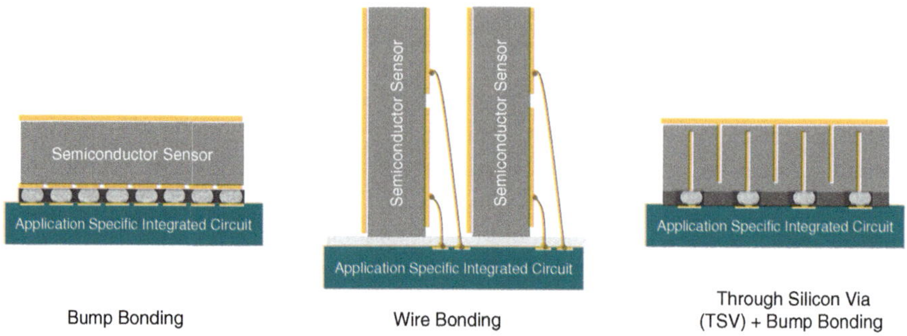

Fig. 3.11 Three electrical interconnection methods can be used to interface a semiconductor sensor with an application-specific integrated circuit. Among these, the through-silicon via approach is typically implemented in combination with the bump-bonding technique

in pixelated PCDs. Using this approach, analytical expressions for the induced charge and current pulses can be derived **without directly solving the weighting potentials numerically**.

3.3.1 Pixelated Detector Structure

When an external bias voltage V_0 is applied between the cathode and pixelated anodes, an electric field $\mathcal{E}_0$ is established along the $-z$ direction. As shown in Sect. 3.2.1, the surface charge densities on the inner surfaces of the anode and cathode planes are $\varepsilon\mathcal{E}_0$ and $-\varepsilon\mathcal{E}_0$, respectively. At steady state and in the absence of external radiation, these surface charges remain constant over time; hence, no current flows between the pixelated anode and the ASIC.

Let $t = 0$ denote the time at which an x-ray photon interacts with the PCD at depth $z = z_0$, depositing energy E and generating n electron–hole pairs. At this instant, since the electrons and holes occupy the same location and $q_n + q_p = 0$, no net charge is induced on the electrodes.

The cross-sectional area of the detector in the xy plane is typically much larger than that of an individual pixelated anode pad. Thus, except for pixels located near the detector edges, the cathode plane can be approximated as infinitely large. The origin of the xy coordinate system can then be aligned with the x-ray interaction site, as illustrated in Fig. 3.13.

The position of each pixelated anode pad can be specified by the coordinates of its center, (x_0, y_0). The pixel closest to $(x = 0, y = 0)$, i.e., the one covering the projected shadow of the initial charge carriers, is referred to as the **central pixel** for a given x-ray interaction event.

For $t > 0$, the electrons and holes separate under the influence of $\mathcal{E}_0$: electrons drift toward the anode, while holes drift toward the cathode. On average, their drift velocities are

$$\boldsymbol{v}_n = -\mu_n \boldsymbol{\mathcal{E}}_0; \tag{3.41}$$

$$\boldsymbol{v}_p = +\mu_p \boldsymbol{\mathcal{E}}_0, \tag{3.42}$$

The time taken by the electrons to reach the anode is

$$t_1 = \frac{L - z_0}{v_n}, \tag{3.43}$$

while the time taken by the holes to reach the cathode is

$$t_2 = \frac{z_0}{v_p}, \tag{3.44}$$

Again, we assume $\mu_n \gg \mu_p$, and thus t_1 usually can be assumed to be smaller than t_2.

The z-positions of the electrons and holes as functions of time are therefore

$$z_n(t) = \begin{cases} z_0 + v_n t & 0 < t \leq t_1 \\ L & t > t_1 \end{cases}$$

$$z_p(t) = \begin{cases} z_0 - v_p t & 0 < t \leq t_2 \\ 0 & t > t_2 \end{cases} \tag{3.45}$$

As shown in Eq. (3.45), electrons and holes occupy different z-positions for $0 < t < t_2$. Consequently, they contribute differently to the induced charge at a given point on the anode plane. Our objective is to determine the charge induced on a pixelated anode by the electrons and holes without using the Shockley–Ramo theorem.

Our strategy is to derive an analytical expression for the electric field produced by the electrons and holes at an arbitrary point on the electrode surface. From this field, we calculate the induced surface charge density at that point. Finally, by integrating the surface charge density over the area of a pixelated electrode, we obtain the total induced charge collected by that electrode.

3.3.2 Principle of Superposition and Method of Images

To derive an expression for electric field produced by space charge on a pixelated electrode, we utilize the following special techniques in electrostatics [4]:

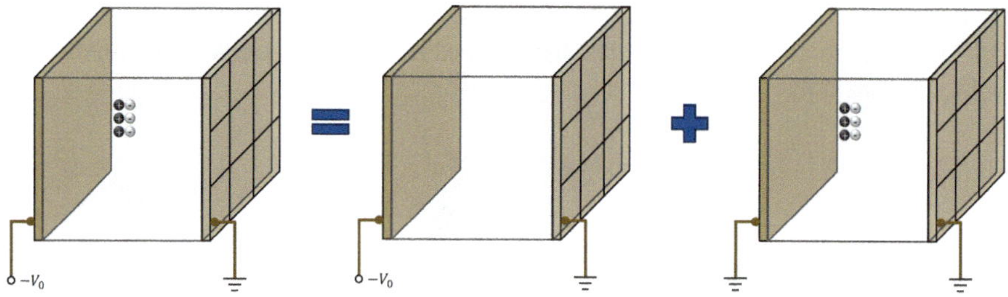

Fig. 3.12 Based on the principle of superposition, the surface charge density on a detector electrode can be determined by decomposing the problem into two subproblems: one with the applied bias voltage V_0 in the absence of space charge and the other one with the space charge present but without the bias voltage

- **Principle of superposition**: The net electric field (or potential) at a point is the vector sum of the fields (or scalar sum of the potentials) produced by individual charges acting independently. This linearity allows us to evaluate the contributions of electrons and holes to the induced charge on the anode *separately* and then sum the results. It also allows us to decouple the electric field generated by the detector bias voltage or semiconductor junction from that produced by radiation-induced space charges. Although the bias voltage (V_0) establishes charges on the electrodes, this is not of interest here. For the purpose of analyzing the electric field from a given space charge, the anode and cathode can therefore be regarded as grounded, as shown in Fig. 3.12. The role of V_0 is to drive the radiation-induced charges in their drift motion, but it does not directly contribute to the detector output signal.
- **Method of images**: For certain simple conductor geometries (e.g., an infinite plane or a sphere), the effect of a conducting boundary can be reproduced by placing fictitious "image" charges outside the physical region. These image charges are chosen so that the required conductor boundary conditions (e.g., a grounded conductor at zero potential) are satisfied exactly. As an example, Fig. 3.13 shows a space charge q_p near a grounded conducting planar electrode and the corresponding image charge $-q_p$. The potential on an arbitrary point (x, y) on the electrode is then

$$V(x, y) = \frac{q_p}{4\pi\varepsilon\sqrt{x^2 + y^2 + (L - z_p)^2}} + \frac{-q_p}{4\pi\varepsilon\sqrt{x^2 + y^2 + (L - z_p)^2}} = 0, \tag{3.46}$$

so the plane is held at zero potential as required. The resulting electric field lines are perpendicular to the plane at the surface, ensuring the other boundary condition to be met (tangential component of the electric field vanishes on the conductor surface).

Based on the principle of superposition and the method of images, we can determine the electric field at the surface of the grounded planar electrode shown in

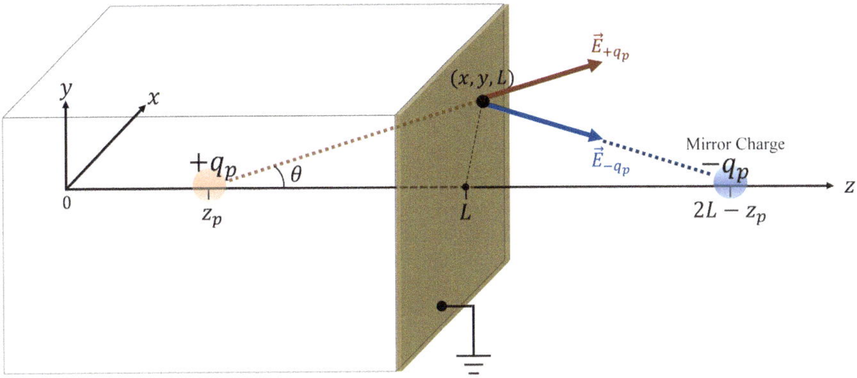

Fig. 3.13 For a given space charge q_p, the electric field at (x, y, L) on the inner surface of the planar electrode at $z = L$ can be calculated using the **method of images**: the electric field at (x, y, L) is given by the sum of the electric fields generated by q_p and its virtual image, $-q_p$

Fig. 3.13. Because the tangential component of the field vanishes on the conductor, it is sufficient to evaluate only the normal component (along the z-axis):

$$\begin{aligned}
\mathcal{E}_p(x, y, L) &= \mathcal{E}_{+q_p}^{\perp} + \mathcal{E}_{-q_p}^{\perp} \\
&= \frac{q_p}{4\pi\varepsilon(x^2 + y^2 + (L - z_p)^2)} \cos\theta - \frac{-q_p}{4\pi\varepsilon(x^2 + y^2 + (L - z_p)^2)} \cos\theta \\
&= 2\frac{q_p}{4\pi\varepsilon(x^2 + y^2 + (L - z_p)^2)} \frac{L - z_p}{\sqrt{x^2 + y^2 + (L - z_p)^2}} \\
&= \frac{q_p}{2\pi\varepsilon} \frac{L - z_p}{(x^2 + y^2 + (L - z_p)^2)^{\frac{3}{2}}}.
\end{aligned} \tag{3.47}$$

3.3.3 Induced Charge Density on Electrode

Based on Gauss's law, the charge density at (x, y, L) is related to $\mathcal{E}_p(x, y, L)$ by

$$\sigma_p(x, y, L) = -\varepsilon\, \mathcal{E}_p(x, y, L). \tag{3.48}$$

At first glance, Eq. (3.47) appears to provide a complete expression for $\mathcal{E}_p(x, y, L)$. However, note that in addition to the grounded electrode plane at $z = L$, there is another grounded electrode plane at $z = 0$, namely, the cathode, as illustrated in Fig. 3.14. To satisfy the boundary condition at the cathode plane ($z = 0$), an image charge of the space charge located at $z = z_p$ must be placed at $z = -z_p$ with opposite polarity. This image is referred to as virtual charge #2.

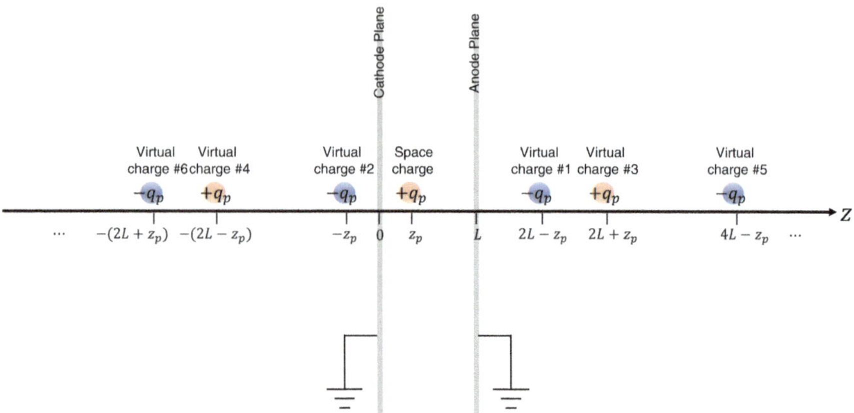

Fig. 3.14 For a space charge sandwiched between two large planar electrodes, the method of images requires an infinite sequence of image charges to simultaneously satisfy the boundary conditions imposed by both electrode planes

Similarly, the virtual "mirror charge" shown in Fig. 3.13 must also be mirrored with respect to the cathode plane. This additional image, referred to as virtual charge #3, is located at $z = -(2L - z_p)$. By continuing this process (Fig. 3.14), one finds that an infinite sequence of image charges is required to simultaneously satisfy the boundary conditions imposed by both electrode planes [3, 17].

The contribution to the electric field at the anode surface from virtual charges #2 and #3 is directed along the $-z$ axis and is given by

$$\mathcal{E}_{\#2,\#3}(x, y, L) = -\frac{q_p}{2\pi\varepsilon}\frac{L + z_p}{\left(x^2 + y^2 + (L - z_p)^2\right)^{3/2}}. \tag{3.49}$$

Likewise, the contribution from virtual charges #4 and #5 in Fig. 3.14 is directed along the $+z$ axis and is given by

$$\mathcal{E}_{\#4,\#5}(x, y, L) = \frac{q_p}{2\pi\varepsilon}\frac{3L - z_p}{\left(x^2 + y^2 + (3L - z_p)^2\right)^{3/2}}. \tag{3.50}$$

Therefore, when all virtual mirror charges are taken into account, the total electric field at the anode surface can be expressed as

$$\mathcal{E}_p(x, y, L) = \frac{q_p}{2\pi\varepsilon}\sum_{m=1,3,5,\ldots}\left\{\frac{mL - z_p}{\left[(mL - z_p)^2 + x^2 + y^2\right]^{3/2}} - \frac{mL + z_p}{\left[(mL + z_p)^2 + x^2 + y^2\right]^{3/2}}\right\}$$

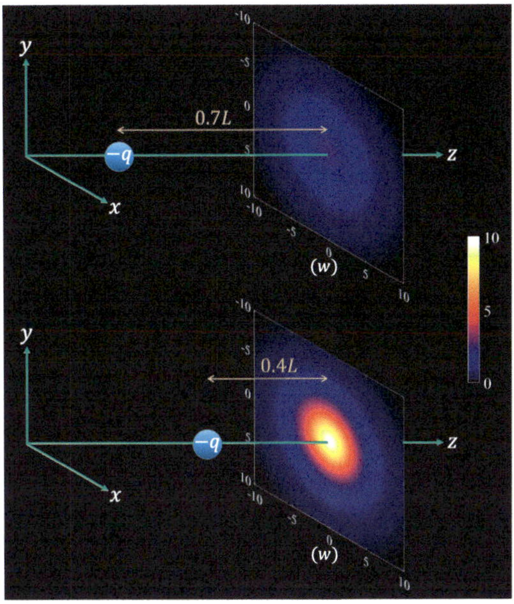

Fig. 3.15 Maps of $\sigma_n(x, y)$ induced by a negative space charge $q_n = -q$. The values of σ_n were normalized by q/L. When $-q$ moved from $z = 0.7L$ to $z = 0.4L$ to be closer to the electrode, the induced surface charges redistributed

$$= \frac{q_p}{2\pi\varepsilon} \sum_{m=1,3,5,\ldots} G(x, y, z_p, mL). \tag{3.51}$$

and

$$G(x, y, z, L) = \frac{L - z}{[(L - z)^2 + x^2 + y^2]^{3/2}} - \frac{L + z}{[(L + z)^2 + x^2 + y^2]^{3/2}}. \tag{3.52}$$

According to Eq. (3.48), the surface charge area density is related to the surface electric field:

$$\begin{aligned} \sigma_p(x, y, L) &= -\varepsilon\, \mathcal{E}_p(x, y, L) \\ &= -\frac{q_p}{2\pi} \sum_m G(x, y, z_p, mL). \end{aligned} \tag{3.53}$$

Using the similar method, it can be shown that the surface density of electron-induced charge on the anode plane is given by

$$\sigma_n(x, y, L) = -\frac{q_n}{2\pi} \sum_m G(x, y, z_n, mL), \tag{3.54}$$

where q_n is negative and z_n denotes the z location of the electrons.

Example maps of $\sigma_n(x, y)$ calculated using Eq. (3.54) are shown in Fig. 3.15. When the negative space charge moves closer to the electrode, the induced surface charges redistributed: locations closer to the z-axis experienced stronger increases in the induced surface charge.

Based on the principle of superposition, the surface density of net induced charge on the anode plane is

$$\begin{aligned}\sigma_{\text{net}}(x, y, L) &= \sigma_p(x, y, L) + \sigma_n(x, y, L)\\ &= -\frac{q_p}{2\pi}\sum_m G(x, y, z_p, mL) - \frac{q_n}{2\pi}\sum_m G(x, y, z_n, mL)\\ &= -\frac{ne}{2\pi}\sum_m \left[G(x, y, z_p, mL) - G(x, y, z_n, mL)\right],\end{aligned} \tag{3.55}$$

where we used $q_p = ne$ and $q_n = -ne$. Because z_p and z_n are time dependent according to Eq. (3.45), σ_{net} is also a function of time.

3.3.4 *Induced Charge and Current for Each Pixelated Electrode*

Based on the formula of $\sigma_{\text{net}}(x, y)$ in Eq. (3.55), the induced charge and induced current in an arbitrary pixel on the anode plane can be calculated: for a pixel with a dimension of $w_x \times w_y$ centered at (x_0, y_0), the induced charge is

$$\begin{aligned}Q &= \int_{-\frac{w_x}{2}+x_0}^{+\frac{w_x}{2}+x_0} \mathrm{d}x \int_{-\frac{w_y}{2}+y_0}^{+\frac{w_y}{2}+y_0} \mathrm{d}y\, \sigma_{\text{net}}(x, y)\\ &= -\frac{ne}{2\pi}\sum_m [\mathcal{F}(x_0, y_0, z_p, mL) - \mathcal{F}(x_0, y_0, z_n, mL)].\end{aligned} \tag{3.56}$$

The formula of function $\mathcal{F}$ is provided in Appendix 2 of this chapter.

The change of Q over time gives the induced current:

$$\begin{aligned}I &= \frac{\partial Q}{\partial t}\\ &= -\frac{ne}{2\pi}\sum_m \left[\frac{\partial}{\partial z_p}\mathcal{F}(x_0, y_0, z_p, mL)\frac{\partial z_p}{\partial t} - \frac{\partial}{\partial z_n}\mathcal{F}(x_0, y_0, z_n, mL)\frac{\partial z_n}{\partial t}\right]\\ &= \frac{ne}{2\pi}\sum_m \Bigg\{[S(x_0, y_0, mL - z_p) + S(x_0, y_0, mL + z_p)]\frac{\partial z_p}{\partial t}\\ &\quad -[S(x_0, y_0, mL - z_n) + S(x_0, y_0, mL + z_n)]\frac{\partial z_n}{\partial t}\Bigg\}.\end{aligned} \tag{3.57}$$

The formula of function S is provided in Appendix 2 of this chapter.

Table 3.1 Detector parameters and physical properties of two semiconductor materials used in the implementation. ϵ: average energy to create an electron–hole pair

w [mm]	0.1, 0.3, 1.0	
L [mm]	0.75, 1.6	
E [keV]	70	
$\mathcal{E}_0$ [V/cm]	5×10^3	
z_0 [mm]	0.5	
Semiconductor	Si	CdTe
ϵ [eV]	3.62	4.43
$n = E/\epsilon$	1.93×10^4	1.58×10^4
μ_n [cm^2/(V·s)]	1400	1100
μ_p [cm^2/(V·s)]	480	100

According to Eq. (3.45),

$$\frac{\partial z_p}{\partial z_t} = \begin{cases} -\mu_p \mathcal{E}_0 & 0 < t \leq t_2 \\ 0 & t > t_2 \end{cases}$$
$$\frac{\partial z_n}{\partial z_t} = \begin{cases} \mu_n \mathcal{E}_0 & 0 < t \leq t_1 \\ 0 & t > t_1 \end{cases} \tag{3.58}$$

Equations (3.57)–(3.58) provide the formula of induced current for an arbitrary pixelated anode. They demonstrate that semiconductors with larger μ_n and μ_p tend to generate a stronger induced current. For a given detector, the peaks of the induced charge and the induced current are linearly proportional to n, namely, linearly proportional to the energy deposited by the x-ray photon. This lays the foundation of estimating x-ray energy based on the height of the induced charge or current pulse.

3.3.5 *Implementation and Applications of Sect. 3.3.4*

The formulas presented in Sect. 3.3.4 were used to calculate the induced charge and current in pixelated PCDs with varying pixel sizes, thicknesses, and semiconductor materials, as listed in Table 3.1. We assumed that the x-ray interaction occurred at $(0, 0, z_0)$, fixed at 0.5 mm regardless of the material. For a given PCD design and a pixel centered at (x_0, y_0), the induced charge was calculated using Eq. (3.56). In the implementation, the summation index m was truncated at 21.

3.3.5.1 Pulse Induced in the Central Pixel

Figure 3.16a–b show the induced charge and current pulses in the central pixel ($x_0 = 0, y_0 = 0$) of a Si-based PCD with thickness $L = 0.75$ mm and a large pixel size ($w = 1.0$ mm, $w/L = 1.3$). Both electron and hole contributions are visible: for

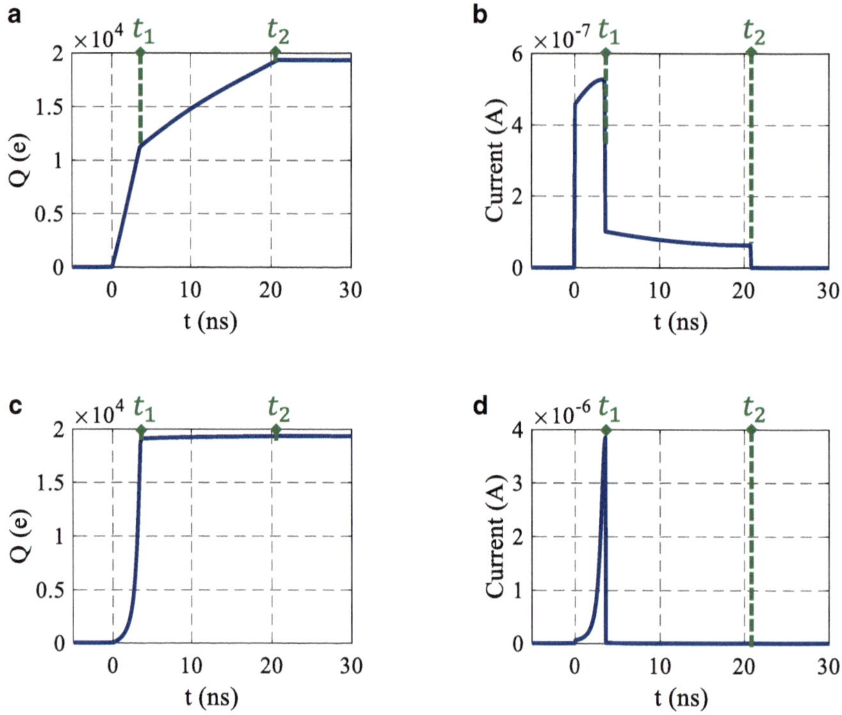

Fig. 3.16 Induced charge (left column) and current (right column) at the central pixel of a Si PCD. (**a**) and (**b**) were calculated under a large pixel condition ($w = 1.0\,\text{mm}$, $L = 0.75\,\text{mm}$); (**c**) and (**d**) were calculated under a small-pixel condition ($w = 0.1\,\text{mm}$, $L = 0.75\,\text{mm}$)

$t < t_1$, the Q–t curve rises steeply as electrons drift to the anode; for $t \in [t_1, t_2]$, electrons have stopped, while holes continue drifting, reducing the slope of $Q(t)$ and the induced current; for $t \geq t_2$, all carriers have reached the electrodes and the current drops to zero. Figure 3.16c–d show the pulses for a smaller anode pixel ($w = 0.1$ mm, $w/L = 0.13$), where $Q(t)$ is nearly flat and the current is zero during $t \in [t_1, t_2]$, indicating that hole contributions are negligible under the small-pixel condition.

Figure 3.17 illustrates the effect of a semiconductor material on the pulse shape. For a given PCD geometry, Si exhibits larger charge and current pulses than CdTe. The charge pulse magnitude is proportional to ne, and Si produces more electron–hole pairs per absorbed energy. The current pulse depends on both ne and electron mobility; silicon's higher mobility and larger n result in stronger and faster-rising pulses compared with CdTe.

3.3.5.2 Pulse Induced in Adjacent Pixel

Figure 3.18 compares the electric signals of the central pixel with an adjacent pixel centered at $(w, 0)$. For the adjacent pixel, the Q–t peak is only 941 e (5% of ne), and

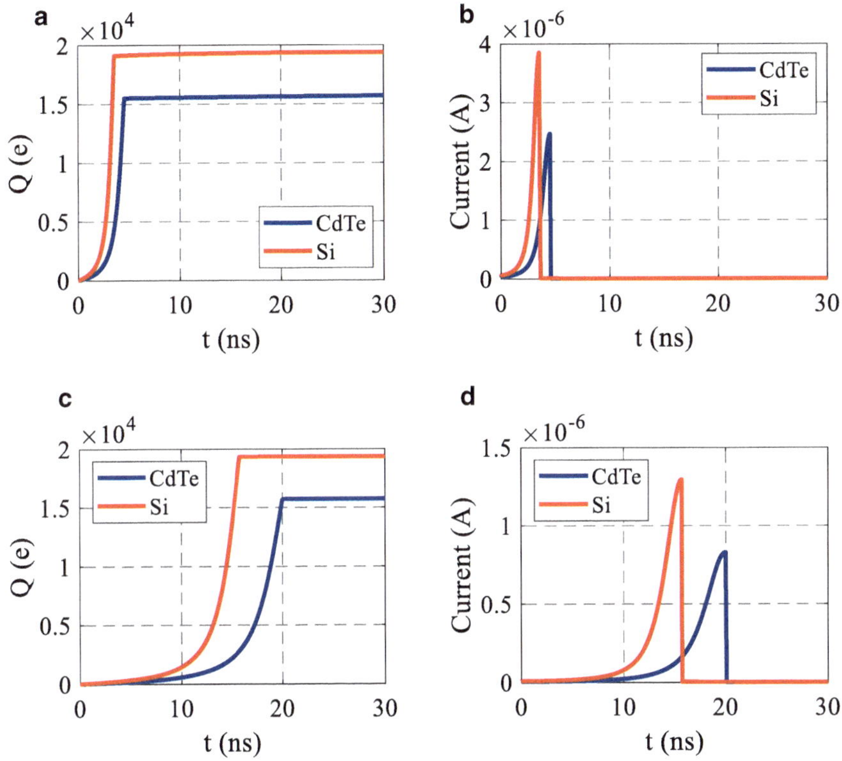

Fig. 3.17 (**a**)–(**b**): Induced charge (left column) and current (right column) at the central pixel of Si- and CdTe-based PCDs. (**a**) and (**b**) were calculated based on the geometry of a PCD in Ref. [8] ($w = 0.1$ mm, $L = 0.75$ mm); (**c**) and (**d**) were calculated based on the geometry of a PCD in Ref. [14] ($w = 0.3$ mm, $L = 1.6$ mm)

the maximum current is 1.67×10^{-8} A (2% of the central peak current). As electrons reach the anode ($t \to t_1$), Q briefly becomes negative because the weighting field is nearly tangential at the adjacent pixel and does not induce charge there. The adjacent pixel's signal is therefore dominated by holes, and the positive-to-negative transition in induced charge produces a brief negative current, as shown in Fig. 3.18b.

3.3.6 Comparison with the Shockley–Ramo Theorem

To validate the induced charge and current calculation method described in Sect. 3.3.4, the Shockley–Ramo theorem was implemented as follows. For a typical PCD geometry ($L = 1.6$ cm, $w_x = w_y = 0.3$ mm), a 3D model was created in SolidWorks™ and exported as a .STL file for use in the MATLAB™ PDE Toolbox. To compute the weighting potential $\varphi_w(x, y, z)$ and weighting field $\mathcal{E}_w(x, y, z)$ of

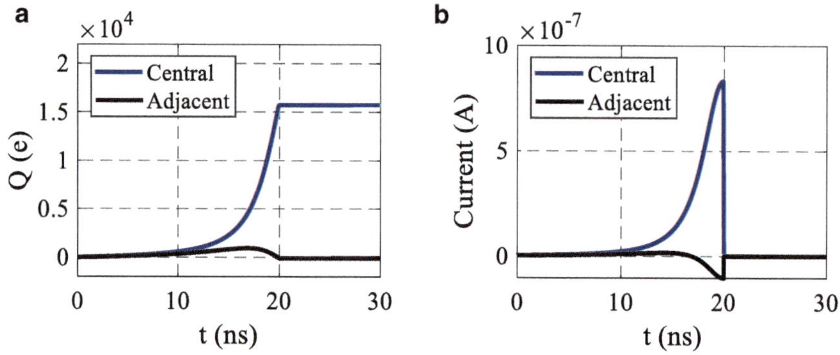

Fig. 3.18 Induced charge (**a**) and induced current (**b**) in a CdTe-based PCD with $w_x = w_y = 0.3$ mm and $L = 1.6$ mm. The central pixel is centered at $(0, 0)$; the adjacent pixel is centered at $(w, 0)$

the central anode pixel, all electrodes were set to zero voltage except the central pixel, which was biased at unit voltage. With these boundary conditions, the 3D model, and assuming zero space charge, the PDE Toolbox solved Poisson's equation to obtain $\varphi_w(x, y, z)$ and $\mathcal{E}_w(x, y, z)$ within the detector domain.

As shown in Fig. 3.19, the method in Sect. 3.3.4 and the Shockley–Ramo theorem produced equivalent initial charge and current pulses in pixelated PCDs.

3.4 Impact of Charge Trapping on Pulse Formation in PCD

3.4.1 Hecht Equation

In the analysis of the initial pulse formation in Sects. 3.2–3.3, we assumed that the number of charge carriers, n, remains constant until carriers reach the electrodes. In practice, this assumption is often violated: as discussed in Sect. 1.9.2 of Chap. 1, n decays in time due to trapping and recombination, typically with a rate of $\tau^{-1}n$, where τ is the mean lifetime of electrons or holes. Because of this time dependence, the signal model for the induced charge/current must be modified.

Consider the planar detector in Fig. 3.7 (electrode spacing L). Based on the Shockley–Ramo theorem, before electrons reach Electrode 1 and holes reach Electrode 2, the incremental change in the induced charge on Electrode 1 between t and $t + \mathrm{d}t$ is

$$\begin{aligned} \mathrm{d}Q_1(t) &= -\,q_n(t)\,\frac{v_n\,\mathrm{d}t}{L} + q_p(t)\,\frac{v_p\,\mathrm{d}t}{L} \\ &= -\big[-e\,n_{t=0}\,\exp\big(-t/\tau_n\big)\big]\frac{v_n\,\mathrm{d}t}{L} + \big[+e\,n_{t=0}\,\exp\big(-t/\tau_p\big)\big]\frac{v_p\,\mathrm{d}t}{L} \end{aligned}$$

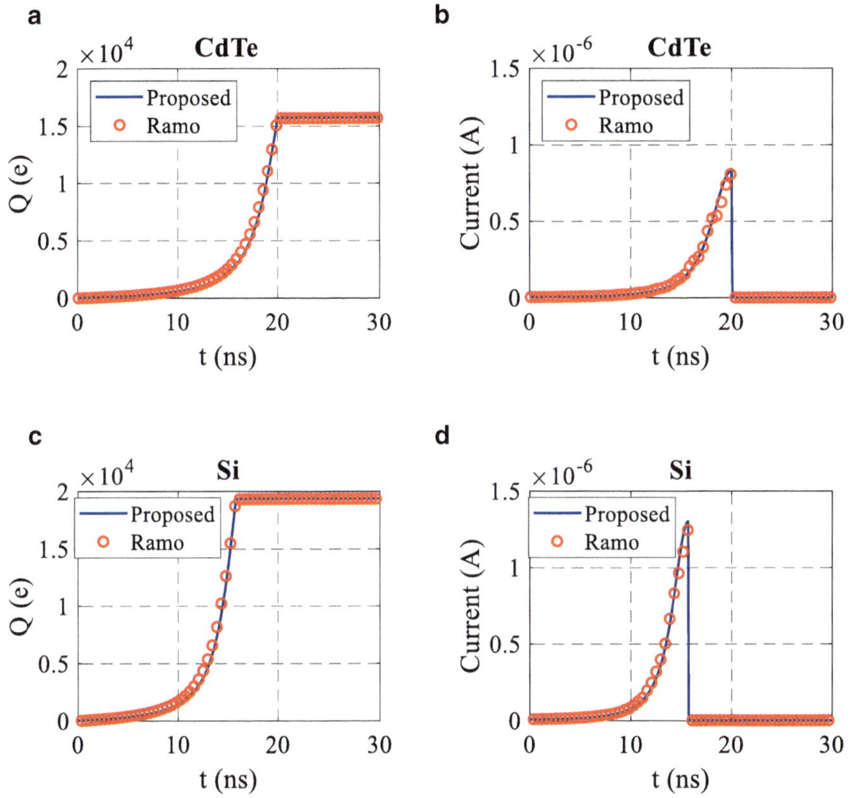

Fig. 3.19 Comparison of charge and current pulse shapes calculated using the Shockley–Ramo theorem with those calculated using the method proposed in Sect. 3.3.4. $w_x = w_y = 0.3$ mm; $L = 1.6$ mm. (**a**) Induced charge in the CdTe detector. (**b**) Induced current in the CdTe detector. (**c**) Induced charge in the Si detector. (**d**) Induced current in the Si detector

$$= e\,n_{t=0}\Big[\frac{v_n}{L}\exp\big(-t/\tau_n\big) + \frac{v_p}{L}\exp\big(-t/\tau_p\big)\Big]\,dt, \tag{3.59}$$

where $q_n(t) = -e\,n_{t=0}\exp(-t/\tau_n)$ and $q_p(t) = +e\,n_{t=0}\exp(-t/\tau_p)$ are the remaining electron and hole charges at time t, v_n and v_p are the respective drift velocities, and $n_{t=0}$ is the initial carrier density created by the interaction.

Integrating from $t' = 0$ to $t' = t$ gives the accumulated induced charge on Electrode 1:

$$Q_1(t) = e\,n_{t=0}\int_0^t \Big[\frac{v_n}{L}\exp\big(-t'/\tau_n\big) + \frac{v_p}{L}\exp\big(-t'/\tau_p\big)\Big]\,dt' \tag{3.60}$$

$$= e\,n_{t=0}\Big\{\frac{v_n\,\tau_n}{L}\Big[1-\exp\big(-t/\tau_n\big)\Big] + \frac{v_p\,\tau_p}{L}\Big[1-\exp\big(-t/\tau_p\big)\Big]\Big\}. \tag{3.61}$$

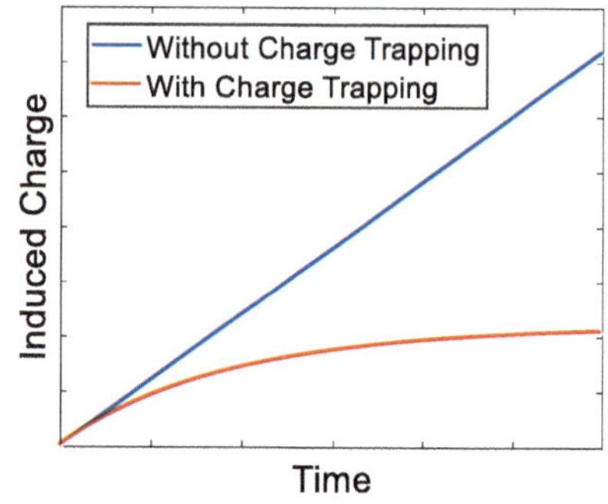

Fig. 3.20 Comparison of $Q_1(t)$ curves with and without charge carrier loss due to trapping

Equation (3.61) is the time-dependent form commonly referred to as the **Hecht equation** for this planar geometry.

In the limit where the carrier lifetimes τ_n and τ_p are much longer than the relevant drift or transit times, the exponential terms can be approximated by $\exp(-t/\tau_{n,p}) \approx 1$, and the above expression reduces to the usual trap-free induced-charge formula previously given in Eq. (3.36).

Figure 3.20 compares the time evolution of the induced charge with and without charge carrier loss due to trapping. When trapping is present, not all charge is collected, and the slope of the $Q_1(t)$ curve decreases over time, indicating a lower induced current. This reduction in collected charge leads to an underestimation of the initial charge created by radiation, a phenomenon known as **ballistic deficit**.

Remark In an ideal "ballistic" measurement, the pulse amplitude reflects the full amount of charge generated and collected. A "deficit" arises when the measured amplitude is lower than the true total charge. Ballistic deficit can be caused not only by charge trapping but also by the response of amplifiers and pulse shapers in the readout electronics. For example, if the shaping time constant is too short compared to the charge collection time, the electronics respond before the charge collection is complete, and part of the input signal does not contribute to the output pulse amplitude. We will discuss the time constant of pulse shapers in Sect. 3.5.

3.4.2 Polarization of Semiconductors Due to Charge Trapping

Electrons and holes can become trapped in a semiconductor even under a strong applied electric field, due to the presence of crystal defects and impurities that introduce localized energy minima within the bandgap. As illustrated in Fig. 3.21, when a charge carrier encounters such a defect level, it can be captured into this local potential well. Even if the trap energy level lies energetically closer to the valence band, a trapped electron does not simply "fall" spontaneously to the valence band. This is because the defect state is typically not aligned with the valence-band maximum in momentum (k) space. Since photons carry negligible

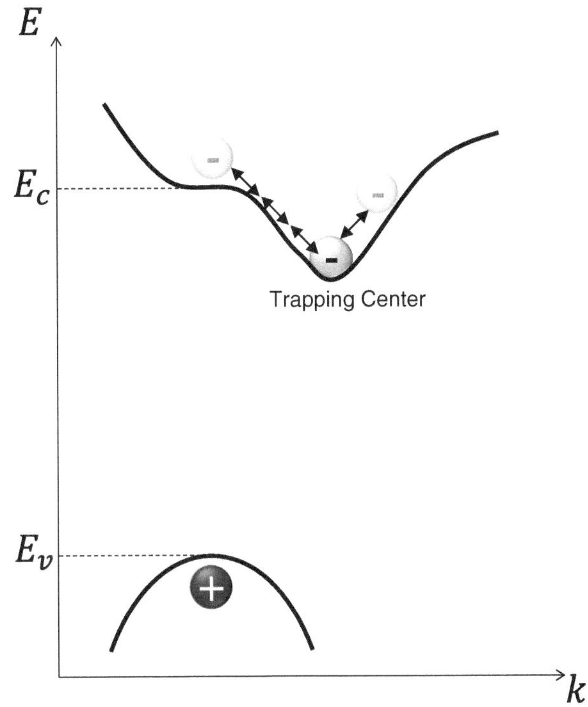

Fig. 3.21 A schematic $E - k$ band diagram of a semiconductor illustrating the presence of a trapping center and the electron trapping effect

momentum, direct radiative recombination is strongly suppressed due to momentum conservation.

Moreover, most trapping centers are not efficient recombination centers; their wave function overlap with extended band states is small, leading to a very low non-radiative transition probability as well. As a result, trapped electrons are more likely to undergo thermal transition back to the conduction band or remain localized near the trap, rather than recombining with holes in the valence band (Fig. 3.21). Consequently, trapped charge carriers are effectively removed from the charge collection process and no longer contribute to the induced current.

Initially, a semiconductor sensor is electrically neutral. When external radiation generates electron–hole pairs, these carriers, if not trapped, are eventually swept out of the detector, restoring the semiconductor to its neutral state. However, if one type of carrier (e.g., holes) becomes trapped while the other type is collected, a net space charge remains within the detector. Even if both electrons and holes are trapped, although the semiconductor remains overall electrically neutral, the spatial separation of charges causes the material to become **polarized**.

In all trapping scenarios, the accumulated space charge generates an internal electric field, which is often opposite to the applied field. This reduces the net electric field, slowing the drift velocity of the remaining mobile carriers (i.e., v_n and v_p in Eq. (3.61)). This prolongs their transit time to the electrodes, which increases the probability of further trapping, leading to a positive feedback or "avalanche" effect: eventually, the net electric field can be nearly canceled by the trapped charges. In this extreme case, very few carriers are collected, and the detector becomes almost unresponsive to incoming radiation.

Therefore, polarization resulting from charge trapping can severely degrade the radiation detection performance of a semiconductor detector. This underscores the need for advanced crystal growth and fabrication techniques to produce PCD-grade semiconductors, minimizing trapping centers and maximizing charge collection efficiency.

3.4.3 *Periodic Cleansing of Trapped Charges*

Over time, when charge carriers become trapped in a semiconductor sensor, the detector's performance can be significantly degraded or even completely lost. To restore the detector's response and maintain consistent pulse formation, the trapped charges must be periodically removed. This can be accomplished by briefly forward-biasing the semiconductor sensor.

When a diode is forward biased, a large number of electrons and holes are injected into the device. In the absence of trapped charges, these carriers travel through the semiconductor and return to the power supply. However, if there is trapped space charge in the diode, the injected electrons or holes can recombine with these trapped charges. Even though the probability of any single injected carrier reaching a trap is small, the sheer number of injected carriers ensures that the trapped charges are effectively neutralized.

As a side note, thermoluminescent dosimeters (TLDs) exploit the charge trapping effect to store information about accumulated radiation energy. In a TLD, electrons and holes can remain trapped for long periods at room temperature. The stored charge is later released by heating the TLD, which provides sufficient thermal energy for the electrons to escape the traps and recombine with holes at luminescence centers, emitting light. While the mechanisms differ, both examples illustrate how trapped charges can be deliberately manipulated to either restore detector function or extract stored information.

3.5 Amplification and Shaping of Initial Pulses

3.5.1 *Preamplification*

As shown in Fig. 3.19, the raw output of a semiconductor detector is extremely weak—typically only tens of thousands of electrons or fewer are generated per x-ray photon. Using such a small signal directly for single-photon counting is problematic.

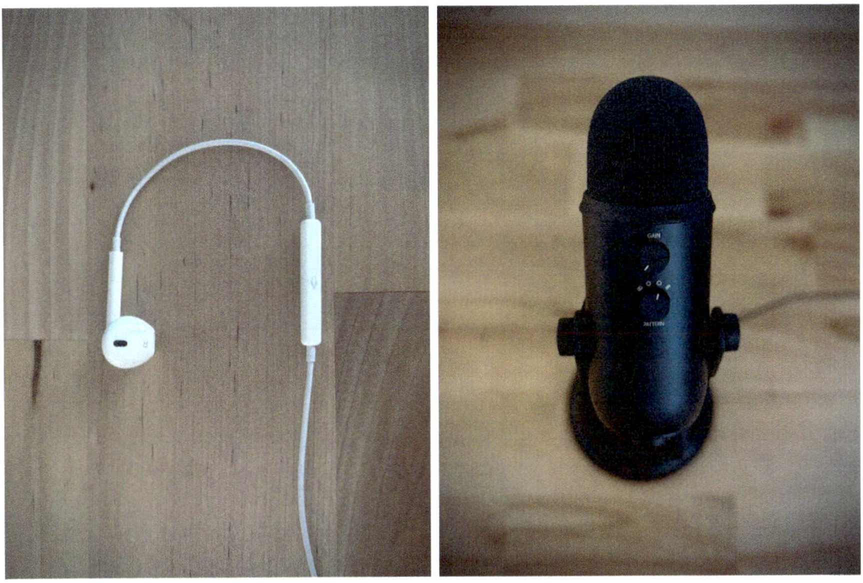

Fig. 3.22 Left: built-in microphone of a wired earphone. Right: studio-quality microphone with an integrated preamplifier

To explain the problem, an analogy can be made with voice recording. Suppose you are a YouTuber recording a guitar cover. If you use the built-in microphone of a wired earphone connected to a computer, the tiny microphone capsule contains only a small field-effect transistor to buffer the transducer's high-impedance signal. Effectively, it has no real preamplification. The resulting audio will sound faint and noisy. Turning up the computer speaker volume only amplifies the noise along with the signal.

Now consider replacing the earphone mic with a studio-quality microphone (Fig. 3.22). Such a microphone has a high-quality preamplifier, pulse shaper, and analog-to-digital converter. The recording in this case is clean, strong, and professional sounding, which may help make it a hit on YouTube.

This analogy highlights the critical role of preamplification: boosting weak signals at the source before electronic noise dominates.

3.5.2 *Quantitative Relationship Between Initial Pulses and Preamp Outputs*

In Sect. 2.9 of Chap. 2, we discussed that a preamplifier can be either voltage sensitive or charge sensitive. For radiation sensors such as semiconductor diodes, the detector capacitance may vary over time due to factors, such as temperature

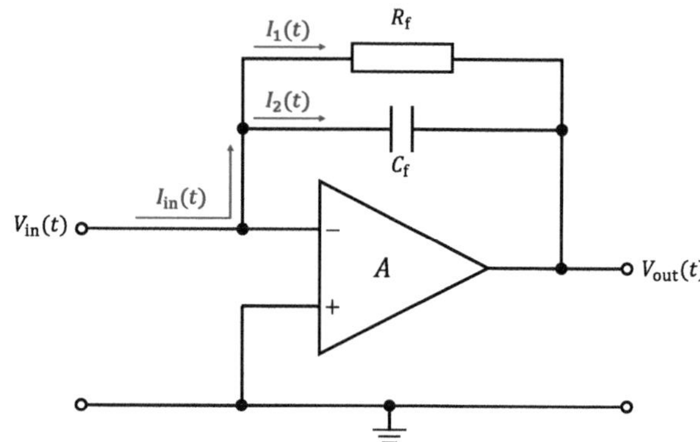

Fig. 3.23 Circuit diagram of a charge-sensitive amplifier. A denotes the open-loop gain of the op amp

fluctuations, charge trapping, and radiation damage. Consequently, the output voltage of the sensor, which is jointly determined by the output charge and capacitance of the sensor, may not reliably reflect the energy of the incident x-ray photon. For this reason, a charge-sensitive preamplifier (CSA) is often preferred.

As described in Sect. 2.9, the CSA employs a feedback capacitor (C_f) to integrate the input charge generated by the sensor. The resulting output voltage is proportional to the total collected charge:

$$\begin{aligned} V_{\text{out}} &\approx -\frac{Q_{\text{in}}}{C_f} \\ &= -\frac{\int I_{\text{in}} \mathrm{d}t}{C_f}, \end{aligned} \tag{3.62}$$

which is independent of the detector capacitance.

If a CSA includes only the feedback capacitor, the stored charge on C_f would continue to build up, ultimately driving the amplifier into saturation. To avoid this, a feedback resistor R_f is often placed in parallel with C_f, allowing the capacitor to discharge gradually and stabilizing the amplifier operation.

To derive the quantitative relationship between the input current and the output voltage of a CSA with both a feedback resistor and capacitor, we consider the circuit shown in Fig. 3.23. From basic circuit principles, the following set of equations can be written:

$$\begin{cases} I_{\text{in}}(t) = I_1(t) + I_2(t) \\ I_1(t) = \frac{V_{\text{in}}(t) - V_{\text{out}}(t)}{R_f} \\ I_2(t) = C_f \frac{\mathrm{d}[V_{\text{in}}(t) - V_{\text{out}}(t)]}{\mathrm{d}t} \\ V_{\text{out}}(t) = -A\, V_{\text{in}}(t) \end{cases} \tag{3.63}$$

Combining these relations yields

$$\frac{d[(A+1)V_{in}(t)]}{dt} + \frac{1}{R_f C_f}(A+1)V_{in}(t) = \frac{1}{C_f} I_{in}(t). \tag{3.64}$$

Since $A \gg 1$, we can approximate $(A+1)V_{in}(t) \approx A\,V_{in}(t) = -V_{out}(t)$. Under this approximation, Eq. (3.64) simplifies to

$$\frac{dV_{out}(t)}{dt} + \frac{1}{R_f C_f} V_{out}(t) = -\frac{1}{C_f} I_{in}(t). \tag{3.65}$$

To solve this partial differential equation (PDE), we multiply both sides by $e^{\frac{t}{R_f C_f}}$:

$$e^{\frac{t}{R_f C_f}} \frac{dV_{out}(t)}{dt} + \frac{1}{R_f C_f} e^{\frac{t}{R_f C_f}} V_{out}(t) = -\frac{1}{C_f} e^{\frac{t}{R_f C_f}} I_{in}(t). \tag{3.66}$$

The left-hand side is the derivative of a product:

$$e^{\frac{t}{R_f C_f}} \frac{dV_{out}(t)}{dt} + \frac{1}{R_f C_f} e^{\frac{t}{R_f C_f}} V_{out}(t) = \frac{d}{dt}\left[e^{\frac{t}{R_f C_f}} V_{out}(t)\right]. \tag{3.67}$$

Therefore,

$$\frac{d}{dt}\left[e^{\frac{t}{R_f C_f}} V_{out}(t)\right] = -\frac{1}{C_f} e^{\frac{t}{R_f C_f}} I_{in}(t). \tag{3.68}$$

Integrating both sides:

$$e^{\frac{t}{R_f C_f}} V_{out}(t) = -\frac{1}{C_f} \int_{-\infty}^{t} e^{\frac{t'}{R_f C_f}} I_{in}(t')dt' + a. \tag{3.69}$$

Multiplying through by $e^{-\frac{t}{R_f C_f}}$ leads to the general solution of $V_{out}(t)$:

$$V_{out}(t) = -\frac{1}{C_f} e^{-\frac{t}{R_f C_f}} \int_{-\infty}^{t} e^{\frac{t'}{R_f C_f}} I_{in}(t')dt' + a e^{-\frac{t}{R_f C_f}}, \tag{3.70}$$

where a is a constant determined by the initial condition.

We typically choose the time origin $t = 0$ to be immediately before the input voltage pulse arrives, so that

$$V_{out}(t \le 0) = 0. \tag{3.71}$$

Evaluating at $t = 0$ gives

$$0 = 0 - \frac{1}{C_f} \int_{-\infty}^{0} 0\, dt' + a. \quad \Rightarrow \quad a = 0. \tag{3.72}$$

Thus, the output voltage of the CSA simplifies to

$$\boxed{V_{\text{out}}(t) = -\frac{1}{C_f} \mathrm{e}^{-\frac{t}{R_f C_f}} \int_0^t \mathrm{e}^{\frac{t'}{R_f C_f}} I_{\text{in}}(t') \mathrm{d}t'}. \tag{3.73}$$

3.5.2.1 Large Time-Constant Limit

As discussed in Sect. 3.3, when an x-ray photon interacts with a semiconductor sensor, the resulting current pulse typically lasts on the order of 10 ns. In many CSAs, the feedback resistor R_f is chosen to be sufficiently large, such that $R_f C_f \gg$ the duration of the input current pulse. Under this condition, the input current can be approximated as a Dirac delta function:

$$I_{\text{in}}(t) \approx Q_0 \, \delta(t). \tag{3.74}$$

According to Eq. (3.73), the corresponding CSA output voltage is

$$\begin{aligned} V_{\text{out}}(t) &\approx -\frac{Q_0}{C_f} \mathrm{e}^{-t/(R_f C_f)} \int_0^t \mathrm{e}^{t'/(R_f C_f)} \delta(t') \, \mathrm{d}t' \\ &= -\frac{Q_0}{C_f} \mathrm{e}^{-t/(R_f C_f)}. \end{aligned} \tag{3.75}$$

As shown in Eq. (3.75), the CSA output exhibits an extremely short **rise time**, and its **peak amplitude** is linearly proportional to the total induced charge Q_0. After reaching the peak, the output **decays exponentially**. The product $R_f C_f$ determines the decay rate: a larger $R_f C_f$ produces a slower decay. This product is therefore a key parameter of the circuit and is commonly referred to as the **time constant**.

In the limit of $R_f \to \infty$, we have $\mathrm{e}^{t'/(R_f C_f)} \to 1$, leading to

$$V_{\text{out}}(t) \to -\frac{Q_0}{C_f}, \tag{3.76}$$

which is identical to Eq. (3.62) for a CSA without a feedback resistor. This result is intuitive, as an extremely large resistor behaves as an open circuit. In this case, the CSA output corresponding to a delta-function input current becomes a **step function**

As shown in Fig. 3.24, for a given input signal, the peak value of the CSA output voltage decreases as the feedback time constant becomes shorter. This behavior indicates that the input charge is not fully integrated when the time constant is small. To ensure high-fidelity measurement of the input photon energy, it is therefore generally desirable to use a large time constant for the CSA.

To achieve a large time constant (typically on the μs level), R_f is typically chosen to be several MΩ or higher while keeping the feedback capacitance C_f small (e.g.,

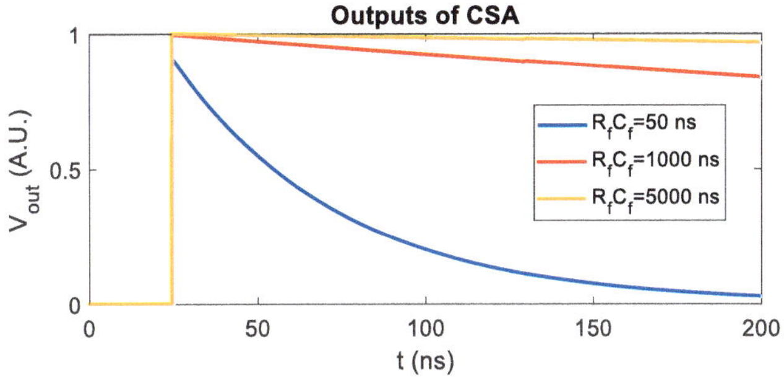

Fig. 3.24 Dependence of CSA output on the time constant. The input is an impulse function

around 0.1 pF) to obtain a high charge-to-voltage conversion gain. A large R_f also helps suppress the so-called Johnson noise, which is the random electrical noise generated by the thermal motion of charge carriers inside a resistor, even when no current is flowing.

However, with a large CSA time constant, the long tail of the output may lead to **pulse pileup**, in which consecutive output pulses overlap. This issue can be mitigated using subsequent high-pass pulse-shaping techniques, as described in Sect. 3.5.4.

Units Analysis: Time Constant

The resistance R is measured in ohms (Ω), defined as

$$\Omega = \frac{[\text{Voltage}]}{[\text{Current}]} = \frac{\text{V}}{\text{A}}. \tag{3.77}$$

The capacitance C is measured in farads (F), defined as

$$\text{F} = \frac{[\text{Charge}]}{[\text{Voltage}]} = \frac{\text{C}}{\text{V}}. \tag{3.78}$$

Therefore, the unit of the product RC is

$$\begin{aligned}[RC] &= \frac{\text{V}}{\text{A}} \cdot \frac{\text{C}}{\text{V}} \\ &= \frac{\text{C}}{\text{A}} \\ &= \text{s}.\end{aligned} \tag{3.79}$$

3.5.3 Periodic Reset of Preamp

As discussed in Sect. 3.5.2.1, for a CSA with a large time constant, the output voltage in response to a sequence of input current pulses resembles a staircase. Each pulse deposits charge onto the feedback capacitor, causing the output voltage to increase incrementally over time as charge accumulates.

Recall that preamplifiers and operational amplifiers are built from transistors, which can operate linearly only within a limited voltage range. When the output voltage V_{out} approaches the supply rail of the circuit (e.g., the positive voltage supply V_{cc} in the configuration shown in Fig. 2.35 of Chap. 2), the transistors inside the amplifier can no longer remain in their forward-active region. In this regime, the base current is cut off, and the device can no longer linearly amplify the input signal.

Consequently, if the CSA output voltage continues to rise due to accumulated charge, it will eventually reach this saturation limit, at which point the amplifier response becomes nonlinear. This saturation compromises both photon-counting accuracy and x-ray energy estimation. To prevent such accumulation, practical CSA designs implement a **periodic reset mechanism** to quickly discharge the feedback capacitor.

Note that the feedback resistor provides a passive pathway for discharging the feedback capacitor, but this discharge is very slow due to the large resistance value. In practice, an active reset circuit is often employed to speed up this process. For example, a MOSFET can be connected across the feedback capacitor. When a periodic reset signal is applied to the MOSFET gate, the transistor temporarily conducts, rapidly discharging the stored charge on C_f. After the reset pulse ends, the MOSFET turns off, and normal charge integration resumes. This periodic reset ensures stable CSA operation and prevents the output from saturating during long acquisition periods.

However, as shown in Fig. 3.25, the reset produces a sharp voltage gradient at the output, which can generate a large signal in the subsequent pulse-shaping circuit. This may cause false counts or overload the shaping circuit. To avoid these issues,

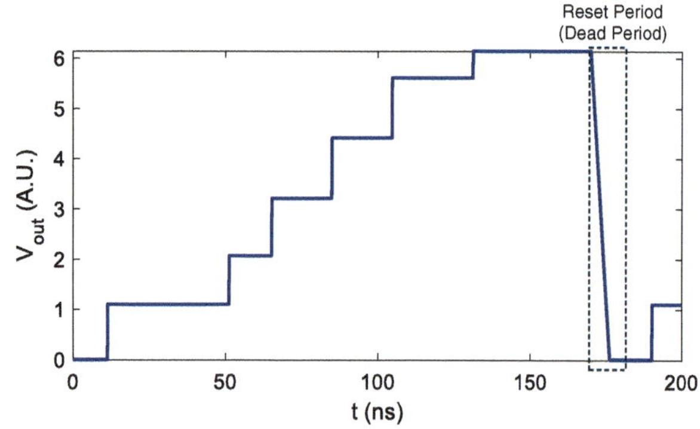

Fig. 3.25 Active reset of the feedback resistor in a CSA creates an abrupt change in the output voltage

counting is typically inhibited during the reset period. This introduces a brief dead-time window, during which information about incident x-rays is lost [10].

3.5.4 High-Pass Pulse-Shaping Circuit

3.5.4.1 Basic Circuit Design

As discussed in Sect. 3.5, the output of the CSA is a staircase-like waveform, with the voltage magnitude progressively increasing until the CSA is reset. If this raw staircase signal were directly used for photon counting or energy estimation, it would introduce severe biases because the signal amplitude depends on time rather than being determined solely by the amount of charges produced by the external radiation.

To address this issue, the CSA output must undergo additional processing, i.e., "shaping." In PCDs, the fundamental components of a pulse shaper are various combinations of resistors (R) and capacitors (C), as illustrated in Fig. 3.26. While many readers may already be familiar with resistors and Ohm's law, capacitors and their electrical behavior can feel less intuitive. The analysis presented in this section is intended to help you build confidence in analyzing capacitors and understanding RC/CR circuits.

Typically, the first step of pulse shaping is to apply a high-pass filter so that only the rising edges of the staircase are preserved. As analyzed in Sect. 3.2, the slope of these edges is proportional to the number of charge carriers n. Therefore, it is sufficient to infer the input photon energy from the edges themselves.

The RC high-pass filter circuit shown in Fig. 3.26a is equivalent to the schematic in Fig. 3.27. In this configuration, **a capacitor C is placed in series with the input voltage signal $V_{in}(t)$**, while **the output voltage $V_{out}(t)$ is taken across a resistor**

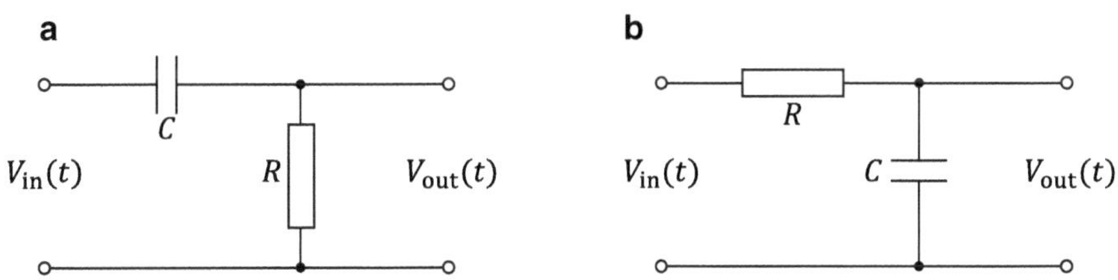

Fig. 3.26 (**a**) C–R pulse shaper (high-pass filter). (**b**) R–C pulse shaper (low-pass filter)

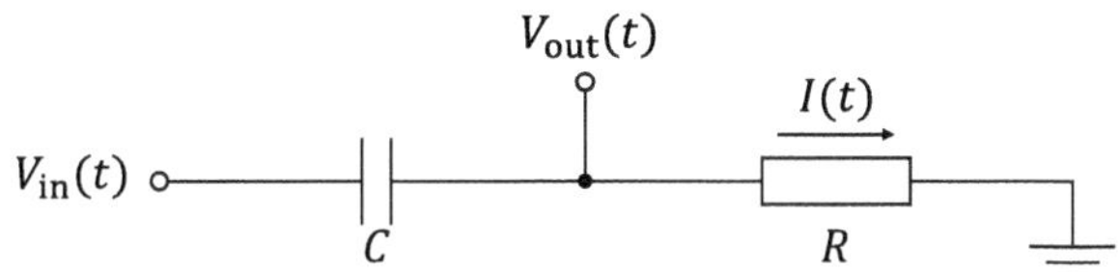

Fig. 3.27 Illustration of a CR high-pass filter

R. At low frequencies, the capacitor impedes current flow, resulting in a negligible voltage drop across the resistor, so that $V_{\text{out}}(t) \approx 0$. At high frequencies, the capacitor effectively acts as a short circuit, allowing the input signal to pass through such that $V_{\text{out}}(t) \approx V_{\text{in}}(t)$.

3.5.4.2 Quantitative Relationship Between Input and Output Voltages

The quantitative relationship between $V_{\text{in}}(t)$ and $V_{\text{out}}(t)$ can be derived from basic circuit principles.

From Fig. 3.27, Ohm's law applied to the resistor gives

$$V_{\text{out}}(t) = I(t)R. \tag{3.80}$$

The charge stored on the capacitor is proportional to its capacitance and the voltage across it, i.e.,

$$Q(t) = [V_{\text{in}}(t) - V_{\text{out}}(t)]C. \tag{3.81}$$

The current into the capacitor toward the resistor direction is the time derivative of this charge:

$$I(t) = \frac{\mathrm{d}Q(t)}{\mathrm{d}t}. \tag{3.82}$$

Combining Eqs. (3.80)–(3.82) yields

$$\frac{V_{\text{out}}(t)}{R} = \frac{\mathrm{d}}{\mathrm{d}t}\big[V_{\text{in}}(t) - V_{\text{out}}(t)\big]C \tag{3.83}$$

or equivalently,

$$\frac{\mathrm{d}V_{\text{out}}(t)}{\mathrm{d}t} + \frac{1}{RC}V_{\text{out}}(t) = \frac{\mathrm{d}V_{\text{in}}(t)}{\mathrm{d}t}. \tag{3.84}$$

To solve this first-order PDE, we apply the same technique used in solving the CSA problem in Eq. (3.65), namely, multiplying through by the integrating factor $\mathrm{e}^{\frac{t}{RC}}$:

$$\mathrm{e}^{\frac{t}{RC}}\frac{\mathrm{d}V_{\text{out}}(t)}{\mathrm{d}t} + \frac{\mathrm{e}^{\frac{t}{RC}}}{RC}V_{\text{out}}(t) = \mathrm{e}^{\frac{t}{RC}}\frac{\mathrm{d}V_{\text{in}}(t)}{\mathrm{d}t}. \tag{3.85}$$

This leads to

$$\frac{\mathrm{d}}{\mathrm{d}t}\left(\mathrm{e}^{\frac{t}{RC}}V_{\text{out}}(t)\right) = \mathrm{e}^{\frac{t}{RC}}\frac{\mathrm{d}V_{\text{in}}(t)}{\mathrm{d}t}. \tag{3.86}$$

Integrating both sides:

$$\begin{aligned} e^{\frac{t}{RC}} V_{out}(t) &= \int_{-\infty}^{t} e^{\frac{t'}{RC}} \, d[V_{in}(t')] + a \\ &= e^{\frac{t}{RC}} V_{in}(t) - \frac{1}{RC} \int_{-\infty}^{t} e^{\frac{t'}{RC}} V_{in}(t') \, dt + a \end{aligned} \tag{3.87}$$

Multiplying through by $e^{-\frac{t}{RC}}$ leads to the general solution of $V_{out}(t)$:

$$V_{out}(t) = V_{in}(t) - \frac{e^{-\frac{t}{RC}}}{RC} \int_{-\infty}^{t} e^{\frac{t'}{RC}} V_{in}(t') \, dt' + a \, e^{-\frac{t}{RC}}, \tag{3.88}$$

where a is a constant determined by the initial condition.

We typically choose the time origin $t = 0$ to be immediately before the input voltage pulse arrives, so that

$$V_{out}(t \leq 0) = 0. \tag{3.89}$$

Evaluating at $t = 0$ gives

$$0 = 0 - \frac{e^{-\frac{t}{RC}}}{RC} \int_{-\infty}^{0} e^{\frac{t'}{RC}} 0 \, dt' + a. \quad \Rightarrow \quad a = 0. \tag{3.90}$$

Thus, the output voltage simplifies to

$$V_{out}(t) = V_{in}(t) - \frac{e^{-\frac{t}{RC}}}{RC} \int_{-\infty}^{t} e^{\frac{t'}{RC}} V_{in}(t') \, dt'. \tag{3.91}$$

3.5.4.3 Example 1: Response to Step Input

Using the Heaviside function, $h(t)$, the input voltage can be expressed as

$$V_{in}(t) = V_0 \cdot h(t) \tag{3.92}$$

According to Eq. (3.91), when $t \geq 0$,

$$\begin{aligned} V_{out}(t) &= V_0 - \frac{V_0}{RC} e^{-\frac{t}{RC}} \int_{0}^{t} e^{\frac{t'}{RC}} \, dt' \\ &= V_0 \, e^{-\frac{t}{RC}}. \end{aligned} \tag{3.93}$$

As shown in Eq. (3.93), the output voltage of the high-pass pulse filter decays exponentially from V_0 to zero. The product RC is the time constant of the high-

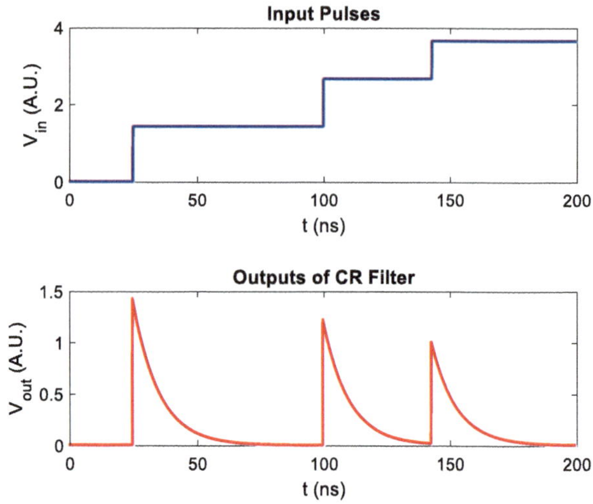

Fig. 3.28 Response of a CR high-pass filter to step inputs, assuming a time constant of 10 ns

pass filter and governs the decay rate: a smaller time constant results in a quicker decay.

As an example, Fig. 3.28 illustrates the response of a CR high-pass filter with a 10 ns time constant to step inputs. The peak amplitude of the output pulses matches the step height of the corresponding input pulses.

3.5.4.4 Example 2: Response to Sinusoidal Input

For a sinusoidal input voltage signal,

$$V_{\text{in}}(t) = V_0 \sin(\omega t), \tag{3.94}$$

where $\omega = 2\pi f$ is the angular frequency of the input signal. According to Eq. (3.91), the output voltage signal is

$$\begin{aligned}
V_{\text{out}}(t) &= V_0 \sin(\omega t) - \frac{V_0}{RC}\,\mathrm{e}^{-\frac{t}{RC}} \int_0^t \mathrm{e}^{\frac{t'}{RC}} \sin(\omega t')\,\mathrm{d}t' \\
&= V_0 \sin(\omega t) - \frac{V_0}{RC}\,\mathrm{e}^{-\frac{t}{RC}}\,\mathrm{Im}\left\{\int_0^t \mathrm{e}^{\left(\frac{1}{RC}+\mathrm{i}\omega\right)t'}\,\mathrm{d}t'\right\} \\
&= V_0 \sin(\omega t) - \frac{V_0}{RC}\,\mathrm{e}^{-\frac{t}{RC}}\,\mathrm{Im}\left\{\left(\frac{1}{RC}+\mathrm{i}\omega\right)^{-1}\left[\mathrm{e}^{\left(\frac{1}{RC}+\mathrm{i}\omega\right)t}-1\right]\right\} \\
&= V_0 \sin(\omega t) - V_0\,\mathrm{Im}\left\{(1+\mathrm{i}\omega RC)^{-1}\left[\mathrm{e}^{\mathrm{i}\omega t}-\mathrm{e}^{-\frac{t}{RC}}\right]\right\}
\end{aligned}$$

$$
\begin{aligned}
&= V_0 \sin(\omega t) - V_0 \,\mathrm{Im}\left\{\frac{1 - \mathrm{i}\omega RC}{1 + (\omega RC)^2}\left[\cos(\omega t) + \mathrm{i}\sin(\omega t) - \mathrm{e}^{-\frac{t}{RC}}\right]\right\} \\
&= V_0 \sin(\omega t) - V_0 \sin(\omega t)\frac{1}{1 + (\omega RC)^2} \\
&\quad + V_0 \cos(\omega t)\frac{\omega RC}{1 + (\omega RC)^2} - V_0 \frac{\omega RC \mathrm{e}^{-\frac{t}{RC}}}{1 + (\omega RC)^2} \\
&= V_0 \sin(\omega t)\frac{(\omega RC)^2}{1 + (\omega RC)^2} + V_0 \cos(\omega t)\frac{\omega RC}{1 + (\omega RC)^2} - V_0 \frac{\omega RC \mathrm{e}^{-\frac{t}{RC}}}{1 + (\omega RC)^2} \\
&= \frac{V_0\, \omega RC}{\sqrt{1 + (\omega RC)^2}}\left[\sin(\omega t)\frac{\omega RC}{\sqrt{1 + (\omega RC)^2}}\right. \\
&\quad \left. + \cos(\omega t)\frac{1}{\sqrt{1 + (\omega RC)^2}} - \frac{\mathrm{e}^{-\frac{t}{RC}}}{\sqrt{1 + (\omega RC)^2}}\right] \\
&= V_\omega \sin(\omega t + \phi) - V_\omega \frac{1}{\sqrt{1 + (\omega RC)^2}}\mathrm{e}^{-\frac{t}{RC}},
\end{aligned}
\tag{3.95}
$$

where

$$
V_\omega = \frac{\omega RC}{\sqrt{1 + (\omega RC)^2}} V_0 = \frac{1}{\sqrt{1 + (\omega RC)^{-2}}} V_0 \tag{3.96}
$$

$$
\phi = \tan^{-1}\left(\frac{1}{\omega RC}\right) \tag{3.97}
$$

3.5.4.5 Cutoff Frequency

When the system reaches a steady state, term $V_\omega \frac{\omega RC}{\sqrt{1+(\omega RC)^2}} \mathrm{e}^{-\frac{t}{RC}}$ in Eq. (3.95) decays to zero. As a result, the output voltage can be approximated by

$$
\begin{aligned}
V_{\mathrm{out}}(t) &= V_\omega \sin(\omega t + \phi) \\
&= \frac{V_0}{\sqrt{1 + (\omega RC)^{-2}}} \sin(\omega t + \phi).
\end{aligned}
\tag{3.98}
$$

The amplitude of the output voltage, V_ω, **increases with increasing frequency**, demonstrating the high-pass behavior of the filter. By convention, the **cutoff frequency** f_c is defined as the frequency at which V_ω reaches $1/\sqrt{2}$ (approximately 70.7%) of V_0. According to Eq. (3.96),

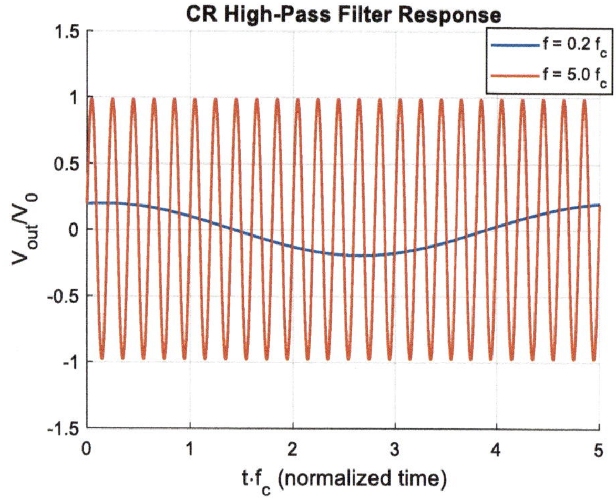

Fig. 3.29 CR high-pass filter response to sinusoidal inputs of varying frequencies

$$\frac{V_0}{\sqrt{2}} = \frac{V_0}{\sqrt{1 + (2\pi f_c RC)^{-2}}} \quad \Rightarrow \quad \boxed{f_c = \frac{1}{2\pi RC}}. \tag{3.99}$$

Using f_c, Eq. (3.98) can be written as

$$V_{\text{out}}(t) = \frac{V_0}{\sqrt{1 + \left(\frac{f_c}{f}\right)^2}} \sin(\omega t + \phi). \tag{3.100}$$

At low frequencies where $f \ll f_c$, the signal is attenuated. At high frequencies where $f \gg f_c$, $|V_{\text{out}}| \approx V_0$, meaning the signal is transmitted with negligible attenuation (Fig. 3.29).

3.5.5 *Low-Pass Pulse-Shaping Circuit*

3.5.5.1 Basic Circuit Design

Although the high-pass pulse-shaping circuit described in Sect. 3.5.4 preserves edges and other high-frequency components of the signal, it also transmits high-frequency electronic noise. Without additional noise suppression, such noise can lead to spurious counts. For this reason, the high-pass filter is typically combined with a low-pass shaper to limit the noise bandwidth.

The RC low-pass filter circuit shown in Fig. 3.26b is equivalent to the schematic in Fig. 3.30. In this configuration, **a resistor** R **is placed in series with the input** voltage signal, $V_{\text{in}}(t)$, while **a capacitor** C **is connected in parallel with the output** voltage signal, $V_{\text{out}}(t)$. At low frequencies, the capacitor behaves almost

Fig. 3.30 Illustration of a RC low-pass filter

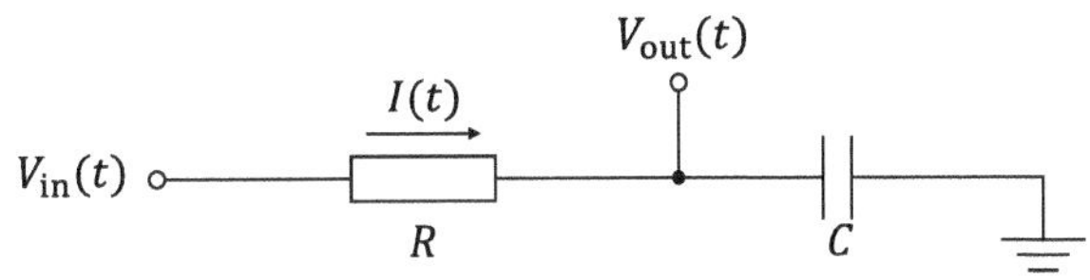

like an open circuit, allowing most of the signal to appear across the output. At high frequencies, however, the capacitor provides a low-impedance path to ground, effectively shorting out the signal and suppressing rapid variations. In this way, the RC circuit selectively attenuates high-frequency noise while passing the slower signal components.

3.5.5.2 Quantitative Relationship Between Input and Output Voltages

The quantitative relationship between $V_{in}(t)$ and $V_{out}(t)$ can be derived from basic circuit principles.

From Fig. 3.30, Ohm's law applied to the resistor gives

$$V_{in}(t) - V_{out}(t) = I(t)R \quad \Rightarrow \quad I(t) = \frac{1}{R}\big[V_{in}(t) - V_{out}(t)\big]. \tag{3.101}$$

The charge stored on the capacitor is proportional to its capacitance and the voltage across it, i.e.,

$$Q(t) = V_{out}(t)C. \tag{3.102}$$

The current into the capacitor is the time derivative of this charge:

$$I(t) = \frac{dQ(t)}{dt}. \tag{3.103}$$

Combining Eqs. (3.101)–(3.103) yields

$$\frac{d}{dt}\big[V_{out}(t)C\big] = \frac{1}{R}\big[V_{in}(t) - V_{out}(t)\big]. \tag{3.104}$$

or equivalently,

$$\frac{dV_{out}(t)}{dt} + \frac{1}{RC}V_{out}(t) = \frac{1}{RC}V_{in}(t). \tag{3.105}$$

To solve this first-order PDE, we apply the same technique used in solving the CSA problem in Eq. (3.65), namely, multiplying through by the integrating factor $e^{\frac{t}{RC}}$:

$$e^{\frac{t}{RC}} \frac{dV_{out}(t)}{dt} + \frac{1}{RC} e^{\frac{t}{RC}} V_{out}(t) = \frac{1}{RC} e^{\frac{t}{RC}} V_{in}(t). \tag{3.106}$$

This leads to

$$\frac{d}{dt}\left(e^{\frac{t}{RC}} V_{out}(t)\right) = \frac{1}{RC} e^{\frac{t}{RC}} V_{in}(t). \tag{3.107}$$

Integrating both sides:

$$e^{\frac{t}{RC}} V_{out}(t) = \frac{1}{RC} \int_{-\infty}^{t} e^{\frac{t'}{RC}} V_{in}(t')\, dt' + a \tag{3.108}$$

Multiplying through by $e^{-\frac{t}{RC}}$ leads to the general solution of $V_{out}(t)$:

$$V_{out}(t) = \frac{e^{-\frac{t}{RC}}}{RC} \int_{-\infty}^{t} e^{\frac{t'}{RC}} V_{in}(t')\, dt' + a\, e^{-\frac{t}{RC}}, \tag{3.109}$$

where a is a constant determined by the initial condition.

By applying the boundary condition of $V_{out}(t \leq 0) = 0$, we get $a = 0$. Thus, the expression for $V_{out}(t)$ simplifies to

$$V_{out}(t) = \frac{e^{-\frac{t}{RC}}}{RC} \int_{-\infty}^{t} e^{\frac{t'}{RC}} V_{in}(t')\, dt'. \tag{3.110}$$

In essence, the capacitor in the RC filter acts as a charge reservoir, storing energy during peaks and releasing it between them to smooth the waveform, while the resistor provides a discharge path.

3.5.5.3 Example 1: Response to Step Input

Using the Heaviside function, $h(t)$, the input voltage can be expressed as

$$V_{in}(t) = V_0 \cdot h(t) \tag{3.111}$$

According to Eq. (3.110), when $t \geq 0$,

$$\begin{aligned} V_{out}(t) &= \frac{V_0}{RC} e^{-\frac{t}{RC}} \int_{0}^{t} e^{\frac{t'}{RC}}\, dt' \\ &= V_0 \left(1 - e^{-\frac{t}{RC}}\right). \end{aligned} \tag{3.112}$$

As shown in Eq. (3.112), the output voltage of the low-pass pulse filter rises exponentially from $t = 0$ to V_0. The product RC is the time constant and governs the rise rate: a smaller time constant results in a slower rise.

As an example, for $R = 1$, kΩ and $C = 1$, pF, the time constant is

$$\begin{aligned} RC &= 1 \times 10^3\ [\Omega] \cdot 1 \times 10^{-12}\ [\mathrm{F}] \\ &= 1 \times 10^{-9}\ [\mathrm{s}] \\ &= 1\ [\mathrm{ns}] \end{aligned} \tag{3.113}$$

This small time constant helps ensure fast signal response and high temporal resolution.

3.5.5.4 Example 2: Response to Sinusoidal Input

For a sinusoidal input voltage signal,

$$V_{\mathrm{in}}(t) = V_0 \sin(\omega t), \tag{3.114}$$

where $\omega = 2\pi f$ is the angular frequency of the input signal. According to Eq. (3.110), the output voltage signal is

$$\begin{aligned} V_{\mathrm{out}}(t) &= \frac{V_0}{RC}\, \mathrm{e}^{-\frac{t}{RC}} \int_0^t \mathrm{e}^{\frac{t'}{RC}} \sin(\omega t')\, \mathrm{d}t' \\ &= \frac{V_0}{RC}\, \mathrm{e}^{-\frac{t}{RC}}\, \mathrm{Im} \left\{ \int_0^t \mathrm{e}^{\left(\frac{1}{RC} + \mathrm{i}\omega\right)t'}\, \mathrm{d}t' \right\} \\ &= \frac{V_0}{RC}\, \mathrm{e}^{-\frac{t}{RC}}\, \mathrm{Im} \left\{ \left(\frac{1}{RC} + \mathrm{i}\omega \right)^{-1} \left[\mathrm{e}^{\left(\frac{1}{RC} + \mathrm{i}\omega\right)t} - 1 \right] \right\} \\ &= V_0\, \mathrm{Im} \left\{ (1 + \mathrm{i}\omega RC)^{-1} \left[\mathrm{e}^{\mathrm{i}\omega t} - \mathrm{e}^{-\frac{t}{RC}} \right] \right\} \\ &= V_0\, \mathrm{Im} \left\{ \frac{1 - \mathrm{i}\omega RC}{1 + (\omega RC)^2} \left[\cos(\omega t) + \mathrm{i} \sin(\omega t) - \mathrm{e}^{-\frac{t}{RC}} \right] \right\} \\ &= \frac{V_0}{\sqrt{1 + (\omega RC)^2}}\, \frac{\sin(\omega t) - \omega RC \cos(\omega t) + \omega RC\, \mathrm{e}^{-\frac{t}{RC}}}{\sqrt{1 + (\omega RC)^2}} \\ &= V_\omega \sin(\omega t - \phi) + V_\omega \frac{\omega RC}{\sqrt{1 + (\omega RC)^2}}\, \mathrm{e}^{-\frac{t}{RC}}, \end{aligned} \tag{3.115}$$

where

$$V_\omega = \frac{V_0}{\sqrt{1 + (\omega RC)^2}} \tag{3.116}$$

$$\phi = \tan^{-1}(\omega RC) \tag{3.117}$$

3.5.5.5 Cutoff Frequency

When the system reaches a steady state, term $V_\omega \frac{\omega RC}{\sqrt{1+(\omega RC)^2}} e^{-\frac{t}{RC}}$ in Eq. (3.115) quickly decays toward zero. As a result, the output voltage is

$$\begin{aligned} V_{\text{out}}(t) &= V_\omega \sin(\omega t - \phi) \\ &= \frac{V_0}{\sqrt{1 + (\omega RC)^2}} \sin(\omega t - \phi). \end{aligned} \tag{3.118}$$

The amplitude of the output voltage, V_ω, decreases with increasing frequency, demonstrating the low-pass behavior of the filter. By convention, the **cutoff frequency** f_c is defined as the frequency at which V_ω falls to $1/\sqrt{2}$ (approximately 70.7%) of V_0. According to Eq. (3.116),

$$\frac{V_0}{\sqrt{2}} = \frac{V_0}{\sqrt{1 + (2\pi f_c RC)^2}} \quad \Rightarrow \quad \boxed{f_c = \frac{1}{2\pi RC}}. \tag{3.119}$$

Using f_c, Eq. (3.118) can be written as

$$V_{\text{out}}(t) = \frac{V_0}{\sqrt{1 + \left(\frac{f}{f_c}\right)^2}} \sin(\omega t - \phi). \tag{3.120}$$

When $f \ll f_c$, $|V_{\text{out}}| \approx V_0$ (passes). This is illustrated in Fig. 3.31, where input signals with frequencies below f_c pass through with minimal attenuation.

3.5.5.6 RC Low-Pass Filter Applied to Detector Voltage Supplier

In semiconductor diode PCDs, RC low-pass filters are used not only for shaping the CSA outputs but also for stabilizing the diode's voltage supplier. Most power supplies, especially rectifier-based ones, do not produce perfectly steady DC voltages. Instead, they contain AC ripple from the mains frequency (e.g., 50/60 Hz) and switching noise from switching regulators. Additional high-frequency noise from other circuit components can further contaminate the supply. An RC filter connected to the output of the power supply removes these unwanted ripple and

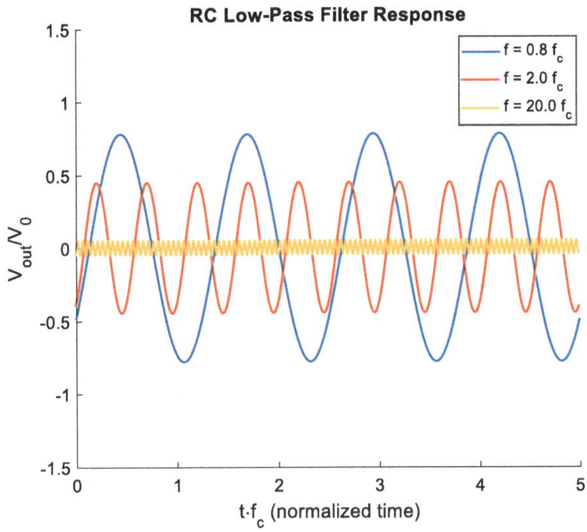

Fig. 3.31 RC low-pass filter response to sinusoidal inputs of varying frequencies

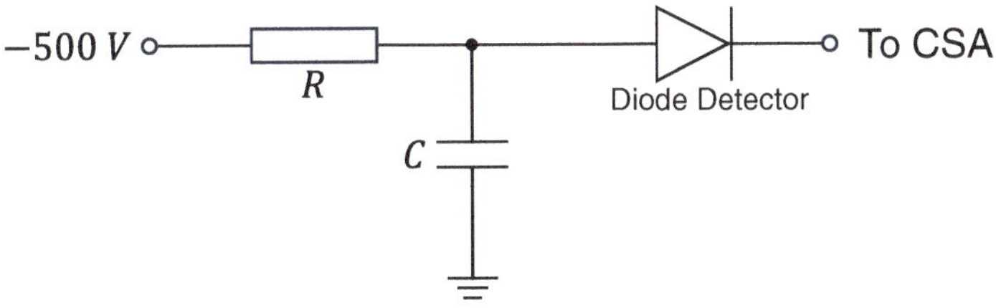

Fig. 3.32 A RC low-pass filter applied to the high-voltage supplier of a diode detector

noise components, ensuring that downstream circuits receive a stable, clean DC voltage.

Since the electric field in a semiconductor is linearly related to the applied bias voltage, stabilizing the supply voltage also stabilizes the drift velocity of charge carriers and, consequently, the output current. Figure 3.32 illustrates an example of an RC filter applied to the voltage supply of a diode detector.

3.5.6 Shaping Amplifier

The pulse shapers discussed in Sects. 3.5.4–3.5.5 do not require additional power supplies. They function purely as passive filters, shaping the voltage signal without increasing its amplitude. In many applications, however, it is desirable to further amplify the preamplifier output signal to improve the signal-to-noise ratio (SNR), energy resolution, or timing resolution. The Ortec Model 451 Spectroscopy Amplifier shown in Fig. 3.2 is an example of a **shaping amplifier**. It simultaneously shapes

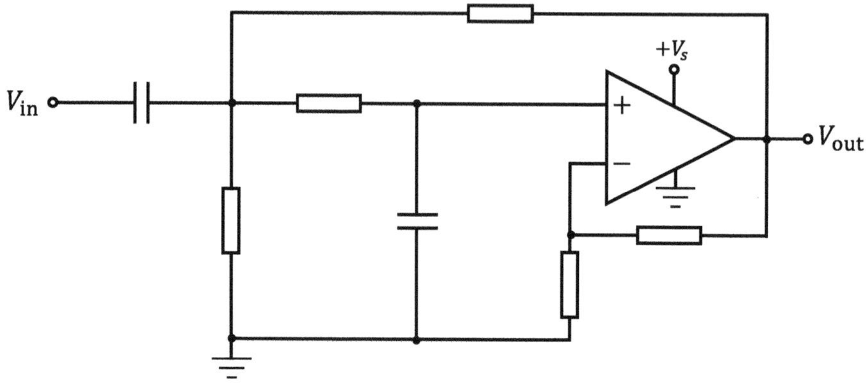

Fig. 3.33 Schematic of a basic shaping amplifier comprising an CR–RC band-pass filter network and an op amp powered by V_s

and amplifies the preamplifier signal while preserving the proportionality between pulse height and the input photon energy. Unlike a preamplifier, which typically has a fixed gain, a shaping amplifier provides adjustable gain control. It cannot replace the preamplifier, however, because its larger physical size prevents it from being mounted in close proximity to the radiation sensor, where short, low-capacitance signal cables are essential for minimizing noise and distortion.

A shaping amplifier achieves pulse shaping by cascading multiple CR–RC filter stages in combination with op amp networks. Amplification is provided by the op amps, which are powered by their own power supplies (Fig. 3.33). By appropriately selecting the shaping time constants (CR and RC) and amplifier configuration, a shaping amplifier can produce a nearly **Gaussian pulse or a bipolar pulse** suitable for compatibility with certain leading edge- or trailing edge-triggered discriminators, described in the next section. For readers interested in more advanced shaping techniques beyond simple passive RC or CR networks, see Reference [10].

3.6 From Analog Pulses to Digital Counts

Once the initial pulse generated by the motion of radiation-induced charge carriers is processed by the preamplifier and pulse shaper, it becomes a stronger and cleaner analog voltage signal. This processed signal is then delivered to a pulse-to-digital counting system, typically consisting of one or more voltage discriminators followed by a digital counter.

3.6.1 Voltage Discriminator and Energy Thresholding

3.6.1.1 Basic Structure

The voltage discriminator, also known as a voltage comparator, can be regarded as a special case of the op amp introduced in Sect. 2.9.1.1 of Chap. 2. According to Eq. (2.54), the output of an open-loop op amp depends on the difference between its two input voltages, V_{in+} and V_{in-}, as

$$V_{out} = \begin{cases} V_{s+}, & \text{if } V_{in+} - V_{in-} > 0, \\ V_{s-}, & \text{if } V_{in+} - V_{in-} < 0. \end{cases} \tag{3.121}$$

Because of its extremely large open-loop gain, the output voltage V_{out} behaves in a binary manner: whenever V_{in+} exceeds V_{in-}, the output rapidly switches to the positive saturation level V_{s+}.

For instance, when the output pulse from the shaper is positive and connected to V_{in+}, while V_{in-} is tied to a user-defined reference voltage V_{ref}, the comparator generates a logic pulse only if the analog pulse amplitude surpasses V_{ref}. Since the pulse amplitude is proportional to the incident photon energy, raising V_{ref} selectively filters out pulses generated by low-energy photons or electronic noise. This voltage comparison process is therefore commonly referred to as the **energy thresholding** process.

As illustrated in Fig. 3.34, the output of the pulse shaper can be connected to multiple voltage discriminators in parallel, each set with a different threshold voltage, V_{ref1}, V_{ref2}, $\cdots$. This configuration enables a coarse, blockwise estimation of the pulse height and thus of the incident photon energy. Because voltage discriminators typically exhibit very high input impedance, connecting several of

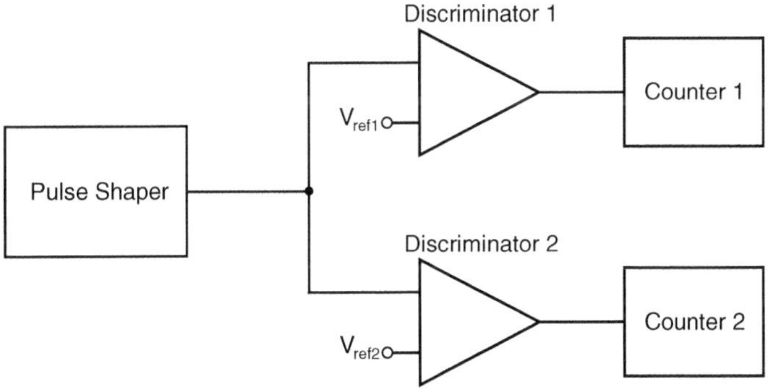

Fig. 3.34 Schematic of a dual-bin photon-counting detector employing two voltage discriminators and two digital counters. If the analog pulse from the pulse shaper exceeds V_{ref1} but not V_{ref2}, the pulse height is considered within the range between V_{ref1} and V_{ref2}. This is known as **energy windowing**

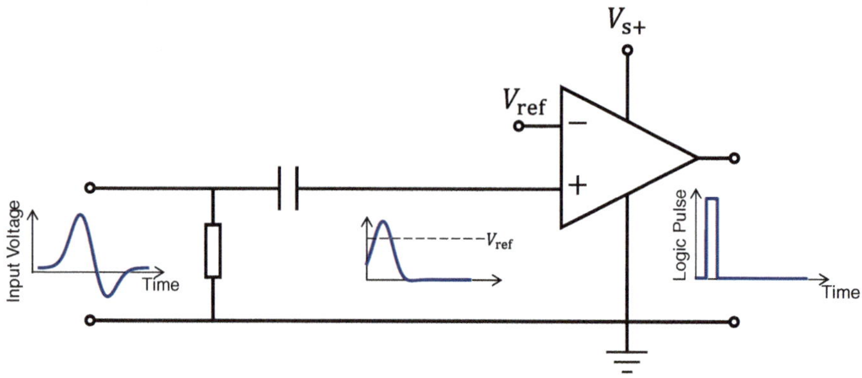

Fig. 3.35 Basic circuit implementation of leading-edge triggering

them in parallel to the pulse shaper output introduces negligible loading and does not significantly distort the analog pulse shape.

In practice, the output of the voltage discriminator is a short digital pulse rather than a step function. In other words, once triggered, the output rises almost instantaneously to V_{s+} and, after a brief and fixed duration, returns almost instantaneously to V_{s-}. An example is shown in Fig. 3.36.

3.6.1.2 Leading Edge-, Trailing Edge-, or Peak-Based Triggering

The logic output pulse of a voltage discriminator can be triggered based on the leading edge, the trailing edge, or the peak of the input analog pulse.

Triggering on the leading (rising) edge occurs when the input signal first exceeds a preset voltage threshold. To detect this edge, a high-pass RC filter, which is essentially a differentiator, can be placed before the positive input of the operational amplifier. As illustrated in Fig. 3.35, by adjusting the RC time constant and, if necessary, cascading additional CR stages, the rising slope of the analog signal produces a stronger positive voltage. When this voltage exceeds the threshold level V_{ref}, the output of the op amp switches from low to high.

Leading-edge triggering is commonly used when fast timing or low-latency response is essential. Because the trigger is produced as soon as the pulse crosses the threshold, subsequent digital operations—such as counting or timing—can respond immediately. Leading-edge triggering is therefore ideal for time-of-arrival measurements or for synchronizing later stages of the signal processing chain.

However, the leading edge of a pulse is often affected by jitter, electronic noise, or variations in pulse rise time, which can introduce timing uncertainty.

Triggering on the trailing (falling) edge can be advantageous when the analog pulses exhibit variable rise times but relatively consistent decay times. Since the trailing edge occurs after the pulse has fully developed, it is often cleaner and more

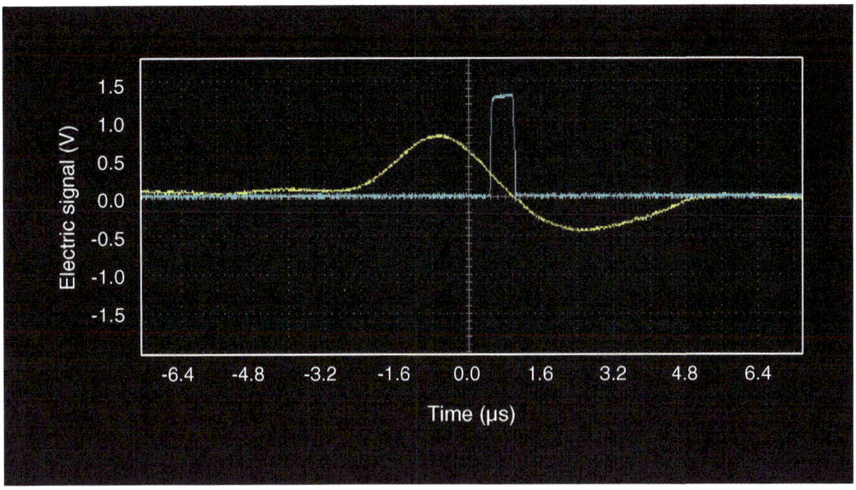

Fig. 3.36 *Yellow curve*: analog pulse from the pulse shaper (Ortec Mode 451 Spectroscopy Amplifier, AMETEK Inc. See Fig. 3.2). *Cyan curve*: digital logic pulse from the voltage discriminator shown in Fig. 3.2 (Ortec 551 Single Channel Analyzer, AMETEK Inc.). The logic pulse is 0.5 μs wide. It is triggered by the trailing edge of the analog pulse produced by the pulse shaper

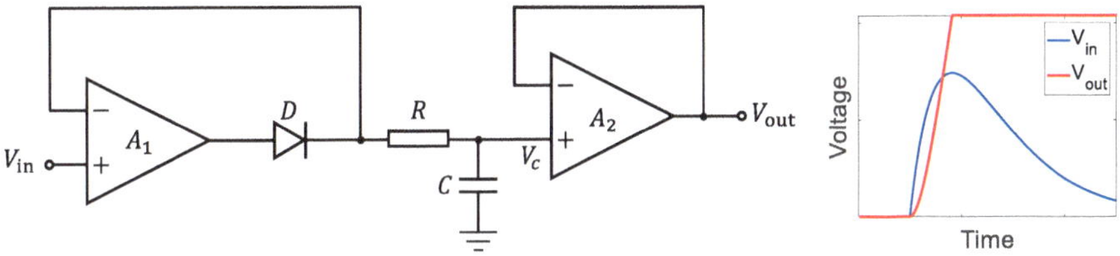

Fig. 3.37 A basic peak detection circuit. The plot on the right compares the input and output signals of the circuit

stable, providing a more reproducible timing reference. Trailing-edge triggering also ensures that charge collection is complete before the signal is registered. Figure 3.36 shows an example of trailing-edge triggering.

If the trailing edges of consecutive pulses overlap, no new trigger can occur until the signal falls below the threshold. This effectively extends the detector dead time and can lead to paralyzable behavior in PCDs.

Peak-based triggering can be implemented using a dedicated peak-detection circuit. Figure 3.37 illustrates a basic design that consists of two op amps (A_1 and A_2), a diode (D), a resistor (R), and a capacitor (C). The input voltage signal $V_{\text{in}}(t)$ is applied to the noninverting input of A_1. At $t = 0$, the capacitor is uncharged, so $V_C = 0$. When the rising edge of $V_{\text{in}}(t)$ arrives, A_1 produces a positive output voltage that forward-biases the diode, allowing the capacitor to charge and thereby increasing V_C. Through feedback, the inverting input of A_1 is driven to match V_{in}, while the positive capacitor voltage appears at the output of A_2 as $V_{\text{out}} = V_c$.

As long as $V_{in}(t)$ continues to rise, the diode remains forward biased, and the capacitor voltage (and thus V_{out}) continues to increase. When $V_{in}(t)$ reaches its peak and begins to fall, the noninverting input of A_1 becomes lower than the inverting input, causing A_1 to output a negative voltage. Because the cathode of the diode remains at a higher potential due to the stored charge on the capacitor, the diode becomes reverse biased, blocking current flow. The charge on the capacitor is then held constant, so V_{out} remains at its peak value until the capacitor is uncharged.

An example of the resulting V_{out} waveform is shown in Fig. 3.37. The turning point of this waveform corresponds to the most negative slope of the input signal. Therefore, by differentiating the signal, the peak location can be identified for triggering, while the magnitude of the plateau in V_{out} can be used to estimate the photon energy.

In the absence of noise, peak triggering provides the most accurate estimation of pulse height. However, even a small amount of jitter or noise in $V_{in}(t)$ can create a transient but significant downward fluctuation in the waveform, potentially leading to a false peak-detection event. As shown by a statistical analysis [11], when a maximum operation is applied, noise in the input signal can induce not only increased output fluctuations but also a statistical bias in the measured peak value. Furthermore, although peak finding appears to be less sensitive to pulse pileup than leading- or trailing-edge triggering, the additional processing required by the peak-detection circuit can increase the pulse processing time, thereby extending the dead time of the PCD system.

3.6.1.3 SCA and MCA

Based on its name, a **single-channel analyzer (SCA)** might seem to be a simple single-threshold discriminator. In fact, an SCA usually incorporates two thresholds and functions as a dual-discriminator system. The lower-level discriminator rejects all pulses below a user-defined threshold, effectively suppressing low-energy noise and unwanted background events. The upper-level discriminator rejects pulses above a second threshold, enabling the selection of a well-defined energy window for signal processing. Most SCAs also allow the upper threshold to be set to positive infinity, in which case the device effectively functions as a single-threshold discriminator.

A **multichannel analyzer (MCA)** is an instrument used to analyze the pulse-height distribution of detector analog signals, where the pulse height corresponds to the energy of the detected event. Unlike an SCA, which selects pulses within a single energy window, an MCA divides the energy range into many windows—typically 1024 or more. Its output is a histogram of pulse heights (or energies), where each bin represents the number of pulses falling within a specific energy interval. An example of an MCA output spectrum is shown in Fig. 3.38.

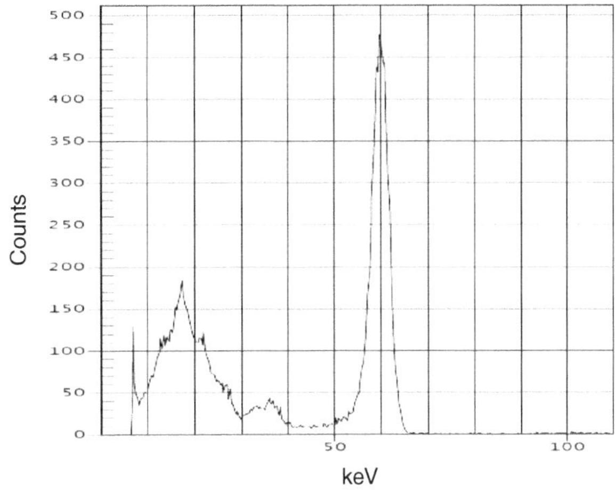

Fig. 3.38 Output of a multichannel analyzer to the CdTe diode detector in Fig. 3.2 to a 59.5 keV γ-ray source

3.6.2 *Digital Counter*

As discussed in Sect. 3.6.1, when an x-ray photon generates an analog pulse in a PCD, the voltage discriminator produces a short-duration logic pulse that triggers a counter to increment by one. The counter output must be **latched** with respect to this transient logic pulse—that is, the stored count value must persist even after the logic pulse has vanished. This behavior contrasts with that of the preamplifier or pulse shaper, whose output signal gradually decays to zero after receiving an input pulse and therefore does not retain any memory of the event once it has passed.

To illustrate the concept of latching, consider two simple analogies. Figure 3.39a shows a backyard gate latch: once the gate closes, it locks automatically and cannot be opened from the outside, no matter how hard one pulls or pushes—a lesson I learned the hard way. The latch preserves the "memory" of the gate-closing action. Similarly, Fig. 3.39b depicts a traditional Chinese counting method, where each event prompts the person to draw one additional stroke on paper. The ink serves as a lasting record of the accumulated counts, even after the person has left.

In a PCD, such latching and counting are implemented electronically. This is typically accomplished using a shift register constructed from a series of flip-flops, which store and update the digital count in response to each logic pulse.

3.6.2.1 Latch Fundamentals

Figure 3.40a shows a circuit consisting of two NPN transistors, two voltage inputs (A and B), a voltage output (V_{out}), and a power supply (V_{cc}). When both A and B are at zero volts, the base–emitter junctions of the transistors are reverse biased. Consequently, no current flows from V_{cc} through the transistors and the resistor to

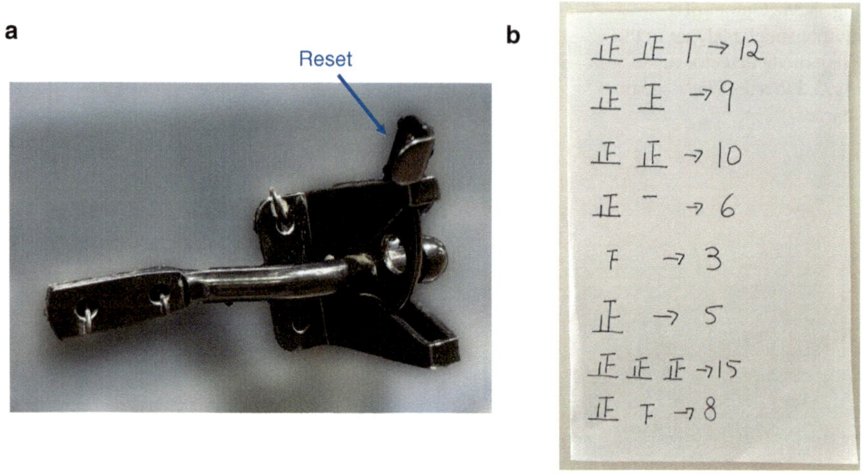

Fig. 3.39 (**a**) A gate latch. (**b**) A traditional tally-mark counting method, in which each Chinese character "正"consists of five strokes. The total number of strokes represents the cumulative count

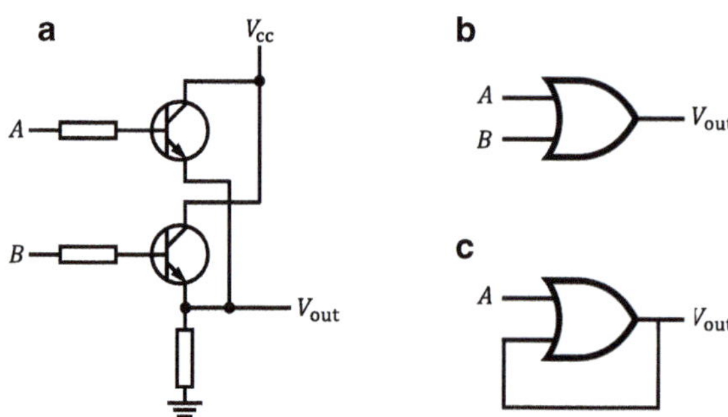

Fig. 3.40 (**a**) Transistor-based OR gate. (**b**) Standardized OR gate symbol defined by the American National Standards Institute (ANSI). (**c**) An OR gate–based latch. A positive signal applied to input *A* produces a positive output, which remains high regardless of any subsequent changes in *A*

ground. As there is no voltage drop across the resistor, V_{out} remains at the same potential as ground.

When either *A* or *B*, or both, receive a positive voltage, at least one base–emitter junction becomes forward biased, establishing a current path from V_{cc} through the resistor to ground. As a result, V_{out} rises to a positive voltage. For simplicity, we assume this positive output corresponds to a logic level of 1. This type of circuit functions as an OR gate, with its standardized symbol shown in Fig. 3.40b.

An OR gate alone does not retain information about past input signals. That is, after a transient positive pulse is applied to *A* or *B*, V_{out} eventually returns to 0, as if the pulse never visited.

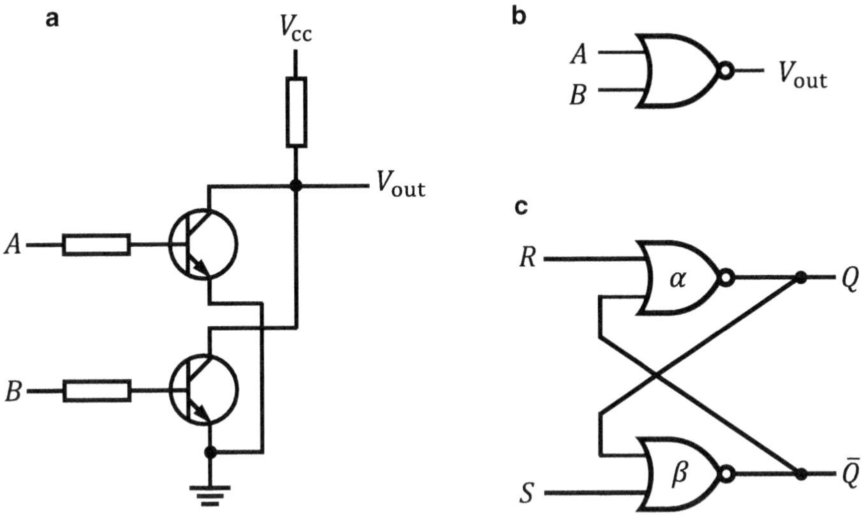

Fig. 3.41 (**a**) Transistor-based NOR gate. (**b**) Standardized NOR gate symbol. (**c**) A set–reset (SR) latch

However, if a feedback loop is established from V_{out} to one of the inputs (e.g., B), the circuit can "remember" a positive input pulse at A. Initially, with both A and B at 0, the output V_{out} is also 0. When the leading edge of a positive logic pulse arrives at A, V_{out} switches to 1. After the pulse ends and A returns to 0, B remains at 1 through the feedback connection shown in Fig. 3.40c, thereby holding V_{out} at 1. In this way, **the circuit effectively latches the input event**.

There is a limitation to the latch shown in Fig. 3.40c: after the first logic pulse, its output is latched at 1 and no longer responds to subsequent input pulses. Clearly, this behavior is unsuitable for counting applications. To overcome this limitation, a NOR gate–based **set–reset (SR) latch** can be employed, as illustrated in Fig. 3.41.

A NOR gate is defined as $\overline{\text{OR}}$: its output is 1 only when both inputs A and B are 0. This behavior can be realized using the circuit shown in Fig. 3.41a. When both A and B are zero, no current flows through the resistor, and V_{out} remains positive. When either A or B, or both, are positive, the output V_{out} drops to 0.

When two NOR gates are connected with feedback loops as shown in Fig. 3.41c, they form an SR latch. Suppose initially both inputs S and R are zero, with one output (Q) at 1 and the other ($\overline{Q}$) at 0. When a positive logic pulse is applied to R, the output of NOR gate α (Q) switches to 0. Consequently, both inputs to NOR gate β become zero, causing $\overline{Q}$ to switch from 0 to 1. When the next positive pulse is applied to S, $\overline{Q}$ switches from 1 to 0. This again drives both inputs of NOR gate α to zero, switching Q back from 0 to 1. In other words, when a train of input logic pulses is alternately applied to S and R, the outputs Q and $\overline{Q}$ toggle between 1 and 0. Conceptually, if Q and $\overline{Q}$ are used as inputs to another SR latch, we can construct a 2-bit counter.

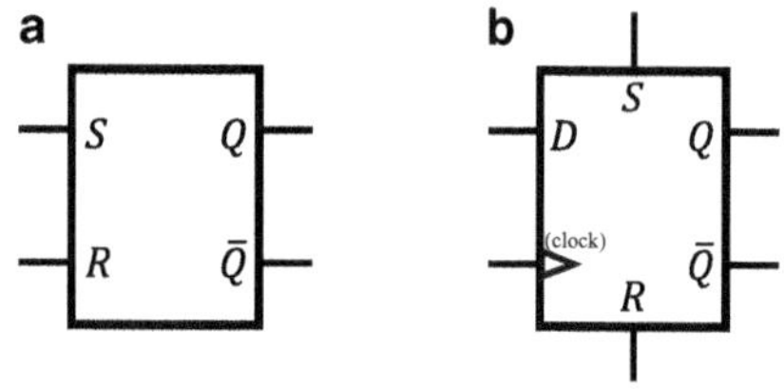

Fig. 3.42 (**a**) Logic symbol of an SR latch. (**b**) Logic symbol of a D flip-flop

Although the SR latch in Fig. 3.41c represents an improvement over the basic latch in Fig. 3.40c, it still has several important limitations. First, its initial state is not well defined: which output (Q or $\overline{Q}$) starts high or low is essentially random. Second, a train of input logic pulses must be applied alternately to S and R, rather than to a single input terminal, which complicates circuit design. Third, when a series of SR latches are connected to form a counter, their outputs are updated asynchronously. That is, the output of any latch can change immediately in response to an input change. If two latches are enabled simultaneously, slight differences in their propagation delays can lead to race conditions, producing incorrect or unpredictable outputs. In addition, short transients or glitches can propagate through the circuit, further degrading counting reliability.

Because of these limitations, practical digital counters and other sequential circuits typically use more advanced latch systems called **flip-flops**, which change state only at a well-defined clock transition. This synchronous operation ensures predictable timing and stable system behavior. Figure 3.42 illustrates the standard symbols for an SR latch and a **D flip-flop**, where "D" stands for *Data*. A D flip-flop includes two additional inputs beyond the S and R of an SR latch: a **clock input**, which controls when the state may change, and the D input, which carries the train of input logic pulses. On the active (e.g., rising) edge of the periodic clock signal, the output Q is updated to match the value present at D:

$$Q = \begin{cases} 0 & \text{if } D = 0, \\ 1 & \text{if } D = 1. \end{cases} \tag{3.122}$$

At all other times, the output Q remains unchanged. This edge-triggered behavior eliminates timing ambiguities and race conditions inherent in asynchronous latch circuits.

In a D flip-flop, the S and R inputs are no longer used for processing photon-induced logic pulses. Instead, they serve as asynchronous control inputs that can override the normal D and clock behavior to clear the memory or reset the counter. Specifically, asserting S forces the outputs to $Q = 1$ and $\overline{Q} = 0$, while asserting R forces $Q = 0$ and $\overline{Q} = 1$, regardless of the state of D or the clock signal. A similar reset capability is also present in the gated latch shown in Fig. 3.39a.

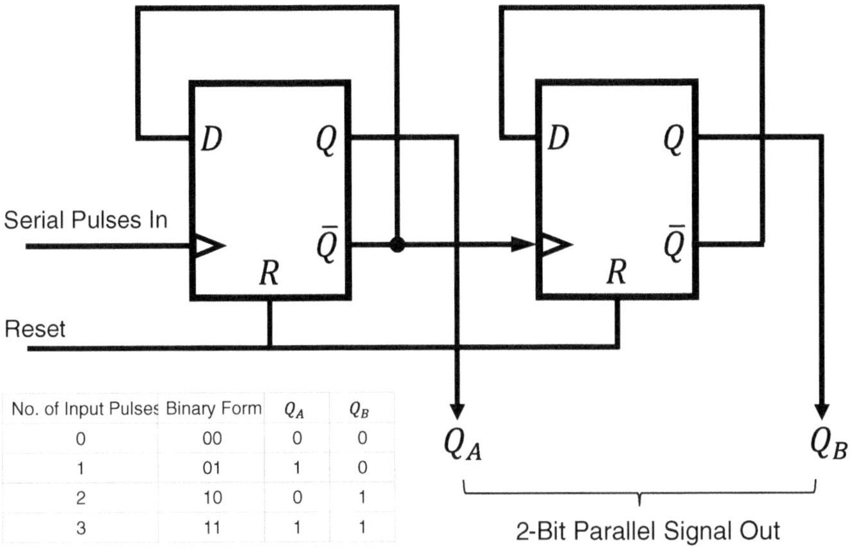

No. of Input Pulses	Binary Form	Q_A	Q_B
0	00	0	0
1	01	1	0
2	10	0	1
3	11	1	1

Fig. 3.43 A 2-bit serial-in, parallel-out shift binary counter implemented using two D flip-flops

3.6.2.2 Counter Fundamentals

A digital counter increments its output by 1 with each input logic pulse produced by a voltage discriminator. A binary digital counter is a commonly used type, which represents the counts in binary form (using 0s and 1s). It can be implemented using D flip-flops or other types of flip-flops. The number of flip-flops in a counter determines the bit depth of the counter and, consequently, the maximum count it can represent.

To illustrate the basic principle of a flip-flop-based counter, we use a simple model shown in Fig. 3.43, consisting of two D flip-flops (A and B) capable of counting from 0 to 3. The input logic pulse is applied only to the clock input of the first flip-flop (A). The clock input of the second flip-flop (B) is driven by the $\overline{Q}$ output of the preceding flip-flop. Each flip-flop's data input D is fed by its own $\overline{Q}$ output. The state of Q represents the state of the corresponding bit in the binary output.

Figure 3.44 shows the counter response to two input pulses. The flip-flops are assumed to change state on the *rising edge* of the clock signal. Before the first input pulse, the counter is initialized such that $Q_A = Q_B = 0$ and $\overline{Q}_A = \overline{Q}_B = 1$.

When the first pulse arrives at the clock input of flip-flop A, the rising edge triggers Q_A and $\overline{Q}_A$ to update according to D_A, which is 1. As a result, Q_A switches to 1 and $\overline{Q}_A$ switches to 0. Due to the feedback loop, D_A changes from 1 to 0, but this occurs after the clock edge and does not affect the current output. Flip-flop B sees a falling edge at its clock input ($\overline{Q}_A$ drops from 1 to 0), which does not trigger

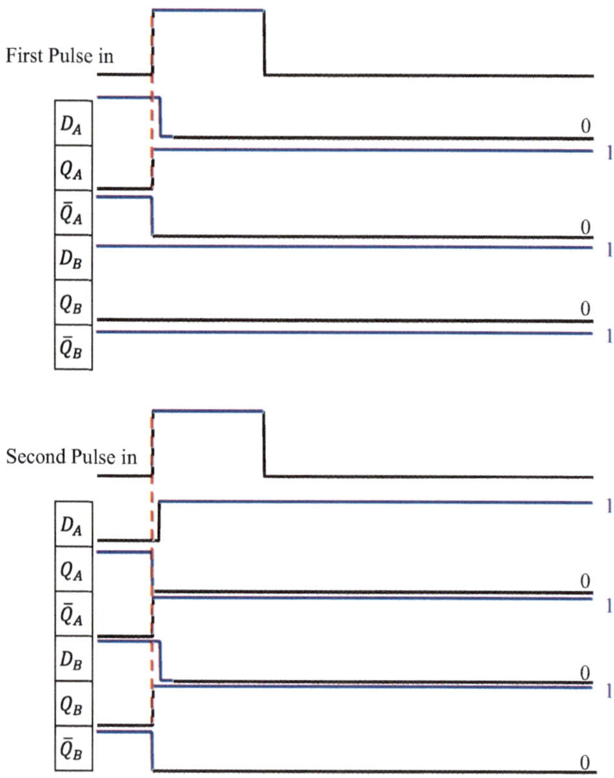

Fig. 3.44 A 2-bit serial-in, parallel-out binary counter implemented using two D flip-flops with rising-edge triggering

a state change. Therefore, after the first pulse, $Q_B = 0$ and $Q_A = 1$. If Q_B is the most significant bit, the binary output is `01`, correctly representing a count of 1.

When the second input pulse arrives, the rising edge at flip-flop A triggers an update based on the current $D_A = 0$, so Q_A switches to 0 and $\overline{Q}_A$ switches from 0 to 1. This rising edge at flip-flop B's clock triggers its output. Since $D_B = 1$, Q_B switches to 1 and $\overline{Q}_B$ switches to 0. After the second pulse, $Q_B = 1$ and $Q_A = 0$, giving a binary output of `10`, corresponding to a count of 2.

This toy-model example illustrates the basic working principle of a digital counter. In photon-counting image detectors, each pixel typically employs more than 10 bits per counter and may use other types of flip-flops to handle a large number of input events before readout. Therefore, adding more energy channels or bins to each detector pixel is not a trivial task, as each channel requires a separate digital counter.

A digital counter consumes power primarily due to the dynamic switching of flip-flop states. The power consumption increases with the count rate and clock frequency, resulting in a higher energy penalty when the PCD counting speed is increased.

3.6.3 Coincidence Detection

In certain medical imaging applications, it is necessary to determine whether pulses produced at two or more detector pixels originated from the same primary event. In positron emission tomography (PET), the primary event is a positron–electron annihilation, which generates two 511 keV γ-ray photons traveling in nearly opposite directions. Identifying two PET detector pixels that simultaneously register these photons enables estimation of the annihilation location using methods such as filtered backprojection or time-of-flight reconstruction [1]. In x-ray photon-counting planar or CT imaging, the primary event is the photoelectric absorption of an x-ray photon in the detector sensor. This can produce signals in multiple detector pixels due to K-fluorescence reabsorption or lateral charge diffusion. Identifying pixels triggered by the same photon improves the accuracy of x-ray energy estimation, enhances spatial resolution, and reduces image noise [6, 7, 9] (Fig. 3.45).

A key characteristic of signals generated by the same primary event is that their corresponding pulses occur almost simultaneously. Consequently, coincidence detection is one of the most effective techniques for identifying such correlated events.

The simplest method for coincidence detection is to use an AND gate to compare the logic pulses produced by two detector pixels. The output of the AND gate is logic 1 only when both inputs are simultaneously at logic 1. In other words, if the

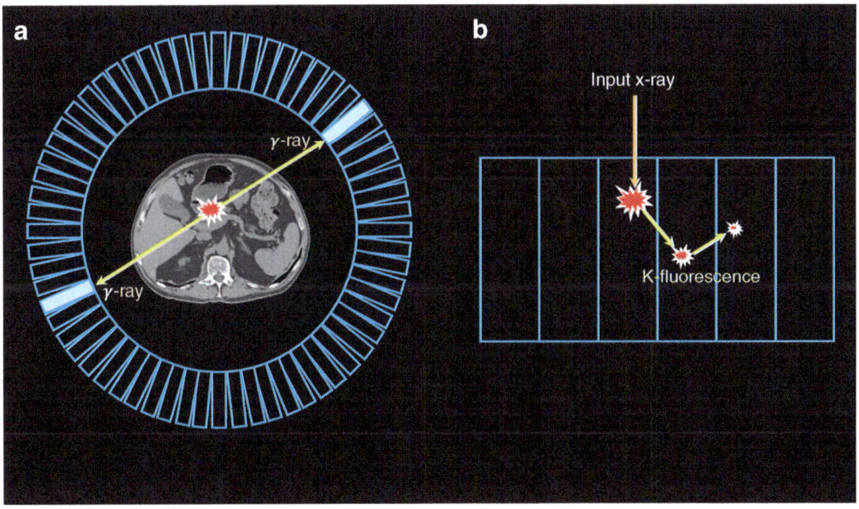

Fig. 3.45 (**a**) Coincidence detection in PET imaging, where the primary event is a positron–electron annihilation within the patient's body, and the secondary events are two γ-ray photons striking two PET detector elements nearly simultaneously. (**b**) Coincidence detection in x-ray PCD imaging, where the primary event is the photoelectric absorption of an incident x-ray photon, and the secondary events are the near-instantaneous reabsorption of K-fluorescence photon(s) in one or more adjacent detector pixels

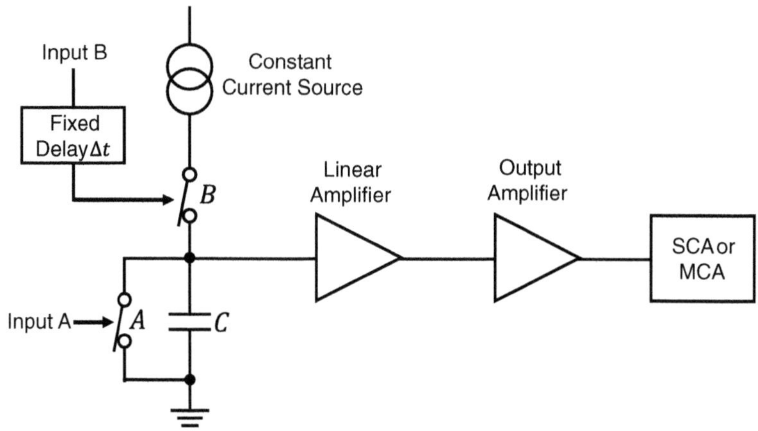

Fig. 3.46 A schematic of a basic time-to-amplitude converter (TAC)

two input pulses overlap in time, a positive logic pulse is generated at the output, indicating a coincident event. Although straightforward to implement, this overlap-based coincidence detection method has several major limitations [12]. First, the pulse processing times of different pixels must be well matched to ensure that true coincident pulses overlap. Second, the input logic pulse width must be sufficiently long to guarantee overlap of coincident pulses. However, increasing the pulse width also raises the probability of false coincidences caused by accidental overlaps. Determining an optimal pulse width is often challenging because it involves a trade-off between sensitivity and specificity.

In practice, coincidence detection is typically implemented using a **time-to-amplitude converter** (TAC). This method involves introducing a controlled time delay, Δt, to one of the two input signals [10, 12]. At first, the use of a time delay in coincidence detection may seem counterintuitive. However, understanding the operation of a TAC clarifies its purpose.

As illustrated in Fig. 3.46, the key components of a TAC include a constant current source, a conversion capacitor C, amplifiers, and two switches controlled by the input logic pulses to be tested for coincidence. Before each measurement, both switches are closed, and the capacitor is uncharged. The leading edge of input pulse A opens switch A, allowing the capacitor to begin charging with a constant current I. When the leading edge of input pulse B, which is intentionally delayed by Δt, arrives, it opens switch B, stopping the charging process. The voltage accumulated on the capacitor is

$$V = \frac{It}{C}, \tag{3.123}$$

where t is the time interval between the leading edges of the two input pulses. This voltage signal is then amplified and converted into a logic pulse whose amplitude is proportional to V and therefore to t. The output is subsequently sent to a SCA. If V falls within a narrow window centered around the voltage corresponding to Δt, the two input pulses are classified as coincident events.

With Eq. (3.123), the benefit of introducing a fixed time delay Δt is now clear. Without the delay, the time interval between two true coincident pulses is typically very small—on the order of a few nanoseconds—determined by factors such as electronic shaping delays, detector response variations, and signal line length differences. Such a short interval would produce a very small voltage V, often below the nonlinear region of the amplifier's dynamic range. This makes it more difficult for the SCA to reliably discriminate true coincident events from electronic noise. By adding a known Δt, the TAC output corresponding to true coincidences is shifted to a well-defined, higher voltage level, improving both the accuracy and precision of coincidence detection.

3.7 Summary of Chap. 3

1. When radiation interacts with a semiconductor sensor, electron–hole pairs are generated. Under an applied electric field, these charge carriers drift toward opposite electrodes. An induced current pulse appears at the electrodes immediately once the carriers start to move.
2. The Shockley–Ramo theorem provides the quantitative relationship between the induced current I, the charge q, the carrier velocity $\mathbf{v}$, and the detector geometry:

$$I = q\,\mathcal{E}_w(\mathbf{x}) \cdot \mathbf{v}(\mathbf{x}), \tag{3.124}$$

   where $\mathcal{E}_w(\mathbf{x})$ is the weighting field, determined entirely by the detector geometry and electrode configuration.
3. For pixelated PCDs, the method of images can be used to analytically calculate the induced charge or current in each pixel, without the need for numerically solving the weighting field.
4. The initial pulse formation mechanism is the same for both PCDs and energy-integrating detectors (EIDs). The key difference lies in the subsequent signal processing: EIDs integrate the induced charge from a sequence of incident photons over a fixed time interval before digitization, while PCDs process individual pulses separately to extract information from each photon interaction.
5. To process individual pulses, the initial signal must be significantly preamplified to minimize the impact of electronic noise. A CSA with a large time constant is typically used to fully collect and amplify the initial pulse signal. The CSA is periodically reset to avoid saturation.
6. The CSA output is then shaped to make it suitable for digital processing. Pulse shapers in PCDs are typically built from combinations of resistors and capacitors. A CR network acts as a high-pass filter that preserves the edges or peaks of the pulse signal, while an RC network acts as a low-pass filter to suppress noise.

7. The shaped signal is sent to a voltage discriminator, which converts analog pulses into digital logic pulses. The discriminator output is binary (high or low), and the transition is triggered by the edge or peak of the shaped analog pulse.
8. A fundamental building block of digital counters is the latch, which retains the logic state of a pulse. By combining multiple latches with feedback, multi-bit binary counters can be constructed for pulse counting.
9. To evaluate whether two pulses from different detector elements originate from the same primary event, a fixed time delay is typically introduced to one of the pulses. The two pulses are then sent to a time-to-amplitude converter (TAC), which produces an output voltage proportional to the time interval between their arrival times. By comparing this output voltage to the expected voltage for a true coincidence, events corresponding to true coincidences can be identified.

Appendix 1: Proof of the Shockley–Ramo Theorem

Green's second identity in vector calculus provides one way to prove the Shockley–Ramo theorem. This was the method originally employed by Ramo in 1939 [15]. An alternative proof, based on energy conservation, is presented in Reference [5].

Let $\phi(\mathbf{x})$ and $\varphi(\mathbf{x})$ be scalar functions with continuous second derivatives in a region V enclosed by a closed surface S. Green's second identity states

$$\int_V \left(\phi\, \nabla^2\varphi - \varphi\, \nabla^2\phi\right) \mathrm{d}V = \oint_S \left(\phi\, \frac{\partial\varphi}{\partial n} - \varphi\, \frac{\partial\phi}{\partial n}\right) \mathrm{d}S, \tag{3.125}$$

where $\frac{\partial}{\partial n}$ denotes the derivative in the outward normal direction on the boundary S and $\mathrm{d}V$ and $\mathrm{d}S$ are the volume and surface elements, respectively.

Figure 3.47 shows a volume containing N conducting electrodes with an arbitrary shape and geometry. Each electrode is biased at an arbitrary potential level of V_i ($i \in [1, N]$). These electrodes are connected through cables, and there is a charge q located at $\mathbf{x}_q$ in the medium between the electrodes.

Let the charge on the i-th electrode be denoted by Q_i. This charge is jointly determined by (a) the applied bias voltage and (b) the presence of space charge q. We decompose Q_i into two components, $Q_i = Q_i^a + Q_i^b$, where Q_i^a corresponds to the charge induced by the bias voltage, and Q_i^b corresponds to the charge induced by q. According to the principle of superposition, the system can be separated into two sub-systems: in the first sub-system, the space charge q is removed, leaving only the bias voltage, which accounts for Q_i^a; in the second sub-system, the bias voltage is removed and all electrodes are grounded, while the space charge q is retained, producing Q_i^b.

Since Q_i^a is time invariant, our focus is on Q_i^b. Accordingly, we restrict our analysis to sub-system (b), as illustrated in Fig. 3.47.

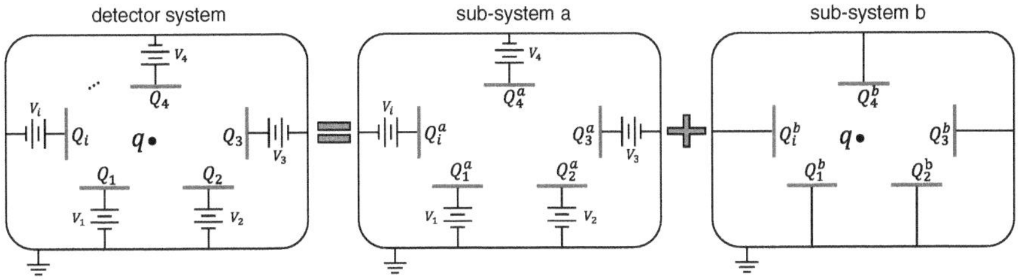

Fig. 3.47 Based on the linear superposition principle, a detector system that contains a space charge can be decomposed into **sub-system a** and **sub-system b**, with sub-system a free of the space charge and sub-system b free of the bias voltage on each electrode

Let $\phi(\mathbf{x})$ denotes the electric potential produced by the space charge q. Based on Poisson's equation,

$$\nabla^2\phi(\mathbf{x}) = -\frac{\rho_q}{\varepsilon}\delta(\mathbf{x} - \mathbf{x}_q). \tag{3.126}$$

In sub-system (b), all electrodes are grounded; hence, $\phi = 0$ on every electrode surface (and on the outer boundary of the volume V).

To use Green's second identity, Ramo handcrafted a scalar function, $\varphi(\mathbf{x})$, as follows:

- Remove the space charge q.
- Set the i-th electrode to a unit potential.
- Ground all other electrodes.

$\varphi(\mathbf{x})$ represents the electric potential for this handcrafted system. It satisfies

$$\nabla^2\varphi(\mathbf{x}) = 0. \tag{3.127}$$

Remember that Green's second identity holds true irrespective of how the system is constructed or implemented. Since $\nabla^2\varphi = 0$, the left-hand side (LHS) of (3.125) reduces to

$$\begin{aligned}
\text{LHS} &= \int_V \left(\phi(\mathbf{x}) \cdot 0 - \varphi(\mathbf{x})\,\nabla^2\phi(\mathbf{x})\right) \mathrm{d}V \\
&= -\int_V \varphi(\mathbf{x})\,\nabla^2\phi(\mathbf{x})\,\mathrm{d}V \\
&= \frac{q}{\varepsilon}\int_V \varphi(\mathbf{x})\,\delta(\mathbf{x} - \mathbf{x}_q)\,\mathrm{d}V \\
&= \frac{q}{\varepsilon}\varphi(\mathbf{x}_q).
\end{aligned} \tag{3.128}$$

Here, we used the sifting property of the Dirac delta function.

For the surface integral on the right-hand side, $\phi = 0$ on every electrode and on the outer boundary, so the $\phi \, \frac{\partial \varphi}{\partial n}$ term vanishes and only the $-\varphi \, \partial_n \phi$ term remains:

$$\begin{aligned}
\text{RHS} &= \oint_S \left(0 \cdot \frac{\partial \varphi}{\partial n} - \varphi \, \frac{\partial \phi}{\partial n} \right) \mathrm{d}S \\
&= - \oint_S \varphi \, \frac{\partial \phi}{\partial n} \, \mathrm{d}S \\
&= - \int_{\text{surface of electrode } i} \varphi \, \frac{\partial \phi}{\partial n} \, \mathrm{d}S \\
&= - \int_{\text{surface of electrode } i} 1 \cdot \frac{\partial \phi}{\partial n} \, \mathrm{d}S \\
&= \int_{\text{surface of electrode } i} (-\frac{\partial \phi}{\partial n}) \, \mathrm{d}S \\
&= \int_{\text{surface of electrode } i} \mathcal{E}_n \, \mathrm{d}S.
\end{aligned} \tag{3.129}$$

Here,

$$\mathcal{E}_n = -\frac{\partial \phi}{\partial n} \tag{3.130}$$

denotes the normal component of the electric field produced by the space charge on the electrode surface. Since $\frac{\partial \phi}{\partial n}$ is defined with respect to the *outward* normal, $\mathcal{E}_n$ points opposite to the inward-facing side of the electrode. Accordingly, by Gauss's law, the induced charge on the inward surface of the electrode is related to $\mathcal{E}_n$ as

$$\int_{\text{surface of electrode } i} \mathcal{E}_n \, \mathrm{d}S = -\frac{Q_i^b}{\varepsilon}. \tag{3.131}$$

By equating Eq. (3.128) with Eq. (3.131), we obtain

$$\frac{q}{\varepsilon} \, \varphi(\mathbf{x}_q) = -\frac{Q_i^b}{\varepsilon}, \tag{3.132}$$

or equivalently,

$$Q_i^b = -q \, \varphi(\mathbf{x}_q). \tag{3.133}$$

Here, $\varphi(\mathbf{x}_q)$ denotes the weighting potential of electrode i evaluated at the position of the space charge $\mathbf{x}_q$. This completes the proof of the Shockley—Ramo theorem given in Eq. (3.23).

Appendix 2: Induced Charge in Pixelated PCDs

Based on Eqs. (3.55) and (3.56), the induced charge in a pixelated anode located at (x_0, y_0) can be expressed as

$$\begin{aligned} Q(x_0, y_0) &= \int_{-\frac{w_x}{2}+x_0}^{+\frac{w_x}{2}+x_0} \mathrm{d}x \int_{-\frac{w_y}{2}+y_0}^{+\frac{w_y}{2}+y_0} \mathrm{d}y\, \sigma_{\text{net}}(x, y) \\ &= -\frac{ne}{2\pi} \sum_{m=1,3,5,\ldots} \int_{-\frac{w_x}{2}+x_0}^{+\frac{w_x}{2}+x_0} \mathrm{d}x \int_{-\frac{w_y}{2}+y_0}^{+\frac{w_y}{2}+y_0} \\ &\quad \times \mathrm{d}y \left[G(x, y, z_p, mL) - G(x, y, z_n, mL)\right]. \end{aligned} \tag{3.134}$$

Taking $m = 1$ as an example: according to the definition of $G(x, y, z, L)$ in Eq. (3.52), we obtain

$$\begin{aligned} &\int_{-\frac{w_x}{2}+x_0}^{+\frac{w_x}{2}+x_0} \mathrm{d}x \int_{-\frac{w_y}{2}+y_0}^{+\frac{w_y}{2}+y_0} \mathrm{d}y\, G(x, y, z, L) \\ &= \int_{-\frac{w_x}{2}+x_0}^{+\frac{w_x}{2}+x_0} \mathrm{d}x \int_{-\frac{w_y}{2}+y_0}^{+\frac{w_y}{2}+y_0} \mathrm{d}y \left[\frac{L-z}{[(L-z)^2+x^2+y^2]^{3/2}} \right. \\ &\quad \left. -\frac{L+z}{[(L+z)^2+x^2+y^2]^{3/2}}\right] \\ &= F(x_0, y_0, L-z) - F(x_0, y_0, L+z) \\ &= \mathcal{F}(x_0, y_0, z, L). \end{aligned} \tag{3.135}$$

where

$$\begin{aligned} F(x_0, y_0, z) &= \arctan\left[\frac{(\frac{w_x}{2}+x_0)(\frac{w_y}{2}+y_0)}{z\sqrt{z^2+(\frac{w_x}{2}+x_0)^2+(\frac{w_y}{2}+y_0)^2}}\right] \\ &\quad - \arctan\left[\frac{(\frac{w_x}{2}+x_0)(-\frac{w_y}{2}+y_0)}{z\sqrt{z^2+(\frac{w_x}{2}+x_0)^2+(-\frac{w_y}{2}+y_0)^2}}\right] \\ &\quad - \arctan\left[\frac{(-\frac{w_x}{2}+x_0)(\frac{w_y}{2}+y_0)}{z\sqrt{z^2+(-\frac{w_x}{2}+x_0)^2+(\frac{w_y}{2}+y_0)^2}}\right] \\ &\quad + \arctan\left[\frac{(-\frac{w_x}{2}+x_0)(-\frac{w_y}{2}+y_0)}{z\sqrt{z^2+(-\frac{w_x}{2}+x_0)^2+(-\frac{w_y}{2}+y_0)^2}}\right]. \end{aligned} \tag{3.136}$$

Substituting Eqs. (3.135)–(3.136) into (3.134), the induced charge becomes

$$Q(x_0, y_0) = -\frac{ne}{2\pi}\sum_m \left[\mathcal{F}(x_0, y_0, z_p, mL) - \mathcal{F}(x_0, y_0, z_n, mL)\right]. \tag{3.137}$$

The time derivative of $Q(x_0, y_0)$ yields the induced current:

$$\begin{aligned}
I(x_0, y_0) &= \frac{\partial Q(x_0, y_0)}{\partial t} \\
&= -\frac{ne}{2\pi}\sum_{m=1,3,5,\ldots}\left[\frac{\partial}{\partial z_p}\mathcal{F}(x_0, y_0, z_p, mL)\frac{\partial z_p}{\partial t}\right. \\
&\quad \left.-\frac{\partial}{\partial z_n}\mathcal{F}(x_0, y_0, z_n, mL)\frac{\partial z_n}{\partial t}\right] \\
&= -\frac{ne}{2\pi}\sum_{m=1,3,5,\ldots}\left\{\frac{\partial}{\partial z_p}[F(x_0, y_0, mL - z_p)\right. \\
&\quad -F(x_0, y_0, mL + z_p)]\frac{\partial z_p}{\partial t} - \frac{\partial}{\partial z_n}[F(x_0, y_0, mL - z_n) \\
&\quad \left.-F(x_0, y_0, mL + z_n)]\frac{\partial z_n}{\partial t}\right\} \\
&= \frac{ne}{2\pi}\sum_{m=1,3,5,\ldots}\left\{[S(x_0, y_0, mL - z_p) + S(x_0, y_0, mL + z_p)]\frac{\partial z_p}{\partial t}\right. \\
&\quad \left.-[S(x_0, y_0, mL - z_n) + S(x_0, y_0, mL + z_n)]\frac{\partial z_n}{\partial t}\right\}
\end{aligned} \tag{3.138}$$

where

$$\begin{aligned}
&S(x_0, y_0, z) \\
&= \frac{\partial F(x_0, y_0, z)}{\partial z} \\
&= -\frac{(\frac{w_x}{2} + x_0)(\frac{w_y}{2} + y_0)[(\frac{w_x}{2} + x_0)^2 + (\frac{w_y}{2} + y_0)^2 + 2z^2]}{[(\frac{w_x}{2} + x_0)^2 + z^2][(\frac{w_y}{2} + y_0)^2 + z^2]\sqrt{z^2 + (\frac{w_x}{2} + x_0)^2 + (\frac{w_y}{2} + y_0)^2}} \\
&\quad + \frac{(\frac{w_x}{2} + x_0)(-\frac{w_y}{2} + y_0)[(\frac{w_x}{2} + x_0)^2 + (-\frac{w_y}{2} + y_0)^2 + 2z^2]}{[(\frac{w_x}{2} + x_0)^2 + z^2][(-\frac{w_y}{2} + y_0)^2 + z^2]\sqrt{z^2 + (\frac{w_x}{2} + x_0)^2 + (-\frac{w_y}{2} + y_0)^2}} \\
&\quad + \frac{(-\frac{w_x}{2} + x_0)(\frac{w_y}{2} + y_0)[(-\frac{w_x}{2} + x_0)^2 + (\frac{w_y}{2} + y_0)^2 + 2z^2]}{[(-\frac{w_x}{2} + x_0)^2 + z^2][(\frac{w_y}{2} + y_0)^2 + z^2]\sqrt{z^2 + (-\frac{w_x}{2} + x_0)^2 + (\frac{w_y}{2} + y_0)^2}}
\end{aligned}$$

$$-\frac{(-\frac{w_x}{2}+x_0)(-\frac{w_y}{2}+y_0)[(-\frac{w_x}{2}+x_0)^2+(-\frac{w_y}{2}+y_0)^2+2z^2]}{[(-\frac{w_x}{2}+x_0)^2+z^2][(-\frac{w_y}{2}+y_0)^2+z^2]\sqrt{z^2+(-\frac{w_x}{2}+x_0)^2+(-\frac{w_y}{2}+y_0)^2}}. \quad (3.139)$$

Using Eqs. (3.134)–(3.139), interested readers can estimate the induced current or charge in a pixelated electrode without the need to numerically solve partial differential equations for the weighting field or weighting potential.

References

1. Bushberg JT, Seibert Jr. JA, Leidholdt EM, Boone JM. The essential physics of medical imaging, 4th ed. London: Wolters Kluwer; 2020.
2. Campanelli RB, Gomes GS, Fernandes MG, Mendes LH, Rosa LB, Reis RC, Antonio EB, Polli JM. Large area hybrid detectors based on Medipix3RX: commissioning and characterization at Sirius beamlines. J Instrum 2023; 18(02):C02008. https://doi.org/10.1088/1748-0221/18/02/C02008.
3. Dick BG. Images and the point charge-capacitor problem. Am J Phys 1973; 41:1289–1290. https://doi.org/10.1119/1.1987548.
4. Griffiths DJ. Introduction to electrodynamics, 3rd ed. New Delhi: Prentice Hall of India; 2009.
5. He Z. Review of the Shockley-Ramo theorem and its application in semiconductor gamma-ray detectors. Nucl Instrum Methods Phys Res Sect A Accel Spectrometers Detect Assoc Equip 2001; 463(1):250–267. ISSN 0168-9002. https://doi.org/10.1016/S0168-9002(01)00223-6.
6. Ji X, Zhang R, Chen G-H, Li K. Impact of anti-charge sharing on the zero-frequency detective quantum efficiency of CdTe-based photon-counting detector system: cascaded systems analysis and experimental validation. Phys Med Biol 2018; 63(9):095003. https://doi.org/10.1088/1361-6560/aab9c9.
7. Ji X, Zhang R, Chen G-H, Li K. Task-driven optimization of the non-spectral mode of photon-counting CT for intracranial hemorrhage assessment. Phys Med Biol 2019; 64(21):215014. https://doi.org/10.1088/1361-6560/ab43a6.
8. Ji X, Feng M, Treb K, Zhang R, Schafer S, Li K. Development of an integrated C-arm interventional imaging system with a strip photon-counting detector and a flat panel detector. IEEE Trans Med Imaging 2021; 40(12):3674–3685. https://doi.org/10.1109/TMI.2021.3095419.
9. Ji X, Feng M, Zhang R, Chen G-H, Li K. An analytical model of energy response function for CdTe-based photon-counting detectors with inter-pixel communication for anti-coincidence corrections. In Zhao W, Yu L, editors. Medical imaging 2022: physics of medical imaging, vol. 12031. International Bellingham: Society for Optics and Photonics, SPIE; 2022. p. 120314N. https://doi.org/10.1117/12.2611848.
10. Knoll GF. Radiation detection and measurement. Hoboken: John Wiley & Sons; 2010.
11. Li K, Chen G-H. Statistical properties of cerebral CT perfusion imaging systems. Part II. Deconvolution-based systems. Med Phys 2019; 46(11):4881–4897. https://doi.org/10.1002/mp.13805.
12. ORTEC. Time-to-amplitude convertersand time calibrator, 2025. https://www.ortec-online.com/.
13. Radtke J, Culberson W, Li K. Abstract ID 6059 seeing is believing: experimental demonstration of basic pulse formation process in photon-counting detectors to medical physics graduate students. AAPM 65th Annual Meeting and Exhibition; 2023.

14. Rajendran K, Petersilka M, Henning A, Shanblatt ER, Schmidt B, Flohr TG, Ferrero A, Baffour F, Diehn FE, Yu L, Rajiah P, Fletcher JG, Leng S, McCollough CH. First clinical photon-counting detector CT system: technical evaluation. Radiology 2022; 303(1):130–138. https://doi.org/10.1148/radiol.212579.
15. Ramo S. Currents induced by electron motion. Proc IRE 1939; 27(9):584–585. https://doi.org/10.1109/JRPROC.1939.228757.
16. Shockley W. Currents to conductors induced by a moving point charge. J Appl Phys 1938; 9(10):635–636. ISSN 0021-8979. https://doi.org/10.1063/1.1710367.
17. Zahn M. Point charge between two parallel grounded planes. Am J Phys 1976; 44:1132–1134. https://doi.org/10.1119/1.10207.

Chapter 4
Basic Physics of Inorganic Scintillators and Scintillator Image Detectors

Contents

K. Li, *Essential Physics of Medical X-ray Sources and Detectors*,
https://doi.org/10.1007/978-3-032-27800-5_4

4.1 Introduction

One of the most classical methods for detecting ionizing radiation is through the use of scintillator-based detectors. In fact, it was a scintillating material—barium platinocyanide [$BaPt(CN)_4$]—that enabled Wilhelm Röntgen to visually discover x-rays emitted from a Crookes tube in 1895. Although Crookes tubes had been operated at high voltages for decades prior, and many scientists may have unknowingly generated x-rays in earlier experiments, it was Röntgen who recognized and documented the phenomenon, earning him the first Nobel Prize in Physics. This highlights the pivotal role of scintillators in radiation detection.

The fundamental principle of a scintillator detector is to convert incoming ionizing radiation into visible light, which is then detected by a light sensor. In modern digital radiation detectors, the sensor transforms lights into an electrical signal that can be interpreted by the human brain or processed by a computer to extract information about the original radiation.

The term "visible light" refers to the portion of the electromagnetic spectrum that the human eye can perceive, typically in the range of 380–740 nm (Table 4.1). A scintillator emits visible light when excited by ionizing radiation, and the light sensor in a scintillator detector is specifically designed to detect this emitted light.

You might wonder: why do we rely on materials that emit *visible* light? Why not use materials that emit ultraviolet (UV) light, paired with UV-sensitive detectors?

In principle, we absolutely can. A scintillator that emits UV light, combined with a UV-sensitive sensor, could be a viable radiation detection system. However, there are historical and practical reasons for the prevalence of visible-light-emitting scintillators. In the early days of x-ray imaging, the "light sensor" was the human eye, which only responds to visible light. Figure 4.1 shows vintage x-ray imaging systems, where operators directly viewed the scintillation lights produced by x-rays. Unfortunately, this setup exposed them to ionizing radiation. Once the health risks of ionizing radiation became apparent, human observation was replaced by films—another detector sensitive to visible light.

While film is now largely obsolete in modern medical imaging, its successors—such as photodiodes and photomultiplier tubes (PMTs)—are also optimized for visible light detection. After over a century of development, the technologies for producing and detecting visible light have become highly advanced and well

Table 4.1 Visible light spectrum with corresponding wavelength and photon energy ranges. Energy values are calculated using $E = \frac{1240}{\lambda}$, where E is in eV and λ is in nm

Light color	Wavelength range (nm)	Energy range (eV)
Violet	380–450	3.26–2.75
Blue	450–495	2.75–2.51
Green	495–570	2.51–2.17
Yellow	570–590	2.17–2.10
Orange	590–620	2.10–2.00
Red	620–750	2.00–1.65

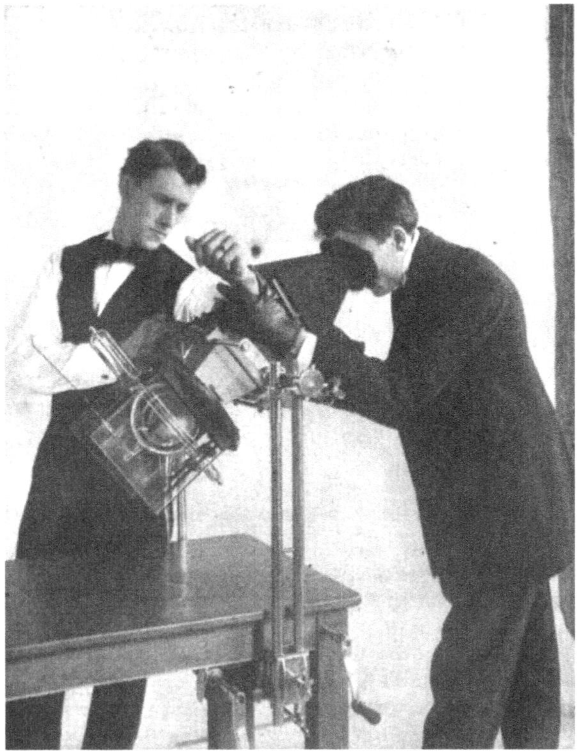

Fig. 4.1 Photo of an early x-ray procedure using a scintillator screen, around 1910. (Source: page 233 of *The Story of Great Inventions* by Elmer Ellsworth Burns, Harper & Brothers, New York, 1910. The copyright of this public-domain book has expired)

integrated into imaging systems. As a result, the practice of converting ionizing radiation into visible light remains the standard in scintillator-based detectors.

For new learners, there may be some misconceptions about how scintillator detectors work. One misunderstanding is that, unlike semiconductors—which directly convert incident photons into electron–hole pairs—scintillators first convert incoming photons into visible light, which then undergoes further conversion in a separate light sensor (e.g., a photodiode).

In reality, the initial stages of energy transfer in both semiconductors and scintillators are quite similar. When an incident x-ray or γ-ray photon interacts with the solid detector material—whether a semiconductor or a scintillator—it deposits its energy (partially or completely) into the material. This energy excites electrons from the valence band into the conduction band, leaving behind holes. Up to this point, the processes in semiconductors and scintillators are fundamentally the same, **except that scintillators typically have wider bandgaps and are insulating, which means they generate fewer electron–hole pairs on average.**

What truly distinguishes a scintillator detector from a semiconductor detector is what happens *after* the electron–hole pairs are created. As illustrated in Fig. 4.2, in semiconductor detectors, the generated electrons and holes are quickly separated and collected due to a strong electric field, producing a direct electrical signal

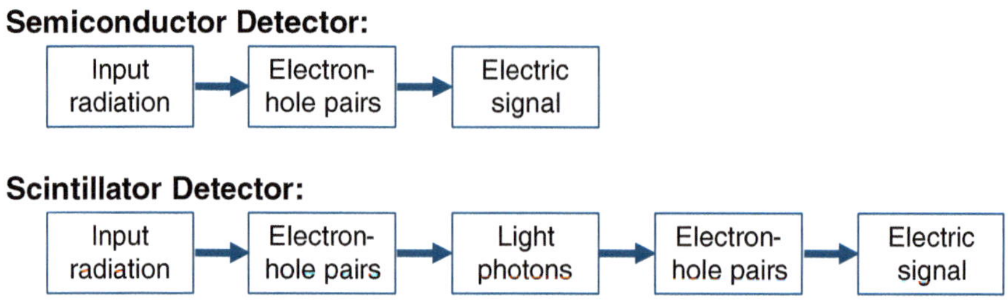

Fig. 4.2 Signal conversion processes involved in semiconductor detectors and scintillator detectors with photodiodes

through their directional drift. In contrast, scintillators lack such an electric field. The electrons and holes thus recombine over time, and a fraction of the recombinations produce light photons. When the scintillator is coupled to a photodiode, these light photons are then absorbed, creating secondary electron–hole pairs. The drift of these secondary charge carriers generates the final electrical signal.

Another misconception, especially among newcomers to x-ray computed tomography (CT) imaging, is the belief that scintillator detectors can only function as energy-integrating detectors (EIDs) and that photon-counting detectors (PCDs) require the use semiconductor sensors. Interestingly, this misunderstanding is far less prevalent in the nuclear medicine imaging community, where many PCDs are in fact scintillator based. One of the key objectives of this chapter is to clarify and reinforce the fundamental distinction between EIDs and PCDs, independent of the sensor material. To this end, the chapter covers both scintillator-based EIDs and scintillator-based PCDs and also briefly touches on semiconductor-based EIDs (Fig. 4.3). Presenting these various combinations of sensor materials and readout modes within the same context helps "connect the dots" and fosters a deeper, more accurate understanding of detector principles.

A fundamental question in scintillator physics is: why do some materials scintillate while others do not? For example, many of you are familiar with cesium iodide (CsI) as a commonly used scintillator, but have you ever wondered why it scintillates, while many other solid materials do not? Answering this question is one of the central objectives of this chapter.

To explain why scintillators emit light, learners were often referred to energy band diagrams like the one shown in Fig. 4.4. These diagrams illustrate the conduction band, the valence band, and additional intermediate energy states introduced by activation centers, which enable the emission of visible light. But this raises another question: why does not an electron excited into the conduction band directly fall back to the valence band and emit a photon, much like what happens in LEDs? Exploring such questions is key to developing a deeper understanding of scintillator behavior.

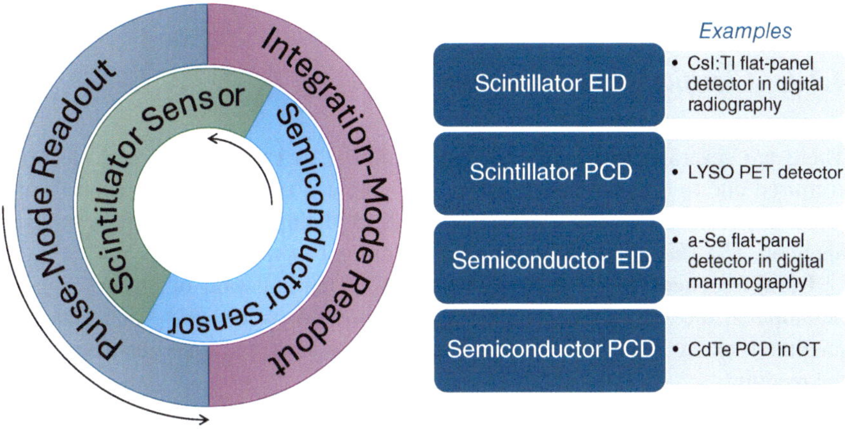

Fig. 4.3 The radiation sensors (scintillator vs. semiconductor) and readout modes (pulse mode vs. charge- or current-integration mode) are represented as two rotating wheels. Each unique combination of sensor and readout mode results in a different type of detector. Several examples of these combinations are illustrated in the chart on the right. *PCD* photon-counting detector, *EID* energy-integrating detector

Fig. 4.4 Energy band diagram of a crystalline scintillator with activators

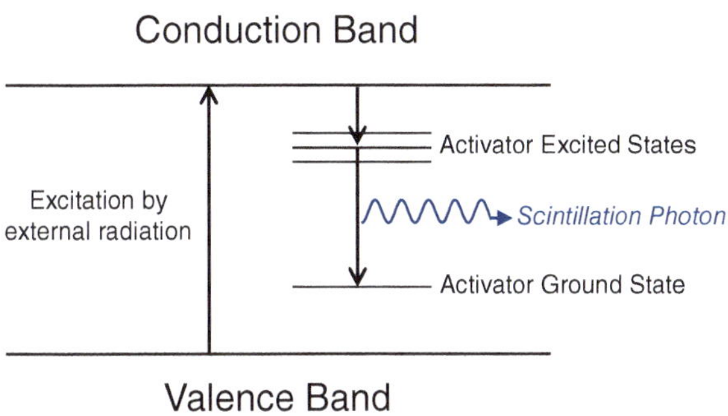

This chapter focuses exclusively on *inorganic scintillators* used in imaging detectors, which are widely employed in medical x-ray and γ-ray imaging. *Organic scintillators*—including plastic and liquid types—represent an alternative to inorganic crystalline scintillators and offer several advantages, such as easier large-scale fabrication, lower cost, and faster light emission with shorter decay times. However, their low mass density and effective atomic numbers substantially reduce their x-ray and γ-ray absorption efficiency, thus limiting their use in medical imaging detectors, where both radiation absorption efficiency and light emission characteristics matter. For information on organic scintillators applied in radiation dosimetry and other fields, readers are referred to References [5, 12].

4.2 Scintillator Material Properties

4.2.1 Clarification of Terminologies

There are several terms that may appear similar to new learners and are often confused due to their subtle differences: luminescence, fluorescence, scintillation, and phosphorescence. In this book, we adopt the definitions used by J. B. Berks in his book, *The Theory and Practice of Scintillation Counting* [3].

Luminescence is a general term describing the emission of electromagnetic radiation in the ultraviolet, visible, or near-infrared regions that occurs when a medium is excited by a process. The resulting emission spectrum is characteristic of the medium.

Based on the means of excitation, luminescence can be classified into the following types:

- **Photoluminescence**: excitation by external electromagnetic waves/photons (e.g., fluorescence)
- **Chemiluminescence and bioluminescence**: light emission resulting from a chemical reaction (e.g., the glow produced by plankton and other microorganisms around the Gulf weed at night, as described in *The Old Man and the Sea*[1]
- **Electroluminescence**: excitation by an electric field or current (e.g., in LEDs)
- **Thermoluminescence**: release of previously stored energy as light upon heating (e.g., thermoluminescent dosimeters)
- **Mechanoluminescence**: excitation by mechanical action
- **Sonoluminescence**: excitation by ultrasound-induced cavitation in liquids

According to Berks [3], photoluminescence can be further classified into the following:

- **Fluorescence** refers to an *immediate* radiative transition (typically within 10^{-9}–10^{-8} s) from an excited quantum state to a lower quantum state.
- **Slow fluorescence**, also known as the β-process, emits light at longer wavelengths than fluorescence. Its emission decays exponentially with a much longer decay time (on the order of 10^{-4} s or more).
- **Delayed fluorescence**, previously referred to as **phosphorescence** or the α-process, has an emission spectrum identical to fluorescence but decays non-exponentially over a longer period. This type of emission can be thermally activated by raising the temperature of the material.

There is some confusion about the definition of phosphorescence, as some literature may attribute it to slow fluorescence rather than delayed fluorescence. To

[1] In *The Old Man and the Sea*, Ernest Hemingway described this glow as "phosphorescence." In modern scientific terms, however, it is more accurately classified as a bioluminescent effect.

avoid ambiguity, we will refrain from using the term "phosphorescence" in this chapter.

There is also uncertainty regarding the definition of **scintillation**, as no universal scientific consensus exists. According to the Merriam-Webster dictionary, "scintillate" means to emit *quick* flashes or sparks, derived from the Latin word scintilla, meaning spark. In the context of this course, we define scintillation as the **immediate emission of light following excitation by ionizing radiation.** In other words, scintillation can be considered *ionizing* radiation-induced fluorescence. Some may argue that scintillation and radiofluorescence differ based on the underlying physical processes, but for learning purposes, this working definition is sufficient as long as it is used consistently.

The next term, **scintillator**, refers to a substance that scintillates. A scintillator does not need to be solid; liquid and even gaseous scintillators exist, such as xenon.

Finally, let us clarify the term **phosphor**, which is sometimes used interchangeably with scintillator in literature. According to Merriam-Webster, a phosphor is a luminescent substance that emits light when excited by radiation. While this is similar to the definition of a scintillator, there are important distinctions: the definition of a phosphor does not require the excitation to be ionizing, nor does it specify that the emission must be immediate. A phosphor's emission can be prompt, slow, or delayed, and it can be excited by nonionizing radiation, such as visible light. For example, glow-in-the-dark watch dials contain phosphor. In this sense, a scintillator is a subtype of phosphor, and it is the primary focus of this chapter because this book concerns with the detection of *ionizing* radiation.

4.2.2 *Energy Band Model*

We will use one of the most common inorganic scintillator materials—cesium iodide (CsI)—as an example to illustrate the energy band model of solid-state scintillators. CsI has a bandgap energy of $E_g = 6.3\,\text{eV}$. According to the classification of solids in Fig. 1.1, CsI is an **insulator**. Therefore, at thermal equilibrium, its conduction band is essentially empty, and its valence band is completely filled. Consequently, there are no electrons in the conduction band, no holes in the valence band, and thus no electronic transitions or photon emissions occur at thermal equilibrium.

CsI has a direct bandgap structure, meaning that the minimum of its conduction band and the maximum of its valence band occur at the same crystal momentum (k) value in the E–k diagram (Fig. 4.5). Many other solid-state scintillators, such as sodium iodide (NaI), also possess direct bandgap structures. But why is a direct bandgap favorable for scintillators?

To understand this, let us first consider what happens after the material absorbs energy from external radiation. The absorbed energy excites an electron from an energy level E_1 in the valence band to an energy level E_2 in the conduction band (Fig. 4.5a). Because the conduction band of an insulator is initially empty, the excited electron tends to relax to the lowest available energy state at the bottom

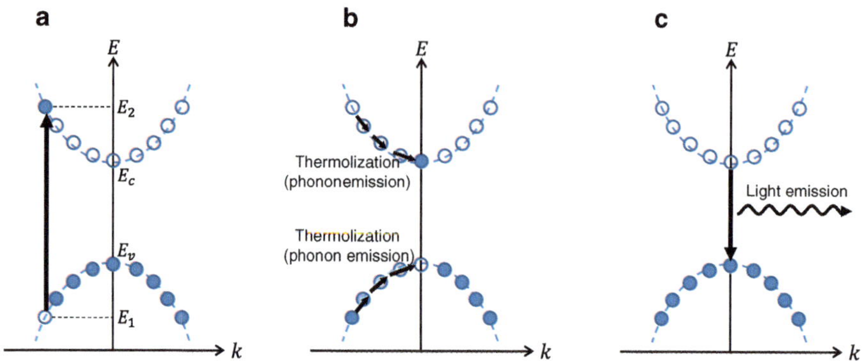

Fig. 4.5 Radiative emission in direct bandgap materials

of the conduction band (E_c). The energy difference between E_2 and E_c, as well as the corresponding momentum (k) difference, must be conserved. These excess quantities are transferred to the crystal lattice, manifesting as acoustic vibrations and dissipating as heat.

This energy and momentum relaxation process is known as **thermalization** (or **phonon emission**), where a phonon represents a quantum of lattice vibration. The thermalization process is extremely fast, typically occurring on the order of 10^{-13}–10^{-12} s. The hole left behind at E_1 undergoes a similar thermalization process, relaxing to the top of the valence band (E_v).

After thermalization, the electron at E_c tends to recombine with the hole at E_v, since an insulator at thermal equilibrium cannot maintain free carriers indefinitely. This recombination can occur via two main mechanisms:

1. **Radiative transition**, in which the electron–hole recombination results in the emission of a photon.
2. **Non-radiative transition**, in which the recombination energy is dissipated through phonon emission.

This raises two key questions: Why do electrons in scintillators primarily undergo radiative transitions? And why do electrons in some materials rarely (or never) emit photons during recombination?

Let us first assume that an electron makes a radiative transition from E_c to E_v, as illustrated in Fig. 4.5c. Based on the conservation of energy and momentum, we have

$$E_c = E_v + h\nu, \tag{4.1}$$

$$\hbar k_c = \hbar k_v + \hbar k_{\text{photon}}, \tag{4.2}$$

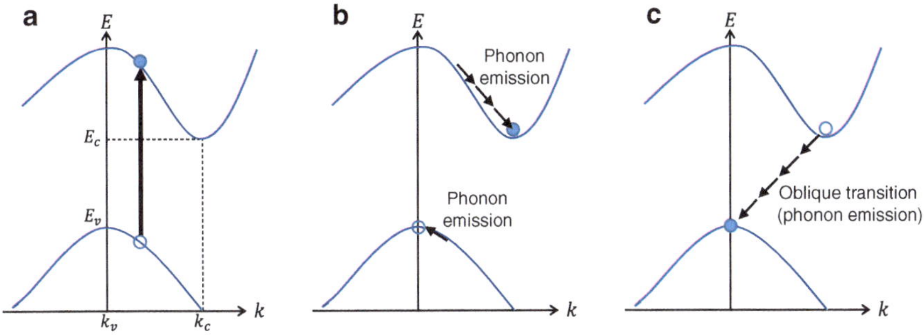

Fig. 4.6 Non-radiative emission in indirect bandgap material

where ν denotes the photon frequency, $h\nu$ is the photon energy, $k_{\text{photon}} = 2\pi/\lambda$ is the photon wave vector, and $\hbar k_{\text{photon}}$ is the photon momentum. For an optical photon, the wavelength λ is typically on the order of 500 nm = 5000 Å.

As shown in Eq. (1.16) of Chap. 1, the wave vector of an electron in a crystal, k_{electron}, is on the order of $\frac{m\pi}{a}$, where a is the lattice period and is typically on the order of 1 Å. Since 5000 Å is much larger than 1 Å, we have

$$\hbar k_{\text{photon}} \ll \hbar k_{\text{electron}}. \tag{4.3}$$

In other words, the photon momentum is negligible compared to that of an electron in a crystal. Therefore, Eq. (4.2) can be approximated as

$$\hbar k_c \approx \hbar k_v. \tag{4.4}$$

Thus, **a radiative transition involving photon emission favors materials with a direct bandgap structure**.

Next, let us examine what happens when an electron makes a transition in an indirect bandgap material. When an electron at energy level E_1 in the valence band is excited to E_2 in the conduction band, it quickly relaxes to the bottom of the conduction band (E_c) through thermalization (phonon emission). Similarly, the corresponding hole at E_1 relaxes to the top of the valence band via the same process.

According to the E–k diagram shown in Fig. 4.6, an electron at (E_c, k_c) in the conduction band must make an *oblique* transition to (E_v, k_v) in the valence band. During this transition process, both energy and momentum conservation laws must be satisfied. However, satisfying the momentum conservation law is difficult for a purely radiative transition, because the photon momentum is much smaller than the difference between k_c and k_v. Therefore, **most transitions in indirect bandgap materials are non-radiative**.[2] These transitions involve the emission of phonons,

[2] In indirect bandgap semiconductors, the recombination of an electron and a hole can emit a photon only with the simultaneous involvement of a phonon to conserve momentum. This

each carrying a small amount of energy but a momentum comparable in magnitude to that of an electron.

In summary, direct bandgap materials tend to produce radiative transitions following excitation by external energy, whereas indirect bandgap materials generally do not emit photons. This leads to a further question: apart from having a direct bandgap structure, what other characteristics make CsI an effective scintillator?

4.2.3 Quasi-Fermi Level and Probability of Emission

The probability of photon emission when an electron transitions from energy level E_2 to E_1 is determined by the joint probability that:

1. The state at E_2 is occupied by an electron
2. The state at E_1 is vacant

Mathematically, this is expressed as

$$P_{\text{emission}} = f(E_2)\,[1 - f(E_1)], \tag{4.5}$$

where $f(E)$ is the Fermi–Dirac distribution function defined in Eq. (1.53).

For an insulator at thermal equilibrium, the Fermi level is uniform throughout the material. As illustrated in Fig. 4.7a, $f(E) \approx 0$ for energies in the conduction band and $f(E) \approx 1$ for energies in the valence band. Consequently, for an insulator at thermal equilibrium,

$$P_{\text{emission}} \approx 0, \tag{4.6}$$

meaning the probability of *spontaneous* emission is essentially zero. No light is emitted under these conditions, which is a desirable property for a scintillator, since it ensures no background emission in the absence of external radiation. In contrast, semiconductors such as GaAs exhibit a small but non-negligible $f(E_2)$ and $[1 - f(E_1)]$ at room temperature. Therefore, it is undesirable to use GaAs as a room-temperature scintillator.

When an insulator absorbs external radiation, it reaches a quasi-equilibrium state characterized by two quasi-Fermi levels, E_{F_c} for the conduction band and E_{F_v} for the valence band. As shown in Fig. 4.7b, both $f(E_2)$ and $[1 - f(E_1)]$ increase, which together enhance the probability of radiative emission.

In Sects. 1.6.3 and 1.7.2, we saw that for a solid in thermal equilibrium, the product of electron and hole concentrations is a fixed value at a given temperature:

process is known as a **phonon-assisted radiative transition**. Because it involves a three-particle interaction (electron, hole, and phonon), its probability is much lower, resulting in significantly weaker light emission compared to direct radiative transitions.

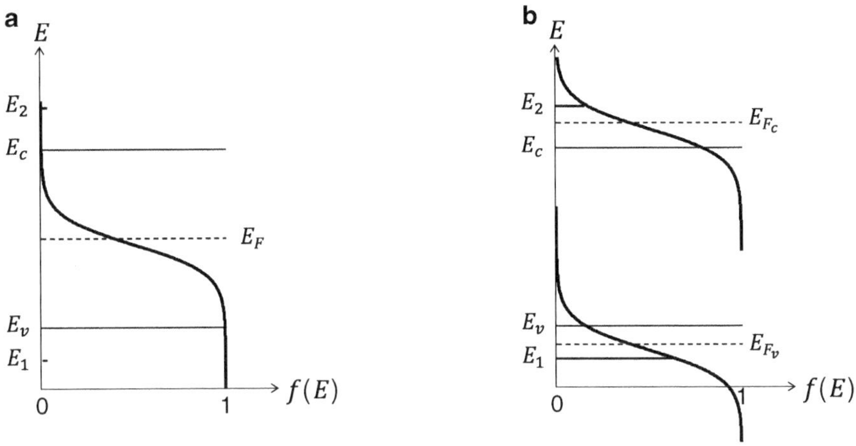

Fig. 4.7 Fermi–Dirac probability distribution function, $f(E)$, for a system in (**a**) thermal equilibrium and (**b**) quasi-equilibrium

$$n_0\, p_0 = n_i^2. \tag{4.7}$$

When radiation is absorbed by the solid, generating Δn electron–hole pairs, the product of electron and hole concentrations no longer equals n_i^2:

$$n\, p = (n_0 + \Delta n)\,(p_0 + \Delta n) > n_0\, p_0, \tag{4.8}$$

or equivalently,

$$n\, p > n_i^2. \tag{4.9}$$

This occurs because, after excitation by external radiation, the system is no longer in thermal equilibrium. In this nonequilibrium state, the concept of **quasi-Fermi levels** becomes necessary. Quasi-Fermi levels describe two separate Fermi energies in a solid under quasi-equilibrium:

- E_{F_c} for the conduction band, describing the occupation probability of electrons
- E_{F_v} for the valence band, describing the occupation probability of holes

The probability that a state at energy E in the conduction band is occupied by an electron is

$$f_e(E) = \frac{1}{1 + \exp\left(\frac{E - E_{F_c}}{k_B T}\right)}, \tag{4.10}$$

and the probability that a state at energy E in the valence band is occupied by an electron is

$$f_v(E) = \frac{1}{1 + \exp\left(\frac{E - E_{F_v}}{k_B T}\right)}. \tag{4.11}$$

Replacing E_F by E_{F_c} in Eq. (1.69) gives the electron concentration in the conduction band:

$$n = N_c \exp\left(-\frac{E_c - E_{F_c}}{k_B T}\right). \tag{4.12}$$

Similarly, the hole concentration in the valence band is

$$p = N_v \exp\left(-\frac{E_{F_v} - E_v}{k_B T}\right). \tag{4.13}$$

The product of n and p is then

$$\begin{aligned} n\,p &= N_c\,N_v \exp\left(-\frac{E_c - E_{F_c} + E_{F_v} - E_v}{k_B T}\right) \\ &= N_c\,N_v \exp\left(-\frac{E_g}{k_B T}\right) \exp\left(\frac{E_{F_c} - E_{F_v}}{k_B T}\right) \\ &= n_i^2 \exp\left(\frac{E_{F_c} - E_{F_v}}{k_B T}\right) \\ &= n_0\,p_0 \exp\left(\frac{E_{F_c} - E_{F_v}}{k_B T}\right), \end{aligned} \tag{4.14}$$

where $E_g = E_c - E_v$ is the bandgap energy. Since $E_{F_c} - E_{F_v}$ can be several eV while $k_B T \approx 0.026\,\text{eV}$ at room temperature, the exponential term in Eq. (4.14) is larger than unity, which explains why $n\,p > n_0\,p_0$ under excitation.

Riding on the concept of quasi-Fermi levels, the probability of emission, P_{emission}, in Eq. (4.5) can be expressed as

$$\begin{aligned} P_{\text{emission}} &= f(E_2) \cdot [1 - f(E_1)] \\ &= \frac{1}{1 + \exp\left(\frac{E_2 - E_{F_c}}{k_B T}\right)} \cdot \frac{\exp\left(\frac{E_1 - E_{F_v}}{k_B T}\right)}{1 + \exp\left(\frac{E_1 - E_{F_v}}{k_B T}\right)} \\ &\approx \exp\left(-\frac{E_2 - E_{F_c}}{k_B T}\right) \cdot \exp\left(\frac{E_1 - E_{F_v}}{k_B T}\right) \\ &= \exp\left(-\frac{E_2 - E_1}{k_B T}\right) \cdot \exp\left(\frac{E_{F_c} - E_{F_v}}{k_B T}\right). \end{aligned} \tag{4.15}$$

According to Eq. (4.14), we have

$$\exp\left(\frac{E_{F_c} - E_{F_v}}{k_B T}\right) = \frac{np}{n_0 p_0}.$$

Therefore, Eq. (4.15) can be rewritten as

$$\boxed{P_{\text{emission}} = \frac{np}{n_0 p_0} \exp\left(-\frac{E_2 - E_1}{k_B T}\right)}. \tag{4.16}$$

This expression shows that **a stronger input radiation intensity increases the probability of emission** because it generates more electron–hole pairs, increasing the value of np. Additionally, **reducing the gap between E_2 and E_1 also increases the probability of emission**.

Since E_2 is in the conduction band and E_1 is in the valence band, $E_2 - E_1 > E_g$. For an insulator such as CsI, the bandgap E_g is far larger than $k_B T$, so the exponential term in Eq. (4.16) is vanishingly small. This makes the probability of emission extremely low, and even if a radiative transition occurs, the photon energy $E_2 - E_1$ is too high to be visible. In short, *pure* CsI is a poor scintillator. Its performance improves dramatically only when a small amount of an additive, known as **activators**, is introduced to engineer its luminescent properties.

4.2.4 The Role of Activators in Scintillators

Activators are small concentrations (typically on the order of 10^{-3} mole fraction) of impurity atoms added to inorganic scintillators to enhance the probability of visible photon emission. These activators modify the energy band structure of the host crystal by introducing localized energy levels within the forbidden gap. Figure 4.8 plots the potential energy of a scintillator system as a function of crystal lattice displacement. The process of radiation-induced light emission can be described in the following steps:

1. An electron is excited from state A in the valence band to state B in the conduction band.
2. The electron relaxes from state B to state C, an energy level introduced by the activator. A collection of such levels is referred to as an **activator center** or **luminescence center**. In the pure crystal, this state does not exist within the forbidden gap. The B→C transition is accompanied by phonon emission (thermalization) rather than photon emission to conserve both energy and momentum.
3. The electron then undergoes a vertical transition from state C to state D, the ground state of the activator. This transition is radiative, and the energy of the emitted photon is given by $h\nu = E_C - E_D$. This radiative transition occurs typically within 10^{-9}–10^{-8} s.

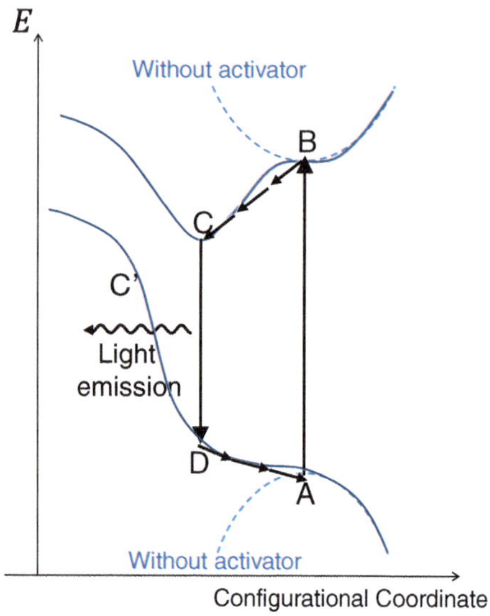

Fig. 4.8 A schematic illustration of the energy band structure and light-emission process in a scintillator containing activators

4. Finally, the electron relaxes from state D back to state A or other unfilled states in the valence band of CsI through phonon emission.

The transitions from *B to C* and *D to A* are non-radiative because they are oblique transitions involving large changes in the crystal momentum. Phonons must therefore participate to conserve momentum. In some cases, an electron may also transition between nearby activator states (e.g., from C to C'). This phonon-mediated process, known as **quenching**, competes with radiative emission and reduces the overall light yield.

Based on the above analysis of the scintillation process, the introduction of activators into a scintillator provides three key advantages:

- **Increased probability of light emission**: Because the activator introduces states C and D within the forbidden gap of the pure crystal,

$$E_C - E_D < E_g < E_B - E_A. \tag{4.17}$$

In practice, the energy difference $E_C - E_D$ can be more than 50% smaller than $E_B - E_A$. Consequently,

$$\exp\left(-\frac{E_C - E_D}{k_B T}\right) \gg \exp\left(-\frac{E_B - E_A}{k_B T}\right). \tag{4.18}$$

According to the expression for P_{emission} in Eq. (4.16), this greatly enhances the probability of photon emission.

- **Reduction of emitted photon energy to the visible range**: In the absence of activators, the bandgap of the pure crystal (e.g., CsI) is approximately 6 eV, which is well above the energy range of visible light (1.6–3.2 eV). Since $h\nu = E_C - E_D$ is significantly smaller than the bandgap, the presence of activators shifts the emitted photon energy into the visible spectrum.
- **Reduced probability of scintillation light reabsorption**: Because $h\nu < E_g$, the emitted photons are unlikely to be reabsorbed by the host crystal to generate new electron–hole pairs. Thus, there is minimal overlap between the absorption and emission spectra. This separation, known as the **Stokes shift**, further improves the collection efficiency of scintillation light.

4.2.5 Charge Carrier Trapping and Delayed Emission in Scintillators

In Chap. 1, we discussed how radiation-induced electrons and holes tend to recombine in a solid material. In semiconductor detectors, such recombinations are undesirable and are suppressed by driving electrons and holes to rapidly drift in opposite directions under an applied bias voltage. In contrast, in scintillator detectors, instantaneous recombination is highly desirable to boost light emission.

We also discussed in Chap. 1 that electrons and holes can be trapped in semiconductors by impurity atoms or crystal defects (trapping centers). Insulating materials, such as inorganic scintillators, are no exception to this trapping effect. Except for latent dosimeters, such as thermoluminescent dosimeters, charge trapping is generally undesirable for both semiconductor and scintillator detectors: in semiconductors, it may reduce the electric field and charge collection efficiency, while in scintillators, it often leads to delayed light emission, or afterglow.

Similar to activators, trapping centers introduce additional metastable energy levels, typically slightly below the conduction band of the intrinsic material. As illustrated in Fig. 4.9, a radiation-excited electron may fall into a trap state. The probability that an electron is captured by a trap, rather than recombining at an activator site or undergoing quenching, depends on the detailed energy-level structure of the material. If the electron falls into a trap, it remains there until thermal energy ($k_B T$) from lattice vibrations or another stimulus releases it back to the conduction band. Once back in the conduction band, the electron can either return to the trap state or transition to a state created by the activators.

Based on the Boltzmann statistics, the probability that an electron possesses enough thermal energy to escape the trap is proportional to

$$P(\text{escape}) \propto \exp\left(-\frac{E_t}{k_B T}\right). \tag{4.19}$$

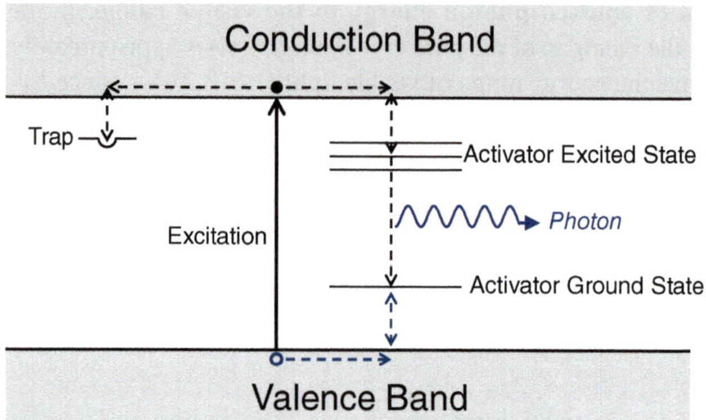

Fig. 4.9 A revised version of Fig. 4.4, i.e., a schematic of the energy band structure of an impurity-activated inorganic scintillator, now including trapping and quenching besides the luminescence process. The probability of each pathway depends on the specific material properties

The rate of electron escape from the trap, s, can be expressed as

$$\kappa = \nu_0 \exp\left(-\frac{E_t}{k_B T}\right), \tag{4.20}$$

where ν_0 denotes the characteristic vibration frequency of the lattice. According to Eq. (1.150) in Chap. 1, the average lifetime (τ) of the trapped electron is related to the rate of escape (κ) by

$$\kappa = \frac{1}{\tau}. \tag{4.21}$$

Therefore,

$$\begin{aligned} \tau &= \frac{1}{\nu_0} \exp\left(\frac{E_t}{k_B T}\right) \\ &= \tau_0 \exp\left(\frac{E_t}{k_B T}\right), \end{aligned} \tag{4.22}$$

where $\tau_0 = \nu_0^{-1}$ is the characteristic (attempt) time. Equation (4.22) indicates that the lifetime of the trapped carrier increases exponentially with the trap depth E_t and decreases exponentially with increasing temperature.

4.2.6 *Common Inorganic Scintillators for Medical Imaging*

For the scintillators discussed in this section, the values of their physical properties, such as light yield, were primarily obtained from the US Department of Energy

National Laboratory Scintillator Library (Reference [32]) and Reference [12]. These values may vary depending on material fabrication conditions, manufacturing technology, and other environmental factors.

4.2.6.1 NaI(Tl)

Thallium-activated sodium iodide (NaI(Tl)) is a classical scintillator material with a light yield of approximately 38,000 photons per MeV of deposited energy. This corresponds to an average radiation energy of about 26 eV required to produce a single emitted photon—significantly higher than the NaI bandgap energy (5.8 eV) due to quenching, trapping, and other non-radiative processes that compete with photon emission.

NaI(Tl) emits blue light with a peak wavelength near 415 nm. Its dominant decay time is about 230 ns, shorter than that of cesium iodide (CsI). Owing to its high light yield, good energy resolution, and moderate cost, NaI(Tl) remains the standard scintillation material for gamma-ray spectroscopy.

A primary drawback of NaI(Tl) is its hygroscopic nature—it deteriorates when exposed to moisture in the atmosphere and must therefore be hermetically sealed for normal operation. The material is also relatively fragile and can be easily damaged by mechanical stress.

4.2.6.2 CsI(Tl) and CsI(Na)

Cesium iodide activated with thallium emits visible scintillation light in different spectral regions: CsI(Tl) emits green light (peaking at 540 nm), while CsI(Na) emits blue light (peaking at 420 nm).

- The green emission from CsI(Tl) is poorly matched to the spectral response of bialkali-based photocathodes used in photomultiplier tubes (PMTs). Therefore, CsI(Na) is generally preferred when PMTs are used as light sensors.
- For detectors employing photodiodes as light sensors, the photosensitivity typically extends well into the red region of the spectrum. In this case, CsI(Tl) is more desirable because its light yield (approximately 60,000 photons/MeV) is significantly higher than that of CsI(Na) (about 39,000 photons/MeV).
- Compared with NaI, CsI is less brittle and less hygroscopic and has both a higher effective atomic number ($Z_{\mathrm{eff}} = 54$) and mass density (4.5 g/cm^3), resulting in superior x-ray and γ-ray absorption efficiency. Moreover, CsI can be grown in a columnar (needle-like) structure, which enhances spatial resolution in imaging detectors. Due to these advantages, CsI(Tl) and CsI(Na) are widely used as scintillation materials in medical imaging applications.
- The main drawback of CsI-based scintillators is their relatively long decay time. The emission from CsI(Tl) typically exhibits two components with decay constants of 0.7 μs and 3.3 μs, along with an afterglow that can interfere with high-frame-rate imaging applications, such as computed tomography (CT).

4.2.6.3 Gadox

Gadox is short for Gd_2O_2S or gadolinium oxysulfide. On its own, Gadox produces little visible light; however, when doped with activator ions, such as Ce^{3+}, Tb^{3+}, or Pr^{3+}, it becomes a highly effective scintillator.

- With a mass density of 6.7 g/cm^3 and a Z_{eff} of 58, Gadox exhibits excellent x-ray absorption efficiency. The presence of high-Z gadolinium contributes strongly to its photoelectric absorption cross-section, making it particularly effective for x-ray and γ-ray detection in the diagnostic energy range.
- The emission spectrum of Gadox depends strongly on the activator ion. Gadox activated by Ce^{3+} emits in the blue region with a peak near 440 nm, Pr^{3+} activation yields blue–green emission around 485 nm, and Tb^{3+} activation produces green emission peaking near 545 nm.
- The scintillation decay time also depends on the activator. Gadox(Ce) exhibits a very fast decay constant of approximately 50–60 ns for 90% of the scintillation lights. Gadox(Pr) has an intermediate decay time on the order of 0.1 ms, while Gadox(Tb) shows a slower decay time of roughly 1 ms.
- A key advantage of Ce-activated Gadox is its fast decay and minimal afterglow. These characteristics enable rapid signal recovery and low image lag, which are essential for high-speed and CT imaging. Although the light yield of Gadox(Ce) ($\sim$9000 photons/MeV) is lower than that of alkali halide scintillators, such as NaI(Tl) or CsI(Tl), its short decay time makes it better suited for time-resolved and high-frame-rate imaging.
- Raw Gadox material is a fine, white, odorless powder. When sintered at high temperature ($\approx$1200–1400 °C), the powder transforms into a dense, polycrystalline ceramic plate with strong mechanical integrity and good optical transparency. Both surfaces of the plate are polished to achieve high optical flatness. The polycrystalline plate is then mechanically scored, laser scribed, or saw diced into an array of narrow columns or pixels (Fig. 4.10). For CT detectors, the typical pixel size ranges from 0.4 mm to 1.5 mm. The cuts extend nearly through the full thickness of the plate but stop short of completely separating the material to maintain mechanical stability. To enhance spatial resolution and reduce optical cross talk between neighboring pixels, the grooves between pixels are filled with a light-reflective material (e.g., TiO_2-loaded epoxy, $BaSO_4$, or a polymer resin). The reflector design and its optical properties will be discussed further in Sect. 4.3.
- Compared with CsI-based scintillators, Gadox(Ce) offers superior mechanical durability, chemical stability, and resistance to humidity, eliminating the need for hermetic sealing. However, its relatively low light output can limit energy resolution in spectroscopic applications.

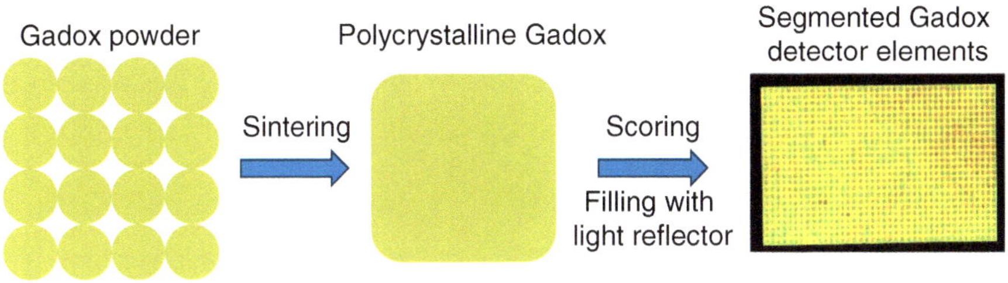

Fig. 4.10 Fabrication of segmented Gadox detector elements from Gadox power

4.2.6.4 BGO

Bismuth germanium oxide or $Bi_4Ge_3O_{12}$ is an inorganic scintillator that does not require any additional activator elements. Its luminescence originates from the optical transitions of Bi^{3+} ions within the crystal lattice. BGO belongs to the class of oxide scintillators and has been extensively studied and used since the 1970s due to its excellent x-ray and γ-ray absorption efficiency and mechanical stability (Fig. 4.11).

- High absorption efficiency: The major advantage of BGO is its large effective atomic number ($Z_{\mathrm{eff}} = 74$) and high density (7.13 g/cm^3), which provide a high probability for photoelectric absorption and Compton scattering. These characteristics make it especially effective for detecting high-energy x-rays and gamma rays for nuclear medicine imaging, high-energy physics experiments, and space-borne gamma-ray telescopes.
- Mechanical and chemical stability: BGO is nonhygroscopic, chemically inert, and mechanically robust. It can be easily cut, polished, and optically coupled to photodetectors without requiring special environmental protection, unlike hygroscopic materials, such as NaI(Tl).
- Scintillation characteristics: The emission peak of BGO is centered around 480 nm (blue light), which matches reasonably well with the sensitivity range of standard PMTs. It has an intermediate scintillation decay time of about 56 ns (90%) or 400 ns (10%), which allows for moderate count-rate capabilities.
- Light yield: The primary limitation of BGO is its relatively low light yield, approximately 8200 photons/MeV—much lower than that of NaI(Tl) or LSO (around 26,000 photons/MeV). This lower light output results in reduced energy resolution.
- Typical usage consideration: Due to its low light yield but high radiation absorption efficiency, BGO is preferred in applications where detective quantum efficiency and compactness are prioritized over energy resolution or timing precision.

Material	Density (g/cm³)	Effective Z	Hygroscopicity	Wavelength of Max Emission (nm	Refractive Index	Decay Time (ns)	Average Light Yield (Photons/MeV)	Averageenergy needed per light photon
NaI(Tl)	3.7	56	Yes	415	1.85	230	38,000	26 eV
CsI(Na)	4.5	54	Yes	420	1.84	460, 4180	39,000	26 eV
CsI(Tl)	4.5	54	Slight	540	1.80	680(64%), 3340(36%)	60,000	17 eV
Pure CsI	4.5	54	Slight	315	1.95	16	2,00	500 eV
GSO(Ce)	6.7	58	No	440	1.85	56 (90%), 400 (10%)	9,000	111 eV
BGO	7.1	74	No	480	2.15	300	8,200	122 eV
YSO (Ce)	4.5	34	No	420	1.80	60-70	9,000–28,000	36–111 eV
LSO (Ce)	7.4	66	No	420	1.83	35–47	26,000–30,000	33–38 eV
LYSO(Ce	7.1	64	No	420	1.82	35–45	27,000–36,000	28–37 eV
$LuBr_3$(Ce)	5.6	60	Yes	500	2.1	30	100,000	10 eV
$LaBr_3$(Ce	5.1	47	Yes	380	2.1	26	63,000	16 eV

Fig. 4.11 Physical properties of some inorganic solid scintillator materials. GSO(Ce): Ce-activated Gd_2O_2S. BGO: $Bi_4Ge_3O_{12}$. LYSO(Ce): Ce-activated $Lu_{2(1-x)}Y_xSiO_5$. LSO(Ce): Ce-activated Lu_2SiO_5. (Data source: References [32] and [12])

Fig. 4.12 A piece of cerium-activated yttrium aluminum garnet [YAG(Ce)] crystal

4.2.6.5 Garnet Scintillators

Garnet is a group of silicate minerals that are commonly used as a gemstone. These minerals share the same cubic crystal structure and general chemical formula, but differ in chemical composition. The general formula is

$$A_3B_2(SiO_4)_3,$$

where A is a divalent cation ($^{2+}$), such as Ca^{2+}, Mg^{2+}, and Fe^{2+} and B is a trivalent cation ($^{3+}$), such as Al^{3+}, Fe^{3+}, and Cr^{3+}.

In garnet scintillators, we generalize the gemstone garnet to include other cations instead of SiO_4 (e.g., AlO_4) and add rare-earth activators (like Ce^{3+}) to facilitate light emission. Examples are

- **YAG(Ce)**. Cerium-activated yttrium aluminum garnet (Fig. 4.12):

$$Y_3Al_5O_{12}:Ce^{3+}$$

- **LuAG(Ce)**. Cerium-activated lutetium aluminum garnet:

$$Lu_3Al_5O_{12}:Ce^{3+}$$

- **GGAG(Ce)**. Cerium-activated gadolinium gallium–aluminum garnet. In this mixed garnet composition, the numbers of Ga^{3+} and Al^{3+} ions occupying the five B-sites in the crystal lattice follow a 3:2 ratio. The chemical formula is

$$Gd_3Ga_3Al_2O_{12}:Ce^{3+}$$

- **GFAG(Ce)**. Cerium-activated gadolinium gallium–aluminum garnet. Although the abbreviation "GFAG" might appear to suggest the presence of fluorine, the letter "F" actually stands for **fully**. It serves as a mnemonic used by researchers and manufacturers to denote a **fully mixed** gallium–aluminum garnet composition, meaning that Ga^{3+} and Al^{3+} ions are homogeneously distributed across all B-sites in the lattice. The chemical formula is typically expressed as

$$Gd_3(Ga, Al)_5O_{12}:Ce^{3+}$$

- **YGAG(Ce)**. Cerium-activated yttrium gallium–aluminum garnet:

$$Y_3(Ga, Al)_5O_{12}:Ce^{3+}$$

Garnet scintillators offer several advantages, including a fast response time, high light yield, excellent mechanical and chemical stability, and nonhygroscopic behavior. Taking YAG(Ce) as an example: its scintillation decay time consists of 88 ns (72%) and 300 ns (28%), which is comparable to that of Gadox. However, YAG produces approximately 17,000 photons/MeV, significantly higher than the light yield of Gadox (9000 photons/MeV). These properties make YAG or other garnet scintillators particularly suitable for high-speed CT detectors.

4.2.6.6 ZnSe

Zinc selenide (ZnSe) is a direct-bandgap semiconductor rather than an insulator, with a bandgap of approximately 2.7 eV. When doped with activator elements, such as tellurium or oxygen, ZnSe can function as an efficient scintillation material. For example, ZnSe(O) emits blue light with a peak wavelength of around 460 nm and a light yield of approximately 20,000 photons/MeV. Its scintillation decay time is about 200 ns, which is comparable to that of Gadox. ZnSe is nonhygroscopic, is chemically stable, and can be easily cut and polished. However, it has a relatively low Z_{eff} of 33, compared to 59 for Gadox. These characteristics make ZnSe a promising candidate for the top layer in a dual-scintillator-layer CT detector design [26], where preferential absorption of low-energy x-rays is desired.

4.2.6.7 Rare Earth Oxyorthosilicate

This includes Ce-activated lutetium oxyorthosilicate [Ce-activated Lu_2SiO_5 or LSO(Ce)], Ce-activated yttrium oxyorthosilicate [Ce-activated Y_2SiO_5 or YSO(Ce), Fig. 4.13], and Ce-activated lutetium–yttrium orthosilicate [Ce-activated $Lu_{1.8}Y_{0.2}SiO_5$ or LYSO(Ce)].

- **YSO(Ce)** is a monoclinic rare-earth orthosilicate crystal that emits light with a peak wavelength of 420 nm. It exhibits relatively fast scintillation kinetics,

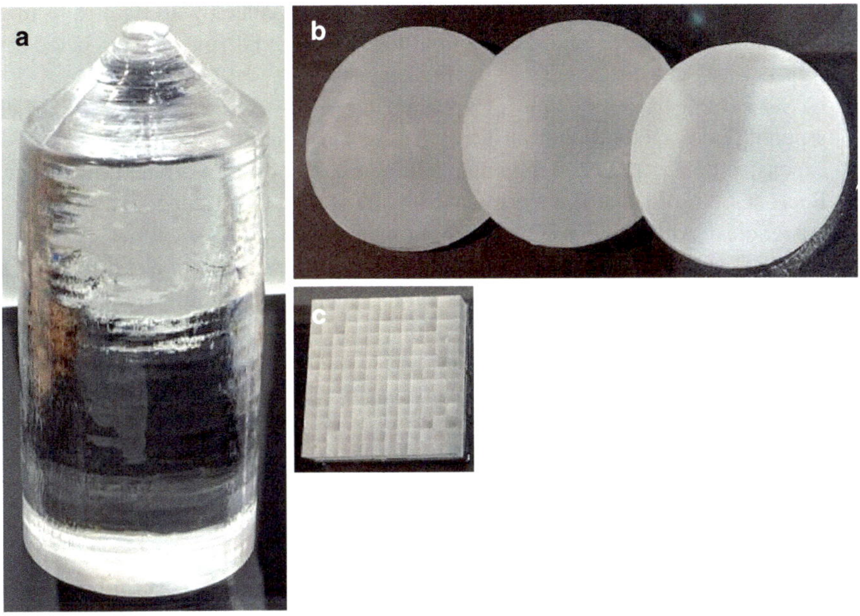

Fig. 4.13 (**a**) A LSO scintillator boule. (**b**) LSO discs cut from the boule and polished. (**c**) A 13 × 13 array of segmented LSO crystal blocks

with a decay time of 50–70 ns. The material is mechanically robust, easily polished, and nonhygroscopic. YSO has a light yield of approximately 9000–28,000 photons/MeV, an intermediate Z_{eff} of 34, and a mass density of 4.5 g/cm^3.

- **LSO(Ce)** is a dense, fast scintillator characterized by a short decay time of approximately 35–47 ns, a high Z_{eff} of 66, and a mass density of 7.4 g/cm^3. It delivers a high light yield of about 26,000–30,000 photons/MeV. However, LSO has some drawbacks: it exhibits intrinsic background radiation of approximately 300 counts/s/cm^3 due to the presence of the naturally radioactive isotope ^{176}Lu, and it is relatively brittle, requiring careful handling and machining. Additionally, lutetium is more expensive than yttrium, increasing material costs.
- **LYSO(Ce)**: By alloying LSO with YSO to form LYSO, the material's band structure is optimized to achieve both improved decay time and enhanced light yield. Monolithic LYSO crystals are typically grown by the Czochralski method, the same process used for many other scintillating crystals. The starting materials are high-purity Lu_2O_3, Y_2O_3, SiO_2, and a small fraction of CeO_2 (0.1–0.2 mol%). The powders are homogeneously mixed and melted in a crucible. A seed LYSO or YSO crystal is dipped into the melt and slowly pulled upward while rotating. Controlled pulling and temperature gradients allow the melt to solidify as a large, transparent monolithic crystal.

  After growth, the crystal is annealed to relieve internal stresses and improve optical clarity. It is then cut and polished to improve optical coupling efficiency to light sensors. The crystal may also be wrapped or coated with light-reflecting materials to enhance light collection for use in PET or gamma detectors (Fig. 4.13c).

LYSO exhibits a decay time of approximately 35–45 ns and a light yield of around 27,000–36,000 photons/MeV. It is easier to grow, more stable, and less expensive than pure LSO while maintaining excellent scintillation performance. Because 10% of Lu atoms are replaced by Y, the intrinsic radioactivity and cost are reduced. Although LYSO has a slightly lower density 7.1 g/cm^3 vs. 7.4 g/cm^3 for LSO) and a marginally lower Z_{eff} (66), its superior scintillation characteristics make it the preferred choice for high timing and energy resolution applications, such as time-of-flight (TOF) PET imaging.

As summarized in Fig. 4.11, LSO and LYSO outperform BGO in nearly all key scintillator properties, making it an increasingly popular material for modern PET detector construction.

- Other rare-earth scintillators used in PET detectors include Ce-activated lutetium bromide [$LuBr_3$(Ce)] and Ce-activated lanthanum bromide [$LaBr_3$(Ce)]. Their key properties are summarized in Fig. 4.11. **There is no single scintillator that is optimal for all applications; the choice of material depends on the specific medical imaging task, performance requirements, cost and material availability, and other practical considerations.**

4.2.6.8 Halide Perovskite Scintillators

Perovskite materials, named after the Russian mineralogist Lev Perovski, comprise a class of compounds characterized by the perovskite crystal structure with the general formula of ABO_3. In this structure, the A-site cation is a larger ion (e.g., Ca^{2+}, Ba^{2+}, or an organic cation, such as CH3NH3$^+$), the B-site cation is a smaller ion, typically a transition metal (e.g., Ti^{4+}, Fe^{3+}, and Mn^{3+}), and the O-site anion may be oxygen or halides, such as Cl^-, Br^-, or I^-.

The first naturally occurring perovskite mineral identified was calcium titanate ($CaTiO_3$). For scintillation applications, the most studied perovskites are halide perovskites, such as $(CH_3NH_3)PbBr_3$ and $(CH_3NH_3PbI_3$. With a bandgap of approximately 2 eV, they are actually semiconductors instead of insulators. While theoretical light yields of halide perovskite scintillators are high due to their narrow band widths, by far there is no experimental evidence that halide perovskite scintillators provide a fundamental advantage in scintillation light yield over conventional inorganic crystalline scintillators, such as LYSO [16, 17]. Moreover, their x-ray and γ-ray absorption efficiencies can be lower due to their reduced mass density and effective atomic number. For example, $(CH_3NH_3)PbBr_3$ has a Z_{eff} of approximately 46 and a mass density of about 3.9 g/cm^3, both lower than those of LSO.

The primary advantages of perovskite scintillators lie in their compositional tunability, bandgap engineering flexibility, and facile fabrication. Unlike conventional inorganic single-crystal scintillators, such as LYSO or BGO, which require high-temperature crystal growth processes (e.g., the Czochralski method), perovskite halides can be synthesized chemically at lower temperatures using simpler solution-based or vapor-phase routes in many research laboratories—one reason for the large volume of publications on this topic. Bulk crystals, thin films, or nanocrystalline powders can thus be produced without complex equipment. Additionally, the mechanical flexibility of perovskite scintillators makes them attractive for constructing bendable or large-area detectors.

However, several challenges remain. Halide perovskites generally suffer from environmental instability under moisture, oxygen, heat, and light exposure, as well as from ion migration [11]. Thick perovskite films also exhibit significant light scattering, which limits optical transport, detective quantum efficiency, and energy resolution [16]. Furthermore, a pronounced trade-off between scintillation light yield and decay time has been reported for halide perovskite scintillators [16].

4.3 Light Transport in Scintillators

4.3.1 Coupling Between Scintillator and Light Sensor

Because scintillation light is emitted in all directions, only a limited fraction can travel directly to the interface between the scintillator and the light sensor (e.g., a photodiode). When the light reaches this interface, total internal reflection occurs if (1) the refractive index of the scintillator (n_0) is greater than that of the sensor (n_1) and (2) the incident angle θ exceeds the critical angle θ_c, given by

$$\theta_c = \sin^{-1}\left(\frac{n_1}{n_0}\right). \tag{4.23}$$

Assume the scintillator surface is flat and polished. Lights that can escape from the scintillator to the sensor are confined within a cone whose apex is at the origin of the emitted light, axis is normal to the slab surface, and vertex angle is θ_c (Fig. 4.14). The total fraction of light collected by the sensor is given by

$$\begin{aligned} F_{\text{collect}} &= \frac{1}{4\pi}\iint \mathrm{d}\Omega \\ &= \frac{1}{4\pi}\int_0^{2\pi} \mathrm{d}\varphi \int_0^{\theta_c} \sin\theta \,\mathrm{d}\theta \\ &= \frac{1}{4\pi} 2\pi(1-\cos\theta_c) \\ &= \frac{1}{2}\left[1-\sqrt{1-\left(\frac{n_1}{n_0}\right)^2}\right]. \end{aligned} \tag{4.24}$$

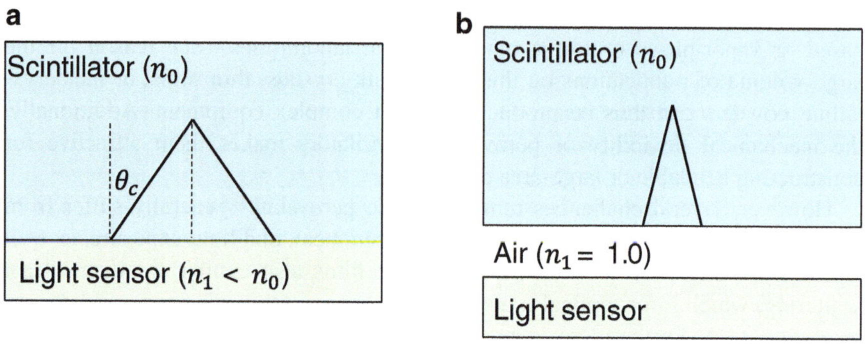

Fig. 4.14 Illustration of the critical angle (θ_c) for total internal reflection. (**a**) Light sensor tightly coupled to the scintillator. (**b**) Presence of an air gap between the sensor and scintillator

For a scintillator with refractive index $n_0 = 1.8$ and a light sensor with refractive index $n_1 = 1.5$, the fraction of emitted light that can be collected by the sensor is

$$\begin{aligned} F_{\text{collect}} &= \frac{1}{2}\left[1 - \sqrt{1 - \left(\frac{1.5}{1.8}\right)^2}\right] \\ &\approx 22\%. \end{aligned} \tag{4.25}$$

However, if the scintillator and sensor are separated by a thin layer of air ($n_1 =$ 1.0), the collection fraction is significantly reduced (Fig. 4.14b):

$$\begin{aligned} F_{\text{collect}} &= \frac{1}{2}\left[1 - \sqrt{1 - \left(\frac{1.0}{1.8}\right)^2}\right] \\ &\approx 8\%. \end{aligned} \tag{4.26}$$

To minimize such losses in light collection efficiency, **a high-viscosity silicone oil or transparent epoxy cement is commonly used to provide an optical coupling layer between the scintillator and the light sensor**. The refractive index of the coupling agent is chosen to match n_0 as closely as possible. In practice, however, achieving a perfect match is difficult because the refractive index depends on wavelength. Even if a good match is achieved at one wavelength, it may not remain optimal across the entire emission spectrum of the scintillator.

Example: Scintillator-to-photodiode coupling

A scintillator's refractive index is 1.8. The scintillator is coupled with a Si photodiode through an optical gel that has a refractive index of 1.5. The refractive index of Si is 4.0.

Question: Ignoring the attenuation and reabsorption of light in the scintillator, calculate the fraction of scintillation light that can be collected by the Si photodiode.

Answer: Since the refractive index of Si exceeds that of the coupling gel, no reflection occurs at the gel–Si interface. In contrast, reflection takes place at the scintillator–gel interface. The fraction of light that can directly transmit through the scintillator–gel interface is

$$\begin{aligned} f_1 &= \frac{1}{2}\left[1-\sqrt{1-\left(\frac{n_{\text{gel}}}{n_{\text{scintillator}}}\right)^2}\right] \\ &= \frac{1}{2}\left[1-\sqrt{1-\left(\frac{1.5}{1.8}\right)^2}\right] \\ &= 0.224. \end{aligned} \tag{4.27}$$

For the scintillator–gel interface, the critical angle $\theta_1 = \sin^{-1}\left(\frac{n_{\text{gel}}}{n_{\text{scintillator}}}\right) = 0.985$. For the other side of the scintillator (interface with air), the critical angle $\theta_1 = \sin^{-1}\left(\frac{n_{\text{air}}}{n_{\text{scintillator}}}\right) = 0.589$. Those lights hitting the top surface within beyond the critical angle θ_2 will be reflected toward the bottom, and a fraction of them can escape to reach the sensor. The fraction of light hitting the top surface that can be eventually collected is given by

$$\begin{aligned} f_2 &= \frac{1}{4\pi}\int_0^{2\pi} \mathrm{d}\phi \int_{\theta_1}^{\theta_2} \sin\theta \mathrm{d}\theta \\ &= \frac{1}{2}(\cos\theta_1 - \cos\theta_2) \\ &= 0.139. \end{aligned} \tag{4.28}$$

Therefore, the total fraction of scintillation light that can be collected by the Si photodiode is

$$f = f_1 + f_2 = 0.363. \tag{4.29}$$

4.3.2 Impact of Reflective Layer and Scintillator Thickness

The formula for the light collection fraction in Eq. (4.24) assumes that light traveling toward the top surface (opposite to the light sensor) is lost. If these rays are within θ_c, they escape from the top surface without being detected. If their θ is larger than θ_c, they are reflected by the top surface and travel toward the bottom surface; however, they still cannot escape from the bottom surface to reach the light sensor due to total internal reflection. Therefore, this assumption is reasonable.

If a light-reflecting material (e.g., TiO_2) is placed on the top surface (Fig. 4.15), the light collection fraction can, in principle, be doubled. However, the main drawback of using a top reflector is the potential degradation of spatial resolution. As shown in Fig. 4.15b, the spatial spreading (L) of the collected light for a reflector-free system is given by

$$L = 2(T - z) \tan \theta_c, \tag{4.30}$$

where z is the depth of light emission and T is the total thickness of the scintillator.

For a system with a top reflector, as shown in Fig. 4.15b, the geometric spreading of the scintillation lights along the lateral direction becomes

$$L = 2(T + z) \tan \theta_c. \tag{4.31}$$

Hence, the spatial light spreading is more pronounced when a reflector is used. If spatial resolution is more critical than light collection efficiency for a given application, it is undesirable to employ a top reflector.

Regardless of whether a reflector is used, Eqs. (4.30) and (4.31) indicate that spatial spreading of light increases with scintillator thickness when the light sensor is attached to the side opposite the radiation incidence. One might ask: why not simply reduce the scintillator thickness T to improve spatial resolution? The reason is that the total fraction of incident X- or γ-rays absorbed by the scintillator is given by

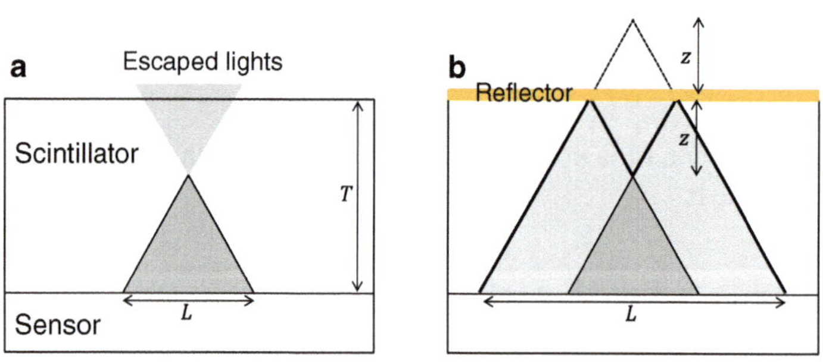

Fig. 4.15 Systems without (**a**) and with (**b**) a top reflective layer

$$F_{\text{absorb}} = 1 - \exp(-\mu T), \tag{4.32}$$

where μ is the linear attenuation coefficient. Therefore, **a trade-off exists between spatial resolution and radiation absorption efficiency in scintillator detectors**.

4.3.3 *Columnar Scintillators*

A columnar scintillator consists of closely packed, micrometer-scale, column-shaped scintillating crystals (Fig. 4.16). Such structures can be fabricated using physical vapor deposition (PVD) methods. The process begins with a cool substrate and scintillator powder placed in a crucible inside a vacuum chamber. When heated, the scintillator material vaporizes and forms a molecular flux that condenses on the substrate surface. Upon arrival, the vaporized molecules possess some kinetic energy and can hop from site to site on the surface. If the substrate is sufficiently warm, the adsorbed molecules can migrate until they find energetically favorable sites, resulting in a relatively dense film. Conversely, when the substrate is cool, the molecules quickly lose their energy and stick where they land. Under these low-mobility conditions, the deposited material self-organizes into a columnar structure due to limited surface diffusion and self-shadowing effects.

The resulting columns are typically 1–10 µm in diameter and 50–600 µm in length, depending on the desired balance between radiation absorption efficiency and spatial resolution. Voids or grain boundaries between columns act as optical barriers. Each column functions similarly to an optical fiber or light pipe, channeling scintillation light toward the light sensor with minimal lateral spread. As illustrated in Fig. 4.17, a light pipe with refractive index n_0 is surrounded by a medium with refractive index $n_1 < n_0$. When scintillation light is generated within the pipe, total internal reflection traps a portion of the light inside the column, guiding it toward the sensor. The fraction of light trapped within the pipe is given by

Fig. 4.16 Columnar CsI. (Image courtesy of Hamamatsu Photonics K.K.)

Fig. 4.17 Light trapping in columnar CsI

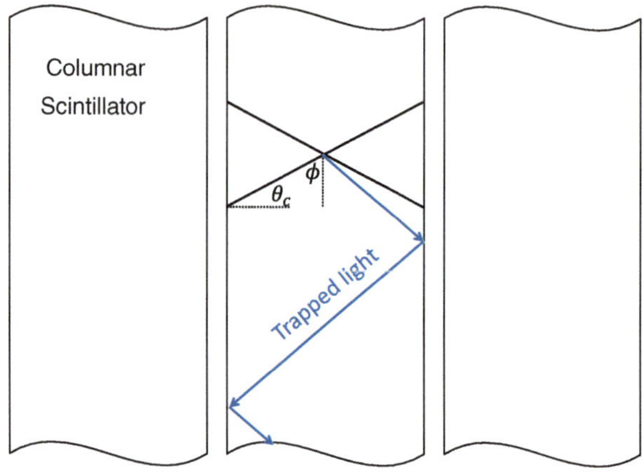

$$
\begin{aligned}
F_{\text{trap}} &= 2 \times \frac{1}{4\pi} \iint \mathrm{d}\Omega \\
&= \frac{1}{2\pi} \int_0^{2\pi} \mathrm{d}\varphi \int_0^{\pi/2-\theta_c} \sin\phi \, \mathrm{d}\phi \\
&= \frac{1}{2\pi} 2\pi [1 - \cos(\pi/2 - \theta_c)] \\
&= 1 - \sin\theta_c \\
&= 1 - \frac{n_1}{n_0}.
\end{aligned}
\tag{4.33}
$$

CsI(Tl) is the most commonly used columnar scintillator in medical imaging detectors. For CsI with a refractive index of $n_0 = 1.8$ and a surrounding void of $n_1 = 1.2$, approximately 33% of the emitted light remains confined within each column, preventing lateral spreading into neighboring columns. This optical guiding effect explains why the columnar structure provides superior spatial resolution for a given scintillator thickness [36].

4.4 Scintillator-Based Energy-Integrating Image Detectors

4.4.1 Scintillator-Based Energy-Integrating Flat-Panel Detectors

4.4.1.1 FPD with Passive Pixels

For digital flat-panel detectors (FPDs) used in radiography, fluoroscopy, and cone-beam CT for image-guided interventional and radiotherapy procedures, the vast

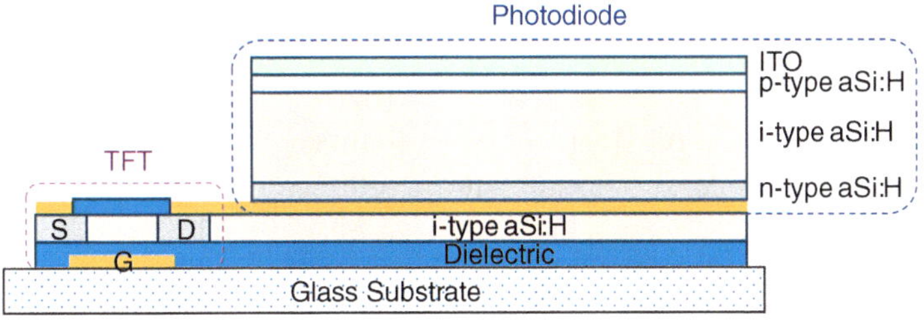

Fig. 4.18 Schematic cross-sectional illustration of an a-Si:H-based photodiode + TFT unit. *S* source, *D* drain, *G* gate, *ITO* indium tin oxide

majority are based on scintillators (e.g., CsI(Tl) or Gadox) that are optically coupled to an array of photodiodes. Because of the inherent trade-off between scintillator thickness and spatial resolution, the scintillator is typically limited to a few hundred microns or less. The spacing between neighboring photodiodes defines the detector pixel pitch, and each photodiode is connected to a thin-film transistor (TFT) for readout.

To fabricate the photodiode–TFT array, different layers of materials shown in Fig. 4.18 are sequentially deposited on a glass substrate as follows:

- Transparent conductive layer (e.g., indium tin oxide, ITO)
- Hydrogenated amorphous silicon (a-Si:H)
- Metal interconnects

Some scintillators, such as CsI:Tl, can be directly deposited onto the photodiode–TFT array. Others, such as Gadox, are bonded to the TFT panel using a thin optical coupling layer (e.g., a silicone-based optical adhesive or index-matching gel). Alternatively, the scintillator can be deposited onto a fiber-optic plate (FOP), which consists of millions of microscopic glass fibers fused together. The bottom surface of the FOP is then optically coupled to the TFT panel [36].

The key advantage of introducing an FOP between the scintillator and the readout panel is that it acts as a radiation shield for the underlying electronics, improving the readout panel's radiation hardness and long-term reliability. The use of an FOP also decouples scintillator growth from TFT fabrication, avoiding thermal or chemical damage to the detector electronics.

As shown in Table 4.2, the charge carrier mobilities of a-Si:H are more than a thousand times lower than those of single-crystal Si. As a result, an FPD made of a-Si:H lacks the speed required for pulse-mode detection. Therefore, a-Si:H-based FPDs operate in the **charge-integrating mode**, with the TFTs functioning as passive switches and charge storage elements. The low charge carrier mobility of a-Si:H also prevents the integration of a preamp within each pixel on the a-Si:H panel.

When a signal readout command is issued via a gate signal, the charge collected over a fixed period of time is transferred over a data line to a CMOS-based CSA located external to the a-Si:H film and typically mounted along the edge of the TFT array (Fig. 4.19). The pixels are **passive**, in the sense that the CSA is not dedicated to

Table 4.2 Comparison of material properties for thin-film flat panels. Poly-Si, polycrystalline silicon; SC-Si, single-crystal silicon

Property	a-Si:H	IGZO	Poly-Si	SC-Si
Electron mobility (μ_e) [cm^2/V · s]	1	~10–50	50–200	1350
Hole mobility (μ_h) [cm^2/V · s]	~10^{-3}–10^{-4}	<10^{-3}	10–100	480
Bandgap (eV)	1.8	3.0–3.6	1.1	1.1
Off-state leakage current [A]	10^{-12}–10^{-10}	10^{-14}–10^{-12}	10^{-12}–10^{-9}	<10^{-14}
Deposition temperature [°C]	200–300	150–300	550–600	>1000
Electrical stability	Moderate	High	High	Excellent
Process complexity	Low	Moderate	High	Very high

each pixel; instead, each row or column of pixels shares the same CSA.[3] Due to the long data line wiring before the initial induced charge is amplified, electronic noise can be significant because of the wiring capacitance and leakage current [29, 30].

Alternatively, the photodiode–TFT array can be fabricated using indium–gallium–zinc oxide (IGZO) [27], which is also widely used to fabricate OLED display panels. As shown in Table 4.2, the electron mobility of IGZO is significantly higher than that of a-Si:H, which increases the readout speed of FPDs and thus their maximum frame rate. However, holes are almost immobile in IGZO, so IGZO TFTs operate exclusively as n-channel devices.

By stacking two scintillator–TFT assemblies, so-called dual-layer FPDs can be implemented [2, 14]. These detectors enable dual-energy measurements from a single x-ray exposure, providing the data needed for dual-energy x-ray imaging applications. The top scintillator is designed to preferentially absorb lower-energy x-rays, while the bottom scintillator preferentially absorbs higher-energy x-rays. This energy "discrimination" can be partially achieved by limiting the thickness of the top scintillator layer (e.g., to 200 μm) or by using a material with a lower Z_{eff}, such as YSO [2].

However, due to the stochastic nature of x-ray interactions, a substantial fraction of high-energy photons are still absorbed in the top layer, and some low-energy photons reach and are absorbed by the bottom layer. To enhance energy separation between the two scintillator layers, a thick copper (Cu) filter (1 mm) has been introduced between them. The Cu filter preferentially absorbs residual low-energy x-rays passing through the top layer, thereby improving energy separation. As shown in Fig. 4.20, however, the Cu filter significantly reduces the overall x-ray absorption efficiency of the detector system, particularly for nonspectral imaging

[3] In the early development of FPDs with TFT arrays, they were sometimes referred to as *active* matrix flat-panel imagers (AMFPIs), even though each pixel does not contain an active amplifier. The term was borrowed from flat-panel display technology, where "active matrix" denotes a matrix-addressed array in which each pixel includes a switching device (the TFT) that actively controls access to that pixel. This is in contrast to a passive matrix, where rows and columns are connected directly and pixel selection relies solely on voltage gradients or resistive isolation, resulting in slower operation and higher cross talk.

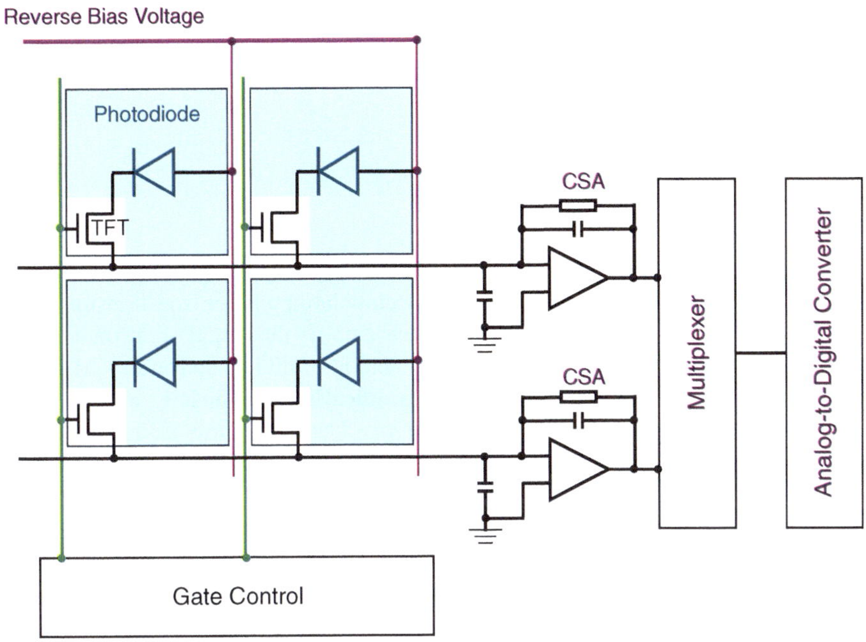

Fig. 4.19 Schematic illustration of a flat-panel detector comprising a 2 × 2 array of passive TFT-based pixels. The gate controller, charge-sensitive amplifiers (CSAs), and analog-to-digital converters (ADCs) are located externally to the TFT panel

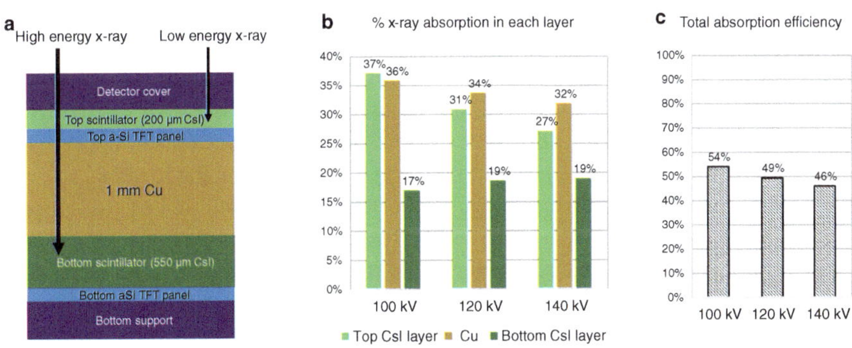

Fig. 4.20 (**a**) Schematic cross-section of a dual-layer flat-panel detector (FPD), with scintillator and Cu layer thicknesses drawn to scale. (**b**) Fraction of incident x-rays absorbed in the top and bottom CsI(Tl) scintillator layers and in the Cu filter. (**c**) Total x-ray absorption fraction in the combined scintillator layers. (Source of figure: Reference [6], licensed under Creative Commons Attribution 4.0 (CC BY 4.0), which permits redistribution and reuse of the material in any medium or format, for any purpose)

tasks, which remain the mainstay of clinical usage of FPDs. Consequently, a removable filter design has been proposed: the filter can be inserted for spectral imaging applications and removed for conventional imaging to maintain detection efficiency [6].

4.4.1.2 FPD with Active Pixels

In more advanced FPDs, each pixel incorporates not only a switching element but also a **local amplifier**. Instead of merely accumulating charge and transferring it passively to the readout bus, these **active pixels** amplify the output of the photodiode locally before the signal is routed to the analog-to-digital converter (ADC) or other readout electronics. This **in-pixel amplification** significantly enhances the detector's SNR by reducing the influence of electronic noise from the wiring capacitance. The function of this local amplifier is conceptually similar to that of an APD operated at high gain, as discussed in Eqs. (2.27)–(2.29). By amplifying the photo-generated signal at the pixel level, the impact of downstream readout noise is greatly mitigated. This capability is particularly valuable in **low-dose imaging** applications, where the number of incident photons is small and the signal level approaches the electronic noise floor.

FPDs that employ per-pixel amplifiers are most commonly realized using CMOS technology. These detectors are fabricated on **single-crystal Si** wafers, in contrast to the a-Si:H technology used in TFT panels. Consequently, detectors based on single-crystal Si are often referred to as **CMOS detectors**, whereas those based on a-Si:H TFTs are more broadly known as flat-panel detectors in general imaging contexts.

As discussed in Chap. 2, the extremely low hole mobility of a-Si:H prevents the efficient fabrication of both n- and p-type transistors on the same substrate, making it difficult to implement full CMOS circuitry [18]. In contrast, single-crystal Si exhibits high carrier mobilities, enabling the integration of true CMOS circuits at the pixel level. This not only allows for in-pixel amplification but also improves the readout speed, reduces image lag, and supports higher readout frame rates.

Another important advantage of CMOS detectors is their ability to support **smaller pixel sizes** (e.g., 50 μm or less). This capability arises from several key factors. First, the high charge carrier mobilities in single-crystal silicon allow transistors to be fabricated with short channel lengths and low parasitic capacitance. In contrast, transistors based on a-Si:H must be relatively large (typically 100–200 μm) to achieve acceptable performance, owing to the low carrier mobility of the amorphous material. Second, the presence of local amplifiers in CMOS detectors enables pixels with smaller photodiode areas to maintain a sufficiently high SNR, even at reduced charge collection areas. Third, the advanced photolithography processes used in single-crystal silicon CMOS fabrication provide far greater precision than the large-area microfabrication techniques used for a-Si:H panels, allowing the integration of denser and smaller circuit elements [4, 18].

Despite these advantages, CMOS-based FPDs also have several limitations. The fabrication cost of single-crystal Si wafers is higher than that of large-area amorphous Si substrates, and wafer size is constrained by semiconductor processing limits. To achieve large detector areas (e.g., 30 cm × 30 cm), multiple CMOS wafers are tiled together, which adds complexity to alignment, calibration, and signal stitching. For this reason, amorphous Si TFT-based FPDs remain dominant in applications, such as image-guided radiation therapy and general radiography, where frame-rate requirements are moderate and the x-ray dose is moderate to high. Conversely, CMOS-based detectors with active amplifiers are increasingly preferred for **low-dose**, **high-speed**, or **high-resolution** applications, such as such as digital mammography [25].

4.4.2 Scintillator-Based Energy-Integrating CT Detectors

In a broader sense, the term CT (computed tomography) encompasses both multi-detector row CT (MDCT), CBCT, and internal-emission tomography, such as PET. In a narrower sense, however, CT usually refers to x-ray MDCT. This section focuses exclusively on the narrower context. Cone-beam CT systems, by contrast, generally employ FPDs, as discussed in Sect. 4.4.1.

4.4.2.1 Technical Requirements of CT Detectors

Modern MDCT detectors must meet the following technical requirements to support a wide range of clinical applications.

- **High frame rate and negligible lag:** In a modern MDCT system, the x-ray tube and detector are mounted on a slip ring, allowing continuous rotation at speeds of 0.25 s per revolution or faster. Unlike CBCT systems, where the x-ray is pulsed, in MDCT the x-ray beam remains continuously on during a scan. While the gantry rotates, the detector must read out 1000 or more frames to generate the sinogram data required for tomographic reconstruction with minimal angular undersampling artifacts, corresponding to a frame rate of at least 4000 frames per second (fps). In addition to high frame rate, the detector must exhibit minimal temporal lag to reduce gantry- and patient-motion artifacts.
- **Wide dynamic range and low electronic noise:** MDCT is a general-purpose modality, scanning patients with a wide range of body habitus and anatomical regions, from the pelvis to the wrist. Some examinations, such as low-dose lung cancer screening CT, operate at very low radiation exposure, whereas others, such as non-contrast head CT for stroke evaluation, require higher exposure levels. Consequently, the detector must have negligible electronic noise to preserve image quality at low dose and a wide dynamic range (1:100,000 or wider) to accommodate high-dose scans. Additionally, the detector should provide a linear response across the full range of x-ray exposures.

- **Intermediate pixel size:** The pixel size of modern CT detectors typically ranges from 0.4 mm to 1.2 mm.[4] Smaller pixels improve spatial resolution for high-resolution imaging tasks; however, they also increase image noise and generate substantially larger data volumes that must be transferred off the slip ring in real time, potentially exceeding the available data bandwidth. Further reductions in pixel size additionally require smaller x-ray focal spots, which can be difficult to achieve given the high tube current requirements of certain CT protocols.
- **Excellent scatter control:** Effective scatter rejection is essential, as scattered x-rays degrade image quality and compromise quantitative accuracy in both diagnostic imaging and radiotherapy planning.
- **High x-ray absorption efficiency:** High x-ray absorption efficiency is critical for maximizing the SNR of CT images. The scintillator must have sufficiently high mass density, Z_{eff}, sufficient thickness to capture a large fraction of incident x-rays without excessive light spread. This property is sometimes loosely referred to as having high "radiation stopping power." However, that term is technically imprecise, as stopping power specifically describes the energy loss per unit path length of a **charged particle** (e.g., an electron or proton) traveling through matter.
- **Excellent uniformity, stability, and radiation hardness:** Detector elements must maintain consistent gain and offset characteristics across the entire array to prevent ring artifacts and preserve quantitative accuracy. Long-term stability under varying temperature, mechanical stress, and radiation exposure is critical for dependable clinical performance. Both the scintillator and semiconductor crystals should exhibit strong resistance to physical and chemical degradation, even after prolonged x-ray irradiation.
- **Other considerations:** The scintillator material should be readily machinable or processable to allow the introduction of optical septa between detector pixels, minimizing optical cross talk and preserving spatial resolution. Its light emission spectrum must be well matched to the spectral response of the photodiode to maximize light collection efficiency. The overall light collection efficiency further depends on the scintillator's intrinsic light yield, its optical transparency, and the transparency and refractive matching of the coupling medium.

4.4.2.2 Scintillators in CT Detectors

Fast scintillators with short decay times are preferred as sensor materials in CT detectors to support high frame rates and minimize image lag. The scintillator must also have a high effective atomic number and mass density to ensure efficient x-ray absorption. As discussed in Sect. 4.2.6, gadolinium oxysulfide (GOS or Gadox), garnet scintillators, and rare-earth oxyorthosilicates are all suitable candidates for

[4] CT manufacturers often report the detector pixel size as the projected value at the scanner isocenter, which equals the physical pixel size divided by the geometric magnification factor.

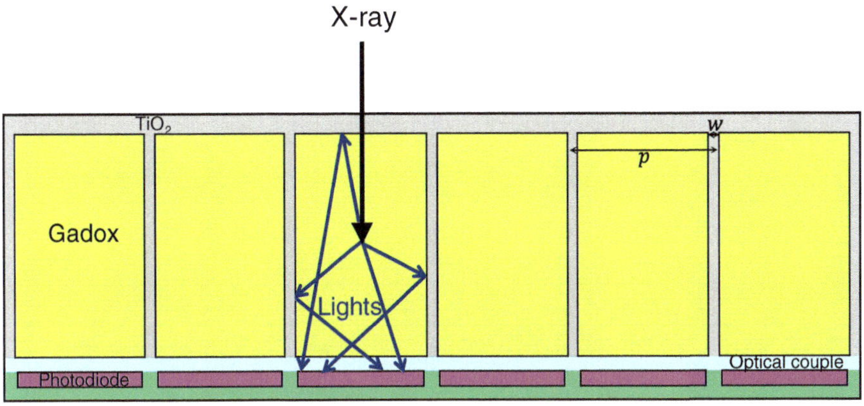

Fig. 4.21 Schematic illustration of a Gadox-based CT detector module

CT detector applications. Among these, Gadox is particularly popular because it does not contain scarce rare-earth elements, making it more cost-effective and widely available. In contrast, CsI(Tl), commonly used in FPDs, is unsuitable for CT due to its relatively long decay time and pronounced afterglow.

The typical scintillator thickness in CT detectors is around 2 mm, more than three times that used in FPDs. This thickness allows absorption of over 90% of x-rays incident on the detector surface. Due to the inherent trade-off between scintillator thickness and spatial resolution, optical septa are used to segment the scintillator for individual pixels and minimize lateral light spread (Fig. 4.21). Each septum, fabricated from light-reflecting materials, such as TiO_2, has a width on the order of 50 μm, largely independent of the detector pixel pitch. While these septa effectively suppress optical cross talk, they introduce "dead" regions, where incident x-rays do not contribute to signal generation. The fraction of the detector area occupied by active scintillator (i.e., the fill factor) is given by

$$\text{Fill Factor} = \left(1 - \frac{w}{p}\right)^2, \tag{4.34}$$

where p is the pixel pitch and w is the width of the optical septum. Since w is fixed to maintain sufficient optical separation, reducing the pixel pitch p reduces the geometric efficiency, imposing a practical limitation on the achievable spatial resolution in scintillator-based CT detectors.

As discussed in Sect. 4.4.2.1, x-rays scattered by the patient must be rejected because a fundamental assumption of CT reconstruction is that once an x-ray photon interacts with the patient—through either photoelectric absorption or scattering—it is effectively removed from the beam (i.e., the x-ray is attenuated). Consequently, nearly all slip-gantry-based MDCT systems employ an anti-scatter grid (ASG) positioned in front of the detector (Fig. 4.22). Since the x-ray focal spot position

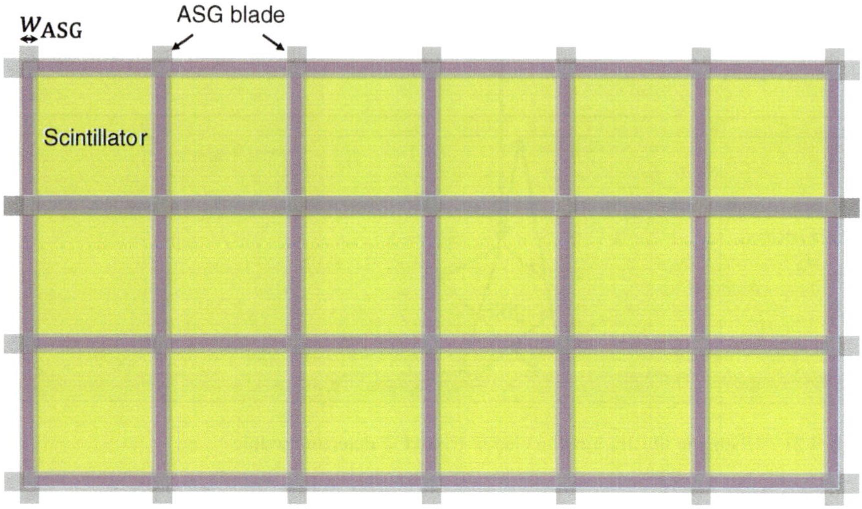

Fig. 4.22 Schematic illustration of a two-dimensional anti-scatter grid positioned in front of a CT detector

relative to the detector is fixed,[5] a single focused ASG is sufficient for a given CT scanner.

The blades of the ASG are typically fabricated from tungsten or other high Z_{eff} materials, with a height on the order of 10 mm to ensure sufficient absorption of scattered x-rays. The width of each ASG blade, w_{ASG}, is on the order of 100 μm, and the inter-blade spacing is approximately 1 mm, which matches the typical pixel pitch of conventional energy-integrating CT detectors. State-of-the-art CT systems generally use two-dimensional (2D) ASGs.[6] In this case, the fill factor of the ASG–detector assembly can be expressed as

$$\text{Fill Factor} = \left(1 - \frac{w_{ASG}}{p}\right)^2, \tag{4.35}$$

where p is the detector pixel pitch. Therefore, when the detector pixel pitch matches the ASG grid period, the optical septa do not introduce *additional* loss in the fill factor. However, if the detector pixel pitch is significantly smaller than the ASG period, the fill factor loss becomes dominated by the optical septa.

Similar to dual-layer flat-panel detectors (FPDs), dual-layer CT detectors have been developed by Philips Healthcare for dual-energy CT imaging [21, 26]. These

[5] State-of-the-art CT systems often incorporate flying-focal-spot technology, in which the focal spot oscillates over a small region to reduce the spatial sampling interval.

[6] Some CT vendors market their ASGs as "3D," which are essentially 2D focused grids.

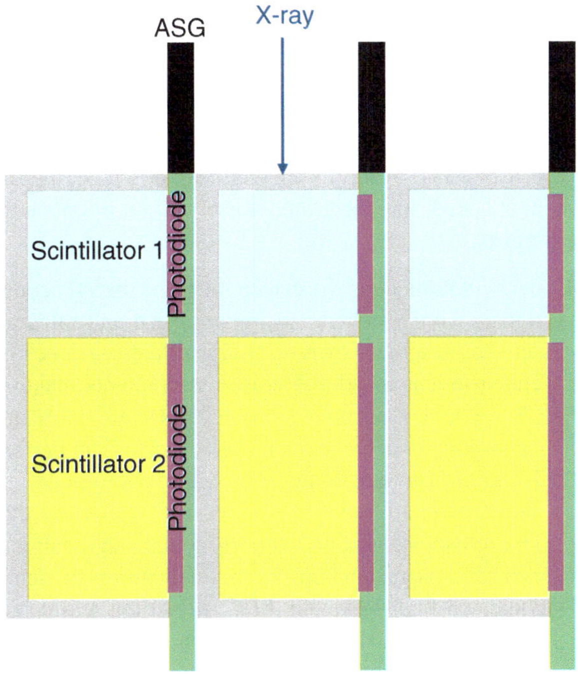

Fig. 4.23 Illustration of a dual-layer CT detector

detectors consist of two stacked scintillator layers, each optimized for preferentially absorbing a different part of the x-ray spectrum. The top scintillator layer (e.g., ZnSe) has a lower Z_{eff} and is relatively thin (e.g., 1 mm) [26]. This layer preferentially absorbs low-energy x-rays while allowing a significant portion of higher-energy photons to pass through. The bottom scintillator layer (e.g., Gadox) has a higher Z_{eff} and is thicker (e.g., 2 mm) [26], ensuring that the remaining higher-energy x-rays are almost completely absorbed.

One notable distinction between dual-layer CT detectors and dual-layer FPDs lies in the photodiode configuration. Instead of being directly laminated beneath the scintillator stack as in FPDs, each single-crystal semiconductor photodiode is positioned edge-on to the incident x-ray beam and laminated to the side of the scintillator layers. As illustrated in Fig. 4.23, the photodiodes are placed behind the ASG. Therefore, these edge-on photodiodes do not create any additional reduction in the fill factor or geometric efficiency of the detector.

The electronic signals generated by the two scintillator layers are processed jointly to produce material-selective CT images. Alternatively, the signals can be combined to reconstruct images equivalent to those from conventional single-energy CT acquisition, providing both spectral and conventional imaging capabilities within a single scan.

4.4.2.3 CT Detectors Readout Electronics

While photon-counting detectors are well-known for using application-specific integrated Circuits (ASICs) as their readout electronics, conventional energy-integrating CT detectors also employ highly sophisticated, custom-designed ASICs (Fig. 4.24). These circuits are engineered to meet the stringent requirements of high readout speed, low electronic noise, and excellent stability under continuous x-ray exposure.

1. Each photodiode in the detector array is directly coupled to an active preamp. The signals from all channels are amplified and integrated in parallel, allowing the detector to achieve extremely high effective frame rates, often exceeding 4000 fps
2. The stringent speed and noise requirements make amorphous Si unsuitable for CT readout electronics. Instead, they are fabricated in single-crystal CMOS technology to achieve the necessary electrical performance.
3. To minimize electronic noise, the photodiode, preamplifier, and ADC are integrated into the same chip that interfaces directly with the scintillator. This close integration reduces signal degradation, parasitic capacitance, and electromagnetic pickup. Each small group of detector pixels is paired with a dedicated high-bit-depth ADC, providing a wide dynamic range and excellent linearity during signal quantization. As a result, in state-of-the-art CT systems, the contribution of electronic noise is effectively negligible, even in those using energy-integrating detectors.
4. Downstream of the ADC, additional digital filtering and signal processing are often applied to further enhance SNR and signal linearity. Field-programmable gate arrays (FPGAs) are commonly employed in CT readout systems as high-

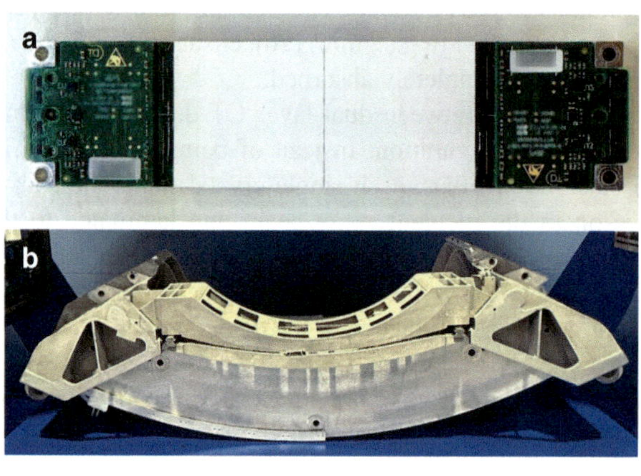

Fig. 4.24 (**a**) A CT detector module manufactured by Detection Technology Plc. (**b**) A 128-row CT detector system (Dunlee CD500) manufactured by Philips Medical Systems DMC GmbH. The system contains 672 × 128 = 86,016 scintillator–photodiode pixel elements. Each pixel has a physical size of 1.4 mm × 1.1 mm. The system also includes a two-dimensional anti-scatter grid

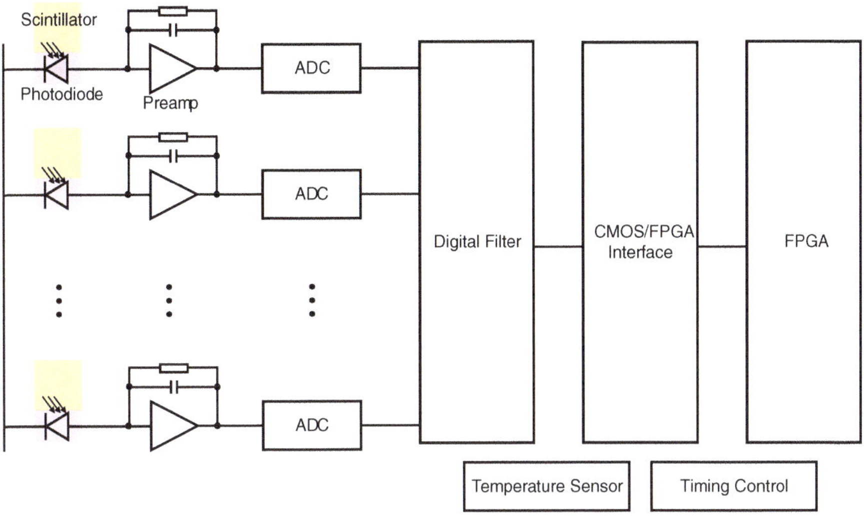

Fig. 4.25 Block diagram of CT detector readout electronics

speed, parallel digital processing and control hubs between the detector ASICs and the reconstruction computer. The FPGA collects the data streams from multiple ASICs, aligns the detector outputs with the recorded gantry angles based on their time stamps, and performs on-the-fly signal preprocessing, such as bad-pixel correction and detector nonuniformity correction, before passing the data to the reconstruction system.

5. During a CT scan, the x-ray source operates in a continuous or near-continuous mode, requiring the readout electronics to perform rapid integration and reset cycles with minimal downtime. The reset times for the amplifiers and ADCs are extremely short to sustain the high frame rates. This continuous high-speed operation significantly increases power dissipation of the ASICs. Consequently, active thermal management systems, including dedicated heat sinks and temperature control circuits, are employed to maintain stable detector performance and prevent thermal drift or gain instability, as even a 0.1% change in the gain may produce noticeable artifacts on CT images due to their excellent low-contrast detectability (Fig. 4.25).

4.5 Scintillator-Based Photon-Counting Image Detectors

4.5.1 Scintillator PCD for Nuclear Medicine Imaging

Scintillator-based photon-counting detectors (PCDs) have been widely used in nuclear medicine imaging since the invention of the Anger gamma camera in the

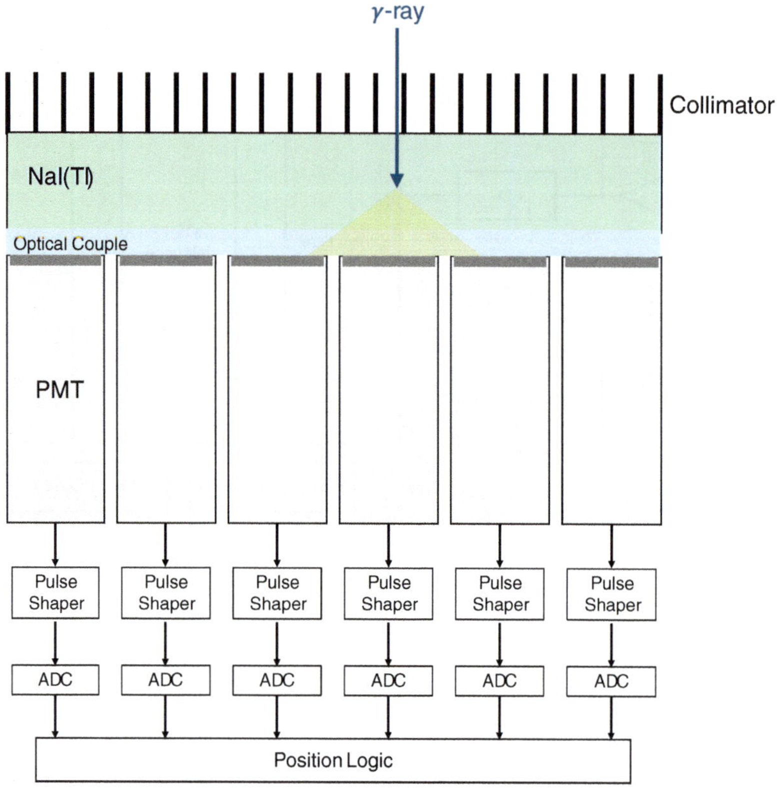

Fig. 4.26 A schematic illustration of the Anger γ-ray photon-counting image detector

1950s [1]. In Hal Anger's original design, a 1/4-in.-thick NaI(Tl) scintillator, 5 in. in diameter, was employed as the γ-ray sensor. The scintillator was optically coupled to seven photomultiplier tubes (PMTs). In modern gamma cameras, a larger array of PMTs with finer spacing is used to improve spatial resolution.

A collimator, placed in front of the scintillator, consists of a lead plate with many parallel holes that act as directional filters. The collimator ensures that only γ-ray photons traveling nearly perpendicular to the detector surface are detected. When a γ-ray photon interacts with the scintillator, it produces a burst of scintillation light photons, which may spread over several PMTs (Fig. 4.26). Each PMT outputs a pulse whose magnitude is proportional to the number of light photons it detects.

To estimate the initial interaction position (x_I, y_I) of the γ-ray photon within the scintillator, the system employs the so-called Anger arithmetic [1, 19]:

$$x_I = \frac{\sum_{V_j \geq V_{\mathrm{TH}}} x_j V_j}{\sum_{V_j \geq V_{\mathrm{TH}}} V_j}, \tag{4.36}$$

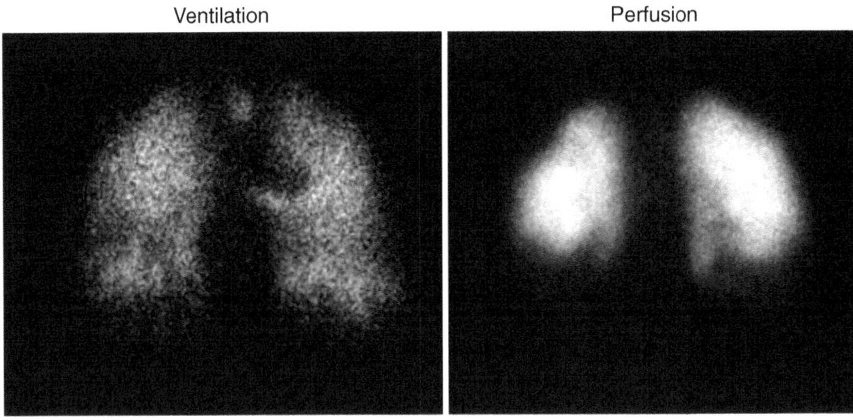

Fig. 4.27 Two-dimensional images obtained using a γ-ray Anger camera from a lung ventilation–perfusion (V/Q) scintigraphy exam. Ventilation imaging assesses airflow in the lungs, with the patient inhaling a ^{99m}Tc-labeled aerosol prior to the scan. Perfusion imaging evaluates pulmonary blood flow, using intravenously injected ^{99m}Tc-macroaggregated albumin. The ventilation and perfusion images are compared to identify mismatched regions, which may indicate pulmonary embolism. In this example, mismatched areas are observed in the lower lobes of both lungs

where V_j denotes the voltage pulse height of the j-th PMT, (x_j, y_j) represents the center position of the j-th PMT, and V_{TH} is a voltage threshold.

As shown in Eq. (4.36), each PMT must be connected to a pulse-mode readout circuit to measure V_j. Modern gamma cameras typically equip each PMT with an ADC, allowing the Anger arithmetic to be performed using digitally processed values of V_j [19]. By mapping the estimated (x_I, y_I) positions of all detected γ-ray events, a two-dimensional image can be reconstructed to represent the spatial distribution of γ-ray-emitting tracers (e.g., ^{99m}Tc) within the patient body (Fig. 4.27).

The shape of the electric pulse produced by a scintillator-based PCD is jointly determined by:

- Incident radiation
- Temporal response of the scintillator
- Time response of the light sensor (PMT, SiPM, etc.)
- Time constant of the readout electronics

The input radiation primarily determines the peak magnitude of the initial signal from the light sensor. A key difference between scintillator-based PCDs and semiconductor-based PCDs lies in the characteristic temporal behavior of the scintillation processes, which typically consist of two general stages. First, there is a finite **rise phase**, during which electrons are "pumped" to the activator excited states. This process is characterized by a rise-time constant τ_1, typically on the order of 10^{-9}–10^{-8} s for fast scintillators. The buildup of electrons in the excited activator centers can be modeled as

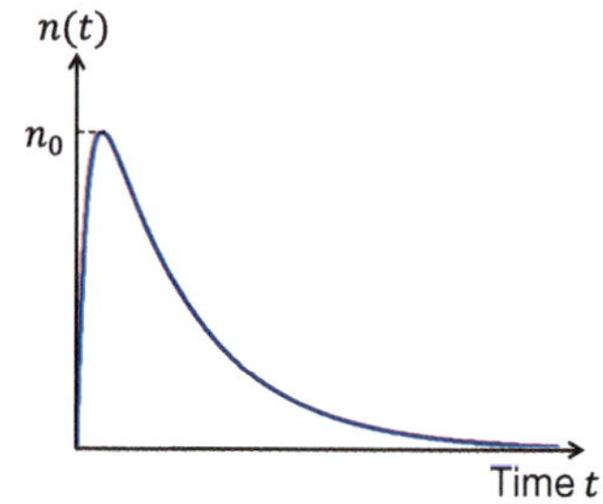

Fig. 4.28 The shape of a scintillation light pulse

$$n(t) = n_0 \left(1 - e^{-\frac{t}{\tau_1}}\right), \tag{4.37}$$

where n_0 denotes the total number of electrons transferred to the activator excited states.

After the rise phase, electrons on the activator excited levels undergo radiative transitions to the activator ground state, while a smaller fraction may undergo nonradiative quenching or thermally excited back to the conduction band. It is reasonable to assume that the radiative decay rate is proportional to the number of electrons in the activator excited states, namely,

$$\frac{dn(t)}{dt} = -\frac{1}{\tau} n(t), \tag{4.38}$$

where τ is the scintillator **decay time** discussed earlier in this chapter. Solving this differential equation yields the standard exponential decay law:

$$n(t) = n_0 \, e^{-\frac{t}{\tau}}. \tag{4.39}$$

Combining the finite rise-time process with the subsequent exponential light decay gives the overall temporal response of the scintillation light pulse:

$$n(t) = n_0 \left(1 - e^{-\frac{t}{\tau_1}}\right) e^{-\frac{t}{\tau}}. \tag{4.40}$$

Figure 4.28 illustrates a typical shape of a scintillation light pulse, showing the rapid rise followed by exponential decay.

As shown earlier in Fig. 4.11, many scintillators, such as CsI(Tl), exhibit more than one (typically two) radiative decay pathway, resulting in multiple exponential decay components with different time constants. In such cases, the temporal response of the scintillation light output can be modeled as a weighted sum of exponential decays:

$$n(t) = n_0 \left(1 - e^{-t/\tau_1}\right) \left(w_f \, e^{-t/\tau_f} + w_s \, e^{-t/\tau_s}\right), \tag{4.41}$$

where w_f and w_s are the relative contributions of the fast and slow components, respectively, and τ_f and τ_s are their corresponding decay constants.

For CsI(Tl), commonly used in flat-panel EIDs, the slow decay component τ_s can be 3 μs or longer. This makes CsI(Tl) poorly suited for photon-counting operation, since pulses from consecutive x-ray or γ-ray interactions are more likely to overlap, leading to distorted energy measurements, and missed counts.

In contrast, NaI(Tl) has a shorter decay time of 230 ns. In a nuclear medicine procedure involving the administration of several hundred MBq of a γ ray-emitting tracer, the event rate per PMT is typically on the order of 10^2–10^3 counts per second, corresponding to a mean inter-event time of 1–10 ms, which is orders of magnitude longer than the 230 ns decay constant of NaI(Tl). This large separation in timescales explains why NaI(Tl) can be successfully used in γ-ray PCDs (e.g., Anger cameras) with minimal pulse pileup.

The pulse shape of a scintillator-based PCD is influenced not only by the scintillation light emission profile, $n(t)$, but also by the temporal response of the photosensor, $p(t)$. The resulting electrical pulse can be modeled as the convolution between $n(t)$ and $p(t)$, so both the scintillator decay characteristics and the sensor response contribute to the overall pulse shape [22]. For applications such as time-of-flight (TOF) PET imaging, both components must exhibit sufficiently fast temporal behavior to achieve the required time resolution [22].

PCD time resolution describes how precisely the detector can measure the arrival time of a γ-ray or x-ray photon. Based on the principles of time-to-amplitube converter presented in Sect. 3.6.3, the timing resolution is fundamentally determined by the sharpness of the electrical pulse generated by each interaction. Time resolution is distinct from **PCD dead time**, which is the period following each detection event during which the detector or its electronics cannot register another event, i.e., the recovery interval of the PCD after recording a photon.

The PCD time resolution (i.e., pulse sharpness) is jointly determined by the sensor time, the light sensor timing jitter, the bandwidth of the front-end electronics, the electronic noise level, and the resulting pulse SNR. In contrast, the dead time is governed by the sensor speed, the time constants of the preamplifier and pulse-shaping electronics, and the reset/recovery characteristics of the detector (especially relevant for SiPMs with finite microcell recharge times). Because these effects arise from different components of the detection chain, improving one metric does not automatically improve or degrade the other.

Nevertheless, indirect links can exist. For example, reducing the scintillator decay time can simultaneously improve timing resolution and shorten the effective dead time by reducing pulse pileup. In contrast, increasing the light yield or reducing electronic noise typically improves only the timing resolution, with little influence on the dead time.

For scintillator-based PCDs used in PET, the coincidence timing resolution is typically on the order of a few hundred ps. However, the detector or channel dead time is generally much longer, commonly in the range of tens to hundreds of ns. In PET, such dead times are not problematic because the photon flux per detector element is relatively low [15]. In contrast, x-ray CT can impose input photon fluxes as high as 10^8 counts/s/pixel [13], which requires detector dead times on the order of 10 ns to be robust against pulse pileups.

4.5.2 Scintillator PCD for CT Imaging

In the same year that Hal Anger invented the NaI(Tl)-based γ-photon–counting camera at UC Berkeley [1], a South African physicist Allan Cormack performed the first proof-of-concept CT experiment at the University of Cape Town. He used a collimated beam from a 7-mCi ^{60}Co γ-ray source and a photon-counting detector (a Geiger counter). The imaging object was a circularly symmetric assembly of aluminum and wood. Cormack demonstrated that the distribution of γ-ray attenuation coefficients within the object could be mathematically reconstructed from the measured projections [7–9]. Based on my estimation, the γ-ray flux in Cormack's setup was approximately 200 counts/s/mm^2, which was sufficiently low to allow a gas-filled PCD (Geiger counter) to operate without significant pulse-pileup effects.

Allan Cormack was eventually awarded the 1979 Nobel Prize in Physiology or Medicine together with the British engineer Godfrey Hounsfield, who developed the first clinical x-ray CT scanner. Hounsfield's CT employed a NaI(Tl) scintillator coupled to a PMT [10, 24]. The output of each PMT was sent to a charge integrator followed by an ADC. A pulse-mode readout was not viable, simply because the x-ray photon flux in x-ray CT was far too high for any PCD technology available at that time.

Since Hounsfield's invention, CT has become one of the most important medical imaging tools, with its detector technology evolving dramatically. It progressed from two NaI(Tl) detector pixels to small linear arrays of NaI(Tl) pixels, then to direct-conversion xenon gas detectors with hundreds of pixels, and ultimately to large curved and multi-row arrays of pixelated ceramic scintillators coupled to semiconductor photodiodes. All of these CT detectors were EIDs, as PCDs with sufficiently short dead time, high x-ray absorption efficiency, long-term stability, and affordable cost were not yet available. It was not until the 2020s that clinical CT scanners equipped with PCDs became commercially and clinically available [20].

At the time of this writing, all clinically available PCD-CTs employ semiconductor-based PCDs. Owing to marketing considerations, some vendors have portrayed the use of semiconductor sensors as a defining hallmark of PCD-CT. Nevertheless, there remains an intriguing question: could scintillator-based PCDs potentially be adopted in future clinical CT scanners?

Scintillator-based CT PCDs offer several advantages:

- Materials such as LYSO and GAGG combine high Z_{eff} and density, providing superior x-ray absorption compared to CdTe or CZT.

- Scintillator-based PCDs do not require high bias voltages, which simplifies detector design and reduces the risk of electrical breakdown. Unlike semiconductors such as CdTe or CZT, scintillators are free from polarization effects caused by charge trapping under prolonged irradiation. This results in more stable and reproducible responses over time, improving both imaging quality and detector reliability in clinical operation.
- The fabrication of detector-grade semiconductors is technically demanding and costly, requiring stringent control over crystal purity, defects, and doping. In contrast, scintillator crystals or ceramics can often be produced more economically and at larger scales, enabling more scalable manufacturing and easier customization of detector geometries. Scintillators can be grown into large areas with uniform response, which reduces ring artifacts in the reconstructed PCD-CT images.
- A wider range of scintillator materials is available. By adjusting the activators, the light yield and speed of scintillators can be engineered to adapt to the specific need.
- Many scintillators are free of hazardous substances, such as lead or cadmium, making them more environmentally friendly and easier to certify as RoHS compliant. This reduces regulatory hurdles for clinical deployment and facilitates safer disposal or recycling at the end of life.
- The response of scintillators is largely insensitive to temperature or humidity fluctuations, owing to their wide bandgaps. In contrast, semiconductor PCDs exhibit significant temperature dependence and therefore require active cooling and thermal stabilization, which increases system complexity and energy consumption.

Taken together, these advantages make scintillators a promising alternative for building PCD sensors for CT applications.

That said, existing scintillator-based PCDs do have several limitations. One major limitation arises from the scintillation light decay time. Although the velocity of light transport within a scintillator is much faster than that of charge carrier transport in semiconductors, the width of the electrical pulse generated by the scintillation process is limited by the temporal decay of the scintillation process, as described in Eq. (4.40). This decay is governed by the stochastic nature of radiative transitions of electrons that have been excited to the conduction band by incoming radiation. As shown in Fig. 4.11, the decay times of some scintillators can reach hundreds of ns, which is too long to temporally resolve closely spaced x-ray photons. Therefore, a key consideration in developing scintillator-based CT PCDs is the identification of scintillators with fast decay times—ideally on the order of 10 ns—to minimize pulse pileup. As indicated in Table 4.3, some Ce-activated scintillators, such as LYSO(Ce), $LaBr_3$(Ce), and GFAG(Ce), exhibit decay times below 40 ns. Nevertheless, further reduction in decay time is still required to meet the stringent speed requirements of CT PCD.

Another major challenge in using scintillators to implement PCDs is the relatively low number of secondary quanta produced per unit of absorbed radiation

Table 4.3 Properties of selected semiconductor and scintillator materials. $\bar{N}$ denotes the average number of secondary quanta (electron–hole pairs in semiconductors or scintillation photons in scintillators) produced per 100 keV of absorbed radiation energy. For semiconductors, "initial pulse width" refers to the average electron transport time across a 1.6-mm-thick layer under an electric field of 6×10^3 V/cm. For scintillators, "initial pulse width" refers to the scintillation light decay time

	Material	Initial pulse width (ns)	$\bar{N}$	Bandgap (eV)	Z_{eff}	Density (g/cm^3)
Semiconductor	Si	20	27,500	1.12	14	2.3
	GaAs	7	23,500	1.44	32	5.3
	CdTe	24	22,600	1.52	50	6.1
	CZT	27	21,600	1.57	49	5.9
Scintillator	GSO(Ce)	56 (90%), 400 (10%)	900	≈6	58	6.7
	LYSO(Ce)	35–45	2700–3600	≈6	64	7.1
	$LaBr_3$(Ce)	26	6300	≈6	47	5.1
	GGAG(Ce)	90	4500–5500	≈6	55	6.6
	GFAG(Ce)	<40	2500–3500	≈6	55–58	6.6

energy. As shown in Table 4.3, the average number of electron–hole pairs generated per 100 keV of absorbed energy in semiconductors is roughly an order of magnitude higher than the average number of scintillation photons produced in typical scintillators. Consequently, when a photodiode is used as the light sensor, the resulting electrical pulse amplitude is approximately one order of magnitude lower than that from a semiconductor PCD, which degrades both energy resolution and image SNR. As emphasized by G. Knoll in his seminal book *Radiation Detection and Measurement, "One of the major limitations of scintillation counters is ... the energy required to produce an information carrier (a photoelectron) is of the order of 100 eV or more ... "* and *"... the use of semiconductor materials as radiation detectors can result in a much larger number of carriers for a given incident radiation event than is possible with any other common detector type"* [12].

Therefore, conventional photodiodes are not suitable for use in scintillator PCDs. Instead, high-gain light sensors are required to generate a sufficiently strong initial electrical pulse from each radiation event for the ASIC to process and register a count. PMTs are also impractical due to their relatively large size compared to the CT detector pitch (typically < 1 mm).

Recent research has therefore focused on the SiPMs in scintillator-based CT PCDs [28, 33]. Modern SiPMs have been significantly miniaturized, with sizes as small as 250 μm, making them compatible with high-resolution CT detector arrays (Fig. 4.29). A typical SiPM may contain up to ~260 SPAD cells, which means it can register at most 260 scintillation photons arriving within the same recovery interval. When the number of scintillation photons produced by an x-ray interaction exceeds the number of available SPAD cells, the SiPM response saturates, causing the output signal to deviate from linear proportionality with the deposited x-ray energy.

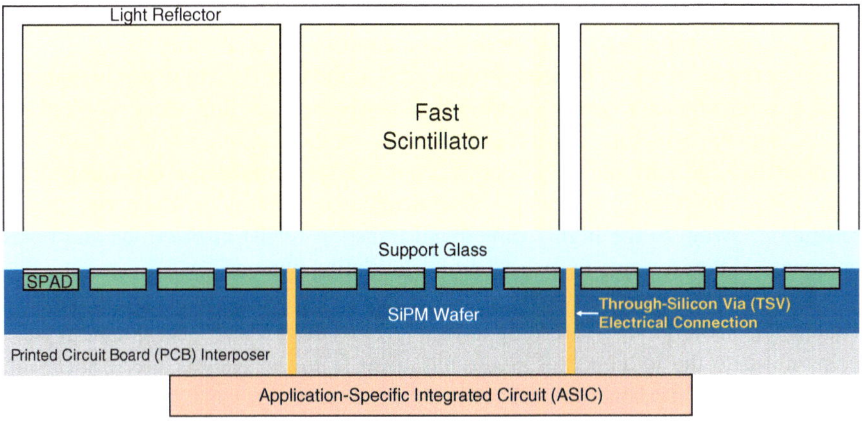

Fig. 4.29 Schematic illustration of a scintillator and SiPM-based PCD designed for PCD-CT applications

Reducing the SPAD cell size increases the number of cells per SiPM and therefore increases the maximum number of photons that can be detected simultaneously, improving linearity across different x-ray energies. Smaller cells also exhibit lower capacitance, leading to faster recharge times after each avalanche event. However, a lower cell capacitance reduces the avalanche charge, which decreases the gain of each SPAD. In addition, reducing the cell size increases the relative amount of nonsensitive dead space (e.g., quenching resistors, guard rings, and isolation structures), thereby reducing the optical fill factor and lowering the overall photon-detection efficiency.

Recent studies have demonstrated that scintillator detectors based on garnet crystals coupled to SiPMs can achieve a pulse width of less than 40 ns [28, 33]. These results highlight the potential of scintillator–SiPM architectures to meet the stringent timing requirements of PCD-CT. Despite the promising progress, significant research and development remain necessary before scintillator-based PCDs can be translated into clinical PCD-CT systems. Key challenges include further reducing scintillation decay times, optimizing light-collection efficiency at submillimeter pixel pitches, and improving SiPM linearity under high photon flux.

4.6 Semiconductor-Based Energy-Integrating Image Detectors

Although this chapter focuses on scintillator detectors, discussing semiconductor detectors may seem like a digression. However, my teaching experience suggests that introducing semiconductor-based EIDs alongside scintillator-based EIDs helps reinforce the concept that the fundamental distinction between EIDs and PCDs lies in the readout circuit design, rather than in the sensor material itself.

For example, in a TFT flat-panel-based EID, if the CsI(Tl) scintillator is replaced with a semiconductor sensor sandwiched between hole- and electron-blocking layers, and the structure is reverse biased to create a strong electric field with minimal leakage current, the system becomes a **semiconductor flat-panel EID** (Fig. 4.30). The primary advantage of semiconductor EIDs is their superior spatial resolution, owing to the highly directional transport (drift) of electrons and holes compared to the isotropic propagation of scintillation photons.

Moreover, semiconductors typically generate a larger number of secondary quanta than scintillators. According to Eqs. (2.27)–(2.29), this higher quantum yield helps mitigate the influence of electronic noise from the readout electronics—an especially important benefit under low-dose conditions.

One popular semiconductor material used to build low-cost direct-conversion EIDs is amorphous selenium (a-Se) [35]. It has been successfully implemented in full-field digital mammography (FFDM) and digital breast tomosynthesis (DBT) systems, where both excellent spatial resolution (for microcalcification detection) and low image noise (for low-contrast mass detection) are crucial [31, 34].

As shown in Table 1.5, a-Se has a relatively wide energy gap, which helps reduce thermal noise. Its intermediate atomic number (34) and mass density (4.3 g/cm^3) provide sufficient x-ray attenuation for the relatively low-energy photons used in FFDM and DBT (~20 keV). On average, it produces approximately 2860 electron–hole pairs per 20 keV of absorbed energy. In contrast, the number of light photons generated per 20 keV absorbed in CsI(Tl) and GSO(Ce) scintillators is only about 120 and 20, respectively.

As also indicated in Table 1.5, in a-Se, the electron mobility is significantly lower than that of holes. Therefore, as shown in Fig. 4.30, **the signal-collecting pixel electrodes in a-Se detectors are cathodes**, rather than anodes as commonly seen in CdTe detectors. The contribution of the slow-moving electrons to the induced signal is suppressed via the small-pixel effect discussed in Sect. 3.2.3.

Assuming an a-Se layer thickness of 300 μm and a bias voltage of 1200 V, the resulting electric field is 4×10^4 V/cm. According to Table 1.5, the hole mobility in a-Se is 0.14 cm^2/V·s. Therefore, the time required for a hole to drift across the layer is given by

$$
\begin{aligned}
t &= \frac{L}{\mu_p \mathcal{E}} \\
&= \frac{0.03\,\text{cm}}{0.14\,\text{cm}^2/(\text{V}\cdot\text{s}) \times 4 \times 10^4\,\text{V/cm}} \\
&= 5\,\mu\text{s}.
\end{aligned}
\tag{4.42}
$$

Although a transit time of 5 μs is too long for photon counting, it is sufficient for energy-integrating FFDM and DBT imaging, where the maximum frame rate is no more than a few frames per second.

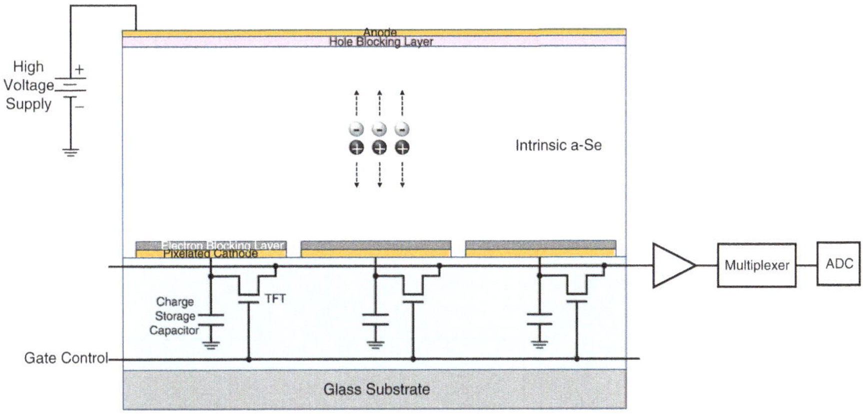

Fig. 4.30 Schematic diagram of a direct-conversion, energy-integrating detector that utilizes amorphous selenium (a-Se) as the x-ray sensor

Another advantage of a-Se is its ability to be deposited uniformly over large areas, making it fully compatible with a-Si TFT readout panels.

Similar to other semiconductor sensors, charge trapping is an important consideration in a-Se detectors. Over time, space charge from trapped electrons or holes can accumulate in the a-Se layer, reducing the internal electric field and charge collection efficiency. During sequential frame acquisition (e.g., in tomosynthesis), detector reset procedures, as described in Sect. 3.4.3, may be applied between exposures to remove the trapped charges and maintain stable performance.

In recent years, high-gain avalanche rushing photoconductor (HARP) detectors have been developed to further enhance the low-dose performance of a-Se EIDs [23]. As illustrated in Fig. 4.31, micro-well structures are fabricated within the a-Se sensor layer. In the bulk region of the a-Se, where most x-rays are absorbed, the electric field is non-avalanche. However, within each well, a much stronger electric field is established between two closely spaced grid electrodes. When photo-generated holes drift into these wells, the intense local field accelerates them enough to ionize additional a-Se atoms, initiating an avalanche multiplication process. This avalanche effect increases the total collected charge, thereby improving signal strength and reducing the relative influence of electronic noise from the readout panel.

As a final note for this chapter, when the primary goal is to enhance the spatial resolution of CT detectors, CdTe or other semiconductor sensors can be used to develop a semiconductor-based energy-integrating CT detector. This approach can achieve ultra–high-resolution performance without the added complexity and cost of a semiconductor PCD, which requires sophisticated pulse-processing electronics and is prone to pulse pileup at high count rates. It is important to recognize that the spatial resolution benefits observed in current PCD-CT systems stem primarily from the use of semiconductor sensor materials, rather than from the photon-counting

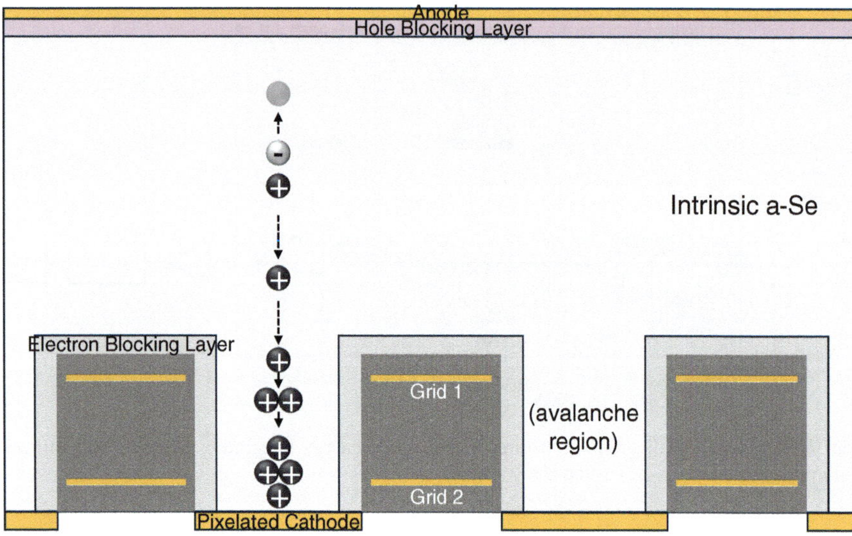

Fig. 4.31 Illustration of the high-gain avalanche a-Se detector. Detailed technical information is available in Reference [23]

mechanism itself. What semiconductor EIDs cannot provide, however, is the ability to simultaneously achieve high spatial resolution and spectral CT measurements.

4.7 Summary of Chap. 4

1. Scintillation is the immediate emission of visible light from a material following excitation by ionizing radiation.
2. To qualify as a scintillator, a material must emit light promptly when irradiated by ionizing radiation and remain non-emissive in the absence of radiation. Therefore, crystalline scintillators are typically insulators with large bandgaps, so that the probability of spontaneous light emission is negligible.
3. The interaction of ionizing radiation with a scintillator generates a large number of electron–hole pairs, splitting the intrinsic Fermi level into two separate quasi-Fermi levels and increasing the probability of light emission.
4. Inorganic scintillators often contain activator atoms that introduce metastable energy levels within the forbidden band. Activators not only enhance the probability of light emission but also ensure that the emitted photons have energies lower than the bandgap energy, thereby minimizing reabsorption.
5. Due to the presence of impurity atoms and crystal defects, electrons and holes may become trapped in a scintillator, leading to delayed light emission or the “afterglow” effect, which is generally undesirable for imaging applications.

6. The choice of scintillator for a specific application depends on multiple factors, such as light yield, x-ray absorption efficiency, light collection efficiency, scintillation decay time, and cost.
7. To improve scintillation light collection efficiency, it is desirable to match the refractive index of the light sensor with that of the scintillator or use an optical coupling medium. Light reflectors can also improve collection efficiency but may degrade spatial resolution. Using columnar scintillators helps mitigate this trade-off.
8. Scintillator-based EIDs are widely used in general x-ray imaging, CT, and cone-beam CT. Flat-panel EIDs differ significantly from CT EIDs in both scintillator materials and readout electronics. CT EIDs require much faster response times, higher X-ray absorption efficiency, and lower electronic noise. Consequently, CT EIDs (particularly their readout electronics) are significantly more complex, costly, and power demanding.
9. Scintillator-based PCDs are widely used in nuclear medicine due to their excellent absorption efficiency for high-energy γ-rays. However, scintillator-based PCDs are less common in CT imaging because of their longer dead time and larger pixel size. Recent advances in scintillator material science and light sensor technology have reduced SiPM size and improved response speed, showing promise for potential clinical CT adoption.
10. EIDs can also be implemented using direct-conversion semiconductor sensors, such as amorphous selenium (a-Se). A notable advantage of semiconductor EIDs is their excellent spatial resolution, making them particularly suitable for breast imaging applications.

References

1. Anger HO. Scintillation camera. Rev Sci Instrum. 1958;29(1):27–33. ISSN 0034-6748. https://doi.org/10.1063/1.1715998.
2. Barnes GT, Sones RA, Tesic MM, Morgan DR, Sanders JN. Detector for dual-energy digital radiography. Radiology. 1985;156(2):537–40. https://doi.org/10.1148/radiology.156.2.4011921. PMID: 4011921.
3. Berks JB. The theory and practice of scintillation counting. Oxford: Pergamon Press; 1964.
4. Bigas M, Cabruja E, Forest J, Salvi J. Review of CMOS image sensors. Microelectron J. 2006;37(5):433–51. ISSN 1879-2391. https://doi.org/10.1016/j.mejo.2005.07.002.
5. Birks JB. The theory and practice of scintillation counting. London: Pergamon Press, 1964.
6. Cai EY, De Caro C, Treb K, Li K. Benefits of using removable filters in dual-layer flat panel detectors. Phys Med Biol. 2023;68(8):085013.
7. Cormack AM. Representation of a function by its line integrals, with some radiological applications. J Appl Phys. 1963;34(9):2722–7. ISSN 0021-8979. https://doi.org/10.1063/1.1729798.
8. Cormack AM. Representation of a function by its line integrals, with some radiological applications. II. J Appl Phys. 1964;35(10):2908–13. ISSN 0021-8979. https://doi.org/10.1063/1.1713127.
9. Di Chiro G, Brooks RA. The 1979 Nobel Prize in Physiology or Medicine. Science. 1979;206(4422):1060–62. ISSN 1095-9203. https://doi.org/10.1126/science.386516.

10. Hounsfield GN. Computerized transverse axial scanning (tomography): part 1. Description of system. Br J Radiol. 1973;46(552):1016–22. ISSN 0007-1285. https://doi.org/10.1259/0007-1285-46-552-1016.
11. Khan MT, Ahmad S. The mercurial rise in research of halide perovskites: what is next. Emergent Mater. 2025;8:3425–32. https://doi.org/10.1007/s42247-024-00834-7.
12. Knoll GF. Radiation detection and measurement. Hoboken: Wiley; 2010.
13. Li K, Liu X, Jacobsen MC, Rong J, Jensen CT, Tamm EP, Dong F. User evaluation of detector performance in clinical photon-counting and energy-integrating CT scanners using DICOM images. Med Phys. 2025;52(10):e70045. https://doi.org/10.1002/mp.70045. https://aapm.onlinelibrary.wiley.com/doi/abs/10.1002/mp.70045.
14. Lu M, Wang A, Shapiro E, Shiroma A, Zhang J, Steiger J, Star-Lack J. Dual energy imaging with a dual-layer flat panel detector. In: Schmidt TG, Chen G-H, Bosmans H, editors. Medical imaging 2019: physics of medical imaging. vol. 10948. International Society for Optics and Photonics, SPIE; 2019. p. 1094815. https://doi.org/10.1117/12.2513499.
15. MacDonald LR, Harrison RL, Alessio AM, Hunter WCJ, Lewellen TK, Kinahan PE. Effective count rates for PET scanners with reduced and extended axial field of view. Phys Med Biol. 2011;56(12):3629. https://doi.org/10.1088/0031-9155/56/12/011.
16. Moseley ODI, Doherty TAS, Parmee R, Anaya M, Stranks SD. Halide perovskites scintillators: unique promise and current limitations. J Mater Chem C. 2021;9:11588–604. https://doi.org/10.1039/D1TC01595H.
17. Mykhaylyk VB, Kraus H, Kapustianyk V, Kim HJ, Mercere P, Rudko M, Da Silva P, Antonyak O, Dendebera M. Bright and fast scintillations of an inorganic halide perovskite CsPbBr3 crystal at cryogenic temperatures. Sci Rep. 2020;10:8601. https://doi.org/10.1038/s41598-020-65672-z.
18. Nathan A, Park B, Sazonov A, Tao S, Chan I, Servati P, Karim K, Charania T, Striakhilev D, Ma Q, Murthy RVR. Amorphous silicon detector and thin film transistor technology for large-area imaging of x-rays. Microelectron J. 2000;31(11):883–91. ISSN 1879-2391. https://doi.org/10.1016/S0026-2692(00)00082-3.
19. Peterson TE, Furenlid LR. SPECT detectors: the Anger Camera and beyond. Phys Med Biol. 2011;56(17):R145. https://doi.org/10.1088/0031-9155/56/17/R01.
20. Rajendran K, Petersilka M, Henning A, Shanblatt ER, Schmidt B, Flohr TG, Ferrero A, Baffour F, Diehn FE, Yu L, Rajiah P, Fletcher JG, Leng S, McCollough CH. First clinical photon-counting detector CT system: technical evaluation. Radiology. 2022;303(1):130–38. https://doi.org/10.1148/radiol.212579.
21. Rassouli N, Etesami M, Dhanantwari A, Rajiah P. Detector-based spectral CT with a novel dual-layer technology: principles and applications. Insights Imaging. 2017;8:589–698.
22. Schaart DR. Physics and technology of time-of-flight PET detectors. Phys Med Biol. 2021;66(9):09TR01. https://doi.org/10.1088/1361-6560/abee56.
23. Scheuermann JR, Goldan AH, Tousignant O, Leveille S, Zhao W. Development of solid-state avalanche amorphous selenium for medical imaging. Med Phys. 2015;42(3):1223–6. https://doi.org/10.1118/1.4907971.
24. Schulz RA, Stein JA, Pelc NJ. How CT happened: the early development of medical computed tomography. J Med Imaging. 2021;8(5):052110. https://doi.org/10.1117/1.JMI.8.5.052110.
25. Schulz-Wendtland R, Dilbat G, Bani M, Fasching PA, Lux MP, Wenkel E, Schwab S, Loehberg CR, Jud SM, Rauh C, Bayer CM, Beckmann MW, Uder M, Meier-Meitinger M. Full Field Digital Mammography (FFDM) versus CMOS technology versus tomosynthesis (DBT)—which system increases the quality of intraoperative imaging? Geburtshilfe Frauenheilkd. 2012;72(6):532–8.
26. Shefer E, Altman A, Behling R, Goshen R, Gregorian L, Roterman Y, Uman I, Wainer N, Yagil Y, Zarchin O. State of the art of CT detectors and sources: a literature review. Curr Radiol Rep. 2013;1:76–91. https://doi.org/10.1007/s40134-012-0006-4.
27. Sheth NM, Uneri A, Helm PA, Zbijewski W, Siewerdsen JH. Technical assessment of 2D and 3D imaging performance of an IGZO-based flat-panel X-ray detector. Med Phys. 2022;49(5):3053–66. https://doi.org/10.1002/mp.15605.

28. Shimazoe K, Kim D, Hamdan M, Kobayashi Y, Kamada K, Yoshino M, Shoji Y, Sakamoto K, Acerbi F, Gola A. An energy-resolving photon-counting x-ray detector for computed tomography combining silicon-photomultiplier arrays and scintillation crystals. Commun Eng. 2024;3:167.
29. Siewerdsen JH. Signal, noise, and detective quantum efficiency of a-Si:H flat-panel imagers. PhD Thesis, University of Michigan (1998).
30. Siewerdsen JH, Antonuk LE, El-Mohri Y, Yorkston J, Huang W, Boudry JM, Cunningham IA. Empirical and theoretical investigation of the noise performance of indirect detection, active matrix flat-panel imagers (AMFPIs) for diagnostic radiology. Med Phys. 1997;24(1):71–89. https://doi.org/10.1118/1.597919.
31. Stone MF, Zhao W, Jacak BV, O'Connor P, Yu B, Rehak P. The x-ray sensitivity of amorphous selenium for mammography. Med Phys. 2002;29(3):319–24. https://doi.org/10.1118/1.1449874.
32. U.S. Department of Energy National Laboratory. Scintillator Library (http://scintillator.lbl.gov). Technical report.
33. van der Sar SJ, Brunner SE, Schaart DR. Silicon photomultiplier-based scintillation detectors for photon-counting CT: a feasibility study. Med Phys. 2021;48(10):6324–38. https://doi.org/10.1002/mp.14886.
34. Zhao B, Zhao W. Imaging performance of an amorphous selenium digital mammography detector in a breast tomosynthesis system. Med Phys. 2008;35(5):1978–87. https://doi.org/10.1118/1.2903425.
35. Zhao W, Rowlands JA. X-ray imaging using amorphous selenium: feasibility of a flat panel self-scanned detector for digital radiology. Med Phys. 1995;22(10):1595–604. https://doi.org/10.1118/1.597628.
36. Zhao W, Ristic G, Rowlands JA. X-ray imaging performance of structured cesium iodide scintillators. Med Phys. 2004;31(9):2594–605. https://doi.org/10.1118/1.1782676.

Chapter 5
Physics in X-Ray Tubes

Contents

K. Li, *Essential Physics of Medical X-ray Sources and Detectors*,
https://doi.org/10.1007/978-3-032-27800-5_5

5.1 Introduction

Unlike gamma rays, which are spontaneously emitted from the atomic nucleus, x-rays are electromagnetic radiation generated when accelerated electrons interact with matter. In the medical domain, the two primary x-ray-producing devices are x-ray tubes and linear accelerators (Linacs). Both are electrical instruments that convert electrical energy into the kinetic energy of electrons and, subsequently, into x-ray energy. Yet x-ray tubes and Linacs are rarely discussed in the same textbook—many outside our field are unaware that a Linac is, in fact, an x-ray-producing machine. Traditionally, x-ray tubes appear in diagnostic imaging texts, while Linacs are covered in radiotherapy texts. Based on our teaching experience in the University of Wisconsin-Madison Medical Physics Graduate Program, we find that connecting these two platforms provides important conceptual benefits.

X-ray tubes and Linacs share several foundational components (Fig. 5.1): an electron emitter, an electron accelerator, and a metallic target where electrons rapidly decelerate to produce x-ray radiation. Because x-ray energy is directly tied to electron kinetic energy, a central technical challenge—particularly in medical applications—is **how to emit electrons from a material and accelerate them to the required energies in the most efficient and economical manner.** Once the electrons interact with the target and decelerate, the overwhelming majority of their kinetic energy is converted into heat, with only a small fraction transformed into x-ray radiation. Thus, a second major challenge is **how to dissipate the resulting heat effectively and economically.**

Despite these similarities, x-ray tubes and Linacs differ substantially in the energies they deliver. Medical x-ray tubes typically generate x-rays below a few hundred kiloelectronvolts (orthovoltage x-rays), whereas Linacs produce megavoltage beams. In x-ray tubes, an electric potential of up to a few hundred kilovolts (kV) is applied between the cathode (electron emitter) and the anode (target). Extending this approach to megavolt energies is impractical because applying megavolt-level potentials across a vacuum tube can lead to catastrophic arcing and electrical

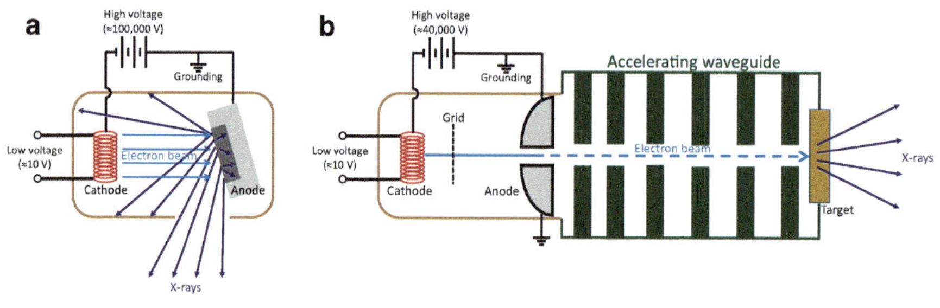

Fig. 5.1 Schematic illustration of two types of x-ray production devices: (**a**) x-ray tube and (**b**) x-ray Linac

breakdown. Linacs therefore adopt a fundamentally different electron acceleration strategy, described in Chap. 6.

The key considerations in designing an x-ray tube include controlled electron emission, effective acceleration to achieve the desired energy and tube current, focal spot formation, and efficient heat management. This chapter focuses on the physics underlying these considerations, rather than engineering implementation. With the exception of the specific acceleration mechanism, many of the physical principles discussed in this chapter for x-ray tubes also apply to Linacs.

5.2 Thermionic Electron Emission

In an x-ray tube or x-ray Linac, a source of free electrons, sometimes referred to as an electron gun, is required. In most existing clinical x-ray systems, electron emission is achieved through the thermionic emission effect, in which electrons are liberated from a material by heating it. Cold electron sources also exist and are discussed in Sect. 5.3.

The thermionic emission effect was first observed in the 1880s by Thomas Edison, who noted that electricity could pass from a heated filament to a metallic plate inside an incandescent lamp [1, 8]. The underlying principle is that sufficient thermal energy must be supplied to electrons so they can overcome the material's **work function**, W. Yes, this is the same W introduced in the discussion of semiconductor-metal contacts in Sect. 2.7 of Chap. 2.

For most metals, thermionic emission becomes significant only at temperatures above approximately 1000 K. Electrons emitted with nonzero initial velocities form an electric current, whose density J increases rapidly with temperature T. The quantitative relationship between J and T was first derived by a British physicist Owen Richardson, who received the 1928 Nobel Prize in Physics [15] "for his work on the thermionic phenomenon and especially for the discovery of the law named after him." Richardson's equation is [20]

$$J = A\,T^2 \exp\left(-\frac{W}{k_B T}\right), \tag{5.1}$$

where A is a numerical constant.

5.2.1 *Derivation of the Richardson Equation*

Back in Chap. 1, when presenting the density of states in solids, Eq. (1.40) showed that in a solid material (e.g., a metal) of volume V, the number of allowable quantum states between (k_x, k_y, k_z) and $(k_x + \mathrm{d}k_x,\ k_y + \mathrm{d}k_y,\ k_z + \mathrm{d}k_z)$ is

$$N(k_x, k_y, k_z)\,\mathrm{d}k_x\mathrm{d}k_y\mathrm{d}k_z = \frac{V}{(2\pi)^3}\,\mathrm{d}k_x\mathrm{d}k_y\mathrm{d}k_z. \tag{5.2}$$

For electrons, the number of quantum states is doubled because of the two possible spin orientations (up and down). Therefore, the number of allowable quantum states *per unit volume* between (k_x, k_y, k_z) and $(k_x + \mathrm{d}k_x,\ k_y + \mathrm{d}k_y,\ k_z + \mathrm{d}k_z)$ is

$$g(k_x, k_y, k_z)\,\mathrm{d}k_x\mathrm{d}k_y\mathrm{d}k_z = 2\frac{1}{(2\pi)^3}\,\mathrm{d}k_x\mathrm{d}k_y\mathrm{d}k_z \tag{5.3}$$

The wave vector $\mathbf{k}$ is related to the momentum $\mathbf{p}$ and velocity $\mathbf{v}$ by

$$\begin{cases} \hbar k_x = p_x = m_e v_x \\ \hbar k_y = p_y = m_e v_y \\ \hbar k_z = p_z = m_e v_z. \end{cases} \tag{5.4}$$

Based on the Boltzmann approximation in Eq. (1.54), the probability that a given quantum state is occupied by an electron is

$$\begin{aligned} f_B(E) &= \exp\left(-\frac{E - E_F}{k_B T}\right) \\ &= \exp\left(\frac{E_F}{k_B T}\right)\exp\left(-\frac{E}{k_B T}\right) \\ &= \exp\left(\frac{E_F}{k_B T}\right)\exp\left(-\frac{E_x + E_y + E_z}{k_B T}\right). \end{aligned} \tag{5.5}$$

Using the density of quantum states from Eq. (5.3) together with the occupation probability in Eq. (5.5), the density of electrons between (k_x, k_y, k_z) and $(k_x + \mathrm{d}k_x,\ k_y + \mathrm{d}k_y,\ k_z + \mathrm{d}k_z)$ is

$$\begin{aligned} n_e(k_x, k_y, k_z)\,\mathrm{d}k_x\mathrm{d}k_y\mathrm{d}k_z &= f_B(E) \times g(k_x, k_y, k_z)\,\mathrm{d}k_x\mathrm{d}k_y\mathrm{d}k_z \\ &= \exp\left(\frac{E_F}{k_B T}\right)\exp\left(-\frac{E_x{+}E_y{+}E_z}{k_B T}\right)\frac{2}{(2\pi)^3}\,\mathrm{d}k_x\mathrm{d}k_y\mathrm{d}k_z. \end{aligned} \tag{5.6}$$

As illustrated in Fig. 5.2, we define the z-axis to be normal to the material surface. Since only electrons with velocity components along z contribute to emission from the surface, the differential emission current density is given by

$$\begin{aligned} \mathrm{d}J(k_x, k_y, k_z) &= -e\, v_z\, n_e(k_x, k_y, k_z)\,\mathrm{d}k_x\mathrm{d}k_y\mathrm{d}k_z \\ &= -e\,\frac{\hbar k_z}{m_e}\exp\left(\frac{E_F}{k_B T}\right)\exp\left(-\frac{E_x + E_y + E_z}{k_B T}\right)\frac{2}{(2\pi)^3}\,\mathrm{d}k_x\mathrm{d}k_y\mathrm{d}k_z \end{aligned}$$

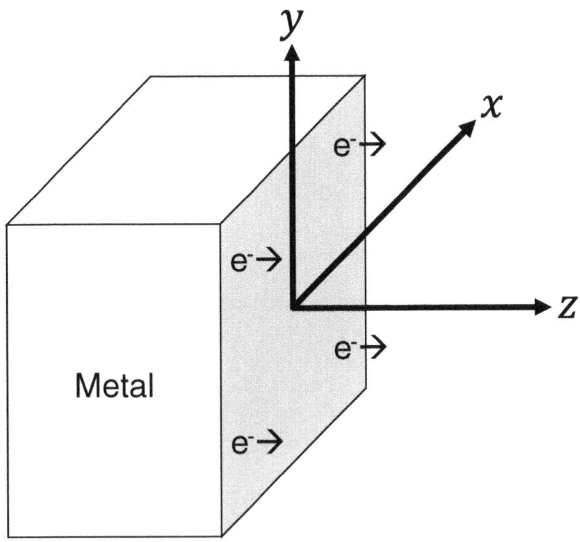

Fig. 5.2 Coordinate system employed in the derivation of the Richardson equation for thermionic emission

$$= -\frac{2e\hbar}{(2\pi)^3 m_e} \exp\left(\frac{E_F}{k_B T}\right) \exp\left(-\frac{E_x + E_y + E_z}{k_B T}\right) k_z \, \mathrm{d}k_x \mathrm{d}k_y \mathrm{d}k_z. \tag{5.7}$$

The total emission current density is

$$\begin{aligned} J &= \iiint \mathrm{d}J(k_x, k_y, k_z)\, \mathrm{d}k_x \mathrm{d}k_y \mathrm{d}k_z \\ &= -\frac{2e\hbar}{(2\pi)^3 m_e} \exp\left(\frac{E_F}{k_B T}\right) \int_{k_u}^{+\infty} \mathrm{d}k_z \, k_z \exp\left(-\frac{E_z}{k_B T}\right) \\ &\quad \iint_{-\infty}^{+\infty} \mathrm{d}k_x \mathrm{d}k_y \exp\left(-\frac{E_x + E_y}{k_B T}\right), \end{aligned} \tag{5.8}$$

where k_u denotes the minimum momentum required to escape the metal surface.

For the double integration in Eq. (5.8), it can be carried out in the polar coordinate system as follows:

$$\begin{aligned} \iint_{-\infty}^{+\infty} \mathrm{d}k_x \mathrm{d}k_y \exp\left(-\frac{E_x + E_y}{k_B T}\right) &= 2\pi \int_0^{+\infty} \mathrm{d}k_{xy} \, k_{xy} \exp\left(-\frac{E_x + E_y}{k_B T}\right) \\ &= \frac{2\pi m_e}{\hbar^2} \int_0^{+\infty} \mathrm{d}\left(\frac{\hbar^2 k_{xy}^2}{2m_e}\right) \exp\left(-\frac{E_{xy}}{k_B T}\right) \end{aligned}$$

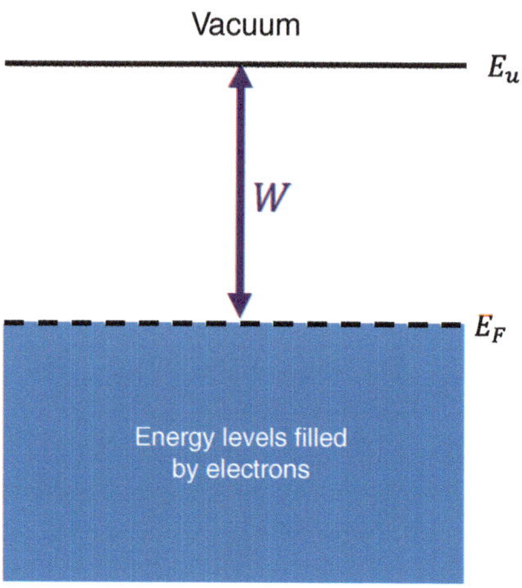

Fig. 5.3 The work function W is defined as the minimum thermodynamic work required to remove an electron from a solid to a point in the vacuum immediately outside the surface. Here, E_u denotes the energy of an electron at rest in the vacuum just outside the metal surface

$$= \frac{2\pi m_e}{\hbar^2} \int_0^{+\infty} \mathrm{d}E_{xy} \exp\left(-\frac{E_{xy}}{k_B T}\right)$$

$$= \frac{2\pi m_e k_B T}{\hbar^2} \tag{5.9}$$

With this result, the J formula in Eq. (5.8) becomes

$$J = -\frac{2e\hbar}{(2\pi)^3 m_e} \exp\left(\frac{E_F}{k_B T}\right) \int_{k_u}^{+\infty} \mathrm{d}k_z\, k_z \exp\left(-\frac{E_z}{k_B T}\right) \frac{2\pi m_e k_B T}{\hbar^2}$$

$$= -\frac{e k_B T}{2\pi^2 \hbar} \exp\left(\frac{E_F}{k_B T}\right) \int_{\frac{\hbar^2 k_u^2}{2m_e}}^{+\infty} \mathrm{d}\left(\frac{\hbar^2 k_z^2}{2m_e}\right) \frac{m_e}{\hbar^2} \exp\left(-\frac{E_z}{k_B T}\right)$$

$$= -\frac{e m_e k_B T}{2\pi^2 \hbar^3} \exp\left(\frac{E_F}{k_B T}\right) \int_{E_u}^{+\infty} \mathrm{d}E_z \exp\left(-\frac{E_z}{k_B T}\right)$$

$$= -\frac{e m_e k_B^2 T^2}{2\pi^2 \hbar^3} \exp\left(\frac{E_F}{k_B T}\right) \exp\left(-\frac{E_u}{k_B T}\right)$$

$$= -\frac{4\pi e m_e k_B^2 T^2}{h^3} \exp\left(-\frac{E_u - E_F}{k_B T}\right). \tag{5.10}$$

In Eq. (5.10), $E_u = \hbar^2 k_u^2/(2m_e)$ denotes the minimum energy required for an electron to escape from the metal. This energy corresponds to that of an electron at rest in the vacuum immediately outside the metal surface. Therefore, the difference $E_u - E_F$ is the work function (W) of the metal (Fig. 5.3), and the emission current density J can be expressed as a function of the work function and the temperature as follows,

$$\boxed{J = A_0 T^2 \exp\left(-\frac{W}{k_B T}\right)}, \tag{5.11}$$

where A_0 is a material- and temperature-independent constant:

$$\begin{aligned} A_0 &= \frac{q_e 4\pi k_B^2 m_e}{h^3} \\ &= -1.2 \times 10^6 \ \mathrm{A\,m^{-2}\,K^{-2}} \end{aligned} \tag{5.12}$$

5.2.2 Discussion of the Richardson Equation

- In practice, an empirical correction factor is applied to A_0 to align the magnitude of J with experimental measurements:

$$A = \kappa A_0 \tag{5.13}$$

  Here, κ is a material-specific factor on the order of 0.5. There is currently no consensus among theoreticians on the exact expression for κ.
- According to the Richardson equation, higher temperatures T enhance thermionic electron emission. Since the cathode must operate at elevated temperatures, it is typically made of high-melting-point metals such as tungsten (3422°C).
- Because thermionic electron emission requires the cathode to be heated to high temperatures and due to the thermal inertia of the cathode, the onset of electron emission is delayed. This limits the speed at which the electron beam can be switched on or off, making thermionic emitters less suitable for applications requiring ultrafast x-ray modulation. This delayed response is a major disadvantage of thermionic emitters compared to cold electron emitters.
- The Richardson equation also shows that a lower work function W facilitates electron emission. Consequently, the surface of tungsten cathodes is often coated with metals such as osmium, iridium, or ruthenium, which have work functions approximately 20% lower. This coating reduces the operating temperature by about 100°C and increases cathode lifetime by roughly a factor of ten.
- To efficiently emit electrons from the metal surface, the cathode filament must provide sufficient surface area. Therefore, it is often constructed from tungsten formed into a tightly coiled wire. This coil design increases the surface area of the tungsten within a compact volume, which is essential for producing a small, focused electron beam and, consequently, a small x-ray focal spot in imaging applications.
- In some electron guns, such as those in Linacs, a control grid is often placed between the cathode and anode. An alternating voltage applied to the grid modu-

lates the potential between the grid and cathode, **forming electron bunches**. As will be described in Sect. 5.4, when the voltage between the cathode and grid is zero, electron flow stops. The grid voltage can also fine-tune the current at the anode. An electron emitter consisting of a cathode, grid, and anode is called a *triode electron gun*, whereas one without a grid is known as a *diode electron gun*.

5.3 Cold Electron Emission

In classical mechanics, a particle must possess kinetic energy at least equal to the height H of a potential barrier in order to overcome it. If electrons in a solid material such as a metal were purely classical particles, then for an electron to escape from the metal into the surrounding vacuum, it would need to acquire sufficient kinetic energy to overcome the material's surface potential barrier, characterized by the work function. As discussed in the previous section, this requirement can be satisfied by heating the metal, leading to thermionic electron emission.

In quantum mechanics, however, electrons exhibit both particle- and wave-like behaviors, and there exists a finite probability for an electron to tunnel through a potential barrier of finite height and width, even when its kinetic energy is lower than the barrier height.

In the absence of an external electric field, the potential barrier at the metal-vacuum interface extends to infinity, and the tunneling probability is vanishingly small. Consequently, electrons remain confined within the metal. When a sufficiently strong electric field is applied normal to the surface, the potential barrier is tilted, and its effective width is reduced. This barrier thinning dramatically increases the tunneling probability, allowing electrons near the Fermi level to escape into the vacuum.

Electron emission under these conditions is known as **field emission** or **cold electron emission**, as it does not rely on thermal excitation. In this regime, the emission current density depends sensitively on the local electric field strength and on the material properties of the emitter. The relationship between the emission current density and the tunneling probability is described quantitatively in this section.

5.3.1 Tunneling Probability

Consider electrons in a conducting material to an electric field $\mathcal{E}$ normal to the surface. The z-axis is perpendicular to the surface, with $z = 0$ at the surface and $z > 0$ in vacuum. Immediately outside the surface, electrons encounter a potential barrier, as shown in Fig. 5.4. For simplicity, we model the barrier as a triangular function,

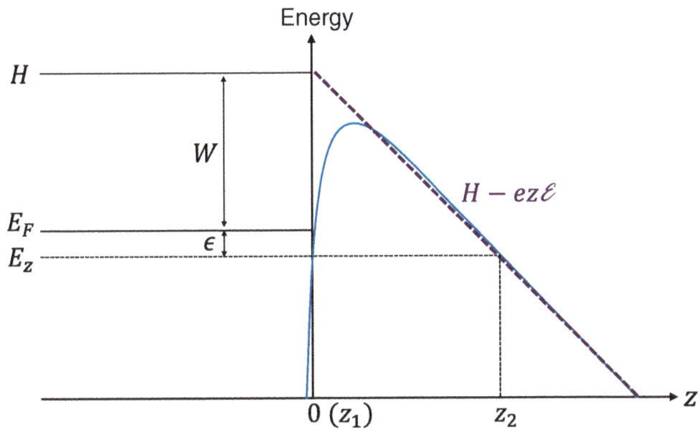

Fig. 5.4 Schematic illustration of the potential barrier at the conductor-vacuum interface. The conductor occupies the region $z < 0$, while the vacuum region corresponds to $z > 0$. An external electric field $\mathcal{E}$, normal to the surface, is applied outside the conductor. W denotes the work function of the conductor. $H - ez\mathcal{E}$ is a triangular approximation of the barrier

$$U(z) = \begin{cases} 0, & z < 0 \quad \text{(inside conductor)}, \\ H - ez\mathcal{E}, & z > 0 \quad \text{(vacuum barrier)}, \end{cases} \tag{5.14}$$

where H is the barrier height and $\mathcal{E}$ is the electric field. The potential inside the conductor is set to zero without loss of generality, since the choice of energy reference does not affect the tunneling probability.

Let E_z denote the electron energy associated with motion normal to the surface. Along the z-direction, there are two classical turning points (Fig. 5.4):

- Inner turning point z_1. It is located near the conductor-vacuum interface ($z = 0$), where electrons have just enough energy ($E_z = U$) to reach the barrier.
- Outer turning point z_2, defined by

$$U(z_2) = E_z \quad \Rightarrow \quad z_2 = \frac{H - E_z}{e\mathcal{E}}.$$

Classically, electrons in the metal with $E_z < H$ cannot escape into the vacuum. Quantum mechanics, however, predicts a different behavior: there is a finite probability T that electrons can **tunnel through** the potential barrier and reach $z > 0$. This tunneling probability is related to the electron wave function $\Psi(z, t)$ as

$$\begin{aligned} T &\propto \Psi^*(z,t)\,\Psi(z,t) \\ &= \psi^*(z)\,e^{+iE_z t/\hbar}\,\psi(z)\,e^{-iE_z t/\hbar} \\ &= |\psi(z)|^2, \end{aligned} \tag{5.15}$$

where $\psi(z)$ and $e^{-iE_z t/\hbar}$ are the spatial (z) and temporal components of the electron wave function, respectively. As shown in Eq. (5.15), the tunneling probability density is independent of time. Consequently, the following **time-independent Schrödinger equation** is used to determine the tunneling probability T:

$$-\frac{\hbar^2}{2m_e}\frac{d^2\psi(z)}{dz^2} + U(z)\psi(z) = E_z\psi(z), \tag{5.16}$$

or

$$\frac{d^2\psi}{dz^2} - \frac{2m_e}{\hbar^2}\left[U(z) - E_z\right]\psi = 0 \tag{5.17}$$

Defining

$$K(z) \equiv \sqrt{\frac{2m_e}{\hbar^2}\left[U(z) - E_z\right]}, \tag{5.18}$$

the Schrödinger equation becomes

$$\frac{d^2\psi}{dz^2} - K^2(z)\psi = 0. \tag{5.19}$$

Within the potential barrier region, $U(z) \geq E_z$, therefore, $K \geq 0$. In this case, the so-called Wentzel-Kramers-Brillouin (WKB) approximation method gives the following approximate solution of $\psi(z)$ [22],

$$\psi(z) \approx \frac{C}{\sqrt{K(z)}} \exp\left[\pm \int_0^z K(z')\,dz'\right], \tag{5.20}$$

where C is a normalization factor.

For an electron incident from the conductor into the potential barrier at $z > 0$, the electron's probability amplitude must decay as it penetrates further into the barrier region, reflecting the decreasing likelihood of finding the particle deeper inside the barrier. Therefore, the physically acceptable (approximate) solution inside the barrier is

$$\psi(z) \approx \frac{C}{\sqrt{K(z)}} \exp\left[- \int_0^z K(z')\,dz'\right]. \tag{5.21}$$

The tunneling probability is given by the ratio of the probability densities at the outer and inner turning points:

$$
\begin{aligned}
T &= \left|\frac{\psi(z_2)}{\psi(z_1)}\right|^2 \\
&\approx \exp\left[-2\int_0^{z_2} K(x)\,\mathrm{d}z\right] \\
&= \exp\left[-2\int_0^{z_2}\sqrt{\frac{2m_e}{\hbar^2}\left(U(z)-E_z\right)}\,\mathrm{d}z\right] \\
&= \exp\left[-\frac{2\sqrt{2m_e}}{\hbar}\int_0^{z_2}\sqrt{H-e\mathcal{E}z-E_z}\,\mathrm{d}z\right] \qquad (5.22)
\end{aligned}
$$

With the substitution $u = H - e\mathcal{E}z - E_z$ and $\mathrm{d}z = -\mathrm{d}u/(e\mathcal{E})$, the integral evaluates to

$$
\int_0^{z_2}\sqrt{H-e\mathcal{E}z-E_z}\,\mathrm{d}x = -\frac{1}{e\mathcal{E}}\int_{H-E_z}^{0} u^{1/2}\,\mathrm{d}u \qquad (5.23)
$$

$$
= \frac{2}{3e\mathcal{E}}(H-E_z)^{3/2}. \qquad (5.24)
$$

Thus, the tunneling probability is

$$
T(E_z) \approx \exp\left[-\frac{4\sqrt{2m_e}}{3\hbar e\mathcal{E}}(H-E_z)^{3/2}\right]. \qquad (5.25)
$$

This result shows that the tunneling probability increases rapidly with increasing E_z and increasing electric field $\mathcal{E}$, which thins the barrier.

5.3.2 *Cold Emission Current: The Fowler-Nordheim Equation*

Similar to derivation of the hot (thermionic) emission current density formula, the cold emission current density is given by the product of charge per electron $(-e)$, the density of electron quantum states $g(k)$, the occupancy probability $f(E)$, and the electron velocity along the z direction as follows:

$$
J = \int \mathrm{d}k_x \int \mathrm{d}k_y \int \mathrm{d}k_z\,(-e)\,v_z\,g(k_x,k_y,k_z)T(E_z)f(E) \qquad (5.26)
$$

One difference from the thermionic emission current density formula is that there is any additional tunneling probability term $T(E)$. This is because only electrons tunneling through the barrier along z contribute to the emission current.

According to Eq. (5.4),

$$v_z = \frac{\hbar k_z}{m_e}. \tag{5.27}$$

According to Eq. (5.3), the density of electron states in the k space is

$$g(k_x, k_y, k_z) = \frac{2}{(2\pi)^3}. \tag{5.28}$$

In conducting materials,

$$f(E) \approx 1 \quad \text{for } E \leq E_F. \tag{5.29}$$

$$\begin{aligned} J &= \frac{-2e}{(2\pi)^3} \int \mathrm{d}k_z \frac{\hbar k_z}{m_e} T(E_z) \int \mathrm{d}k_x \int \mathrm{d}k_y \\ &= -\frac{2e}{(2\pi)^3} \int \mathrm{d}k_z \frac{\hbar k_z}{m_e} T(E_z) \int \mathrm{d}k_{xy}\, 2\pi k_{xy} \\ &= -2\pi \frac{2e}{(2\pi)^3} \int \mathrm{d}\left(\frac{\hbar^2 k_z^2}{2m_e}\right) \frac{1}{\hbar} T(E_z) \int \mathrm{d}\left(\frac{\hbar^2 k_{xy}^2}{2m_e}\right) \frac{m_e}{\hbar^2} \\ &= -2\pi \frac{2e}{(2\pi)^3} \int \mathrm{d}E_z \frac{1}{\hbar} T(E_z) \int \mathrm{d}E_{xy} \frac{m_e}{\hbar^2}, \end{aligned} \tag{5.30}$$

where $k_{xy} = \sqrt{k_x^2 + k_y^2}$, $E_{xy} = \frac{\hbar^2 (k_x^2 + k_y^2)}{2m_e}$, and $E_z = \frac{\hbar^2 k_z^2}{2m_e}$.

Since only occupied states contribute to the emission current, the integration in Eq. (5.30) only need to consider energies within the following range:

$$0 \leq E_x + E_y + E_z \leq E_F. \tag{5.31}$$

Accordingly,

$$\begin{aligned} J &= -\frac{2e}{(2\pi)^2} \frac{m_e}{\hbar^3} \int_0^{E_F} \mathrm{d}E_z\, T(E_z) \int_0^{E_F - E_z} \mathrm{d}E_{xy} \\ &= -\frac{e m_e}{2\pi^2 \hbar^3} \int_0^{E_F} \mathrm{d}E_z\, (E_F - E_z)\, T(E_z) \\ &= -\frac{e m_e}{2\pi^2 \hbar^3} \int_0^{E_F} \mathrm{d}E_z\, (E_F - E_z) \exp\left[-\frac{4\sqrt{2m_e}}{3\hbar e \mathcal{E}} (H - E_z)^{3/2}\right] \\ &= \alpha \int_0^{E_F} \mathrm{d}E_z\, (E_F - E_z) \exp\left[-\beta (H - E_z)^{3/2}\right], \end{aligned} \tag{5.32}$$

where

$$\alpha = -\frac{em_e}{2\pi^2\hbar^3}, \quad \beta = \frac{4\sqrt{2m_e}}{3\hbar e\mathcal{E}}. \tag{5.33}$$

Equation (5.32) shows that the emission current is primarily formed by electrons with energies E_x near the Fermi energy level (E_F), as the tunneling probability decreases exponentially with lowering E_z. Define

$$\epsilon = E_F - E_Z \quad \rightarrow \quad H - E_z = H - E_F + E_F - E_z = W + \epsilon,$$

where $\epsilon \ll W$. With this definition, Eq. (5.32) becomes

$$\begin{aligned}
J &= \alpha \int_{E_F}^{0} \mathrm{d}(-\epsilon)\, \epsilon \, \exp\left[-\beta(W+\epsilon)^{3/2}\right] \\
&= \alpha \int_{0}^{E_F} \mathrm{d}\epsilon \, \epsilon \, \exp\left[-\beta W^{3/2}(1+\epsilon/W)^{3/2}\right] \\
&\approx \alpha \int_{0}^{E_F} \mathrm{d}\epsilon \, \epsilon \, \exp\left[-\beta W^{3/2}(1+\frac{3}{2}\epsilon/W)\right] \\
&= \alpha \exp\left[-\beta W^{3/2}\right] \int_{0}^{E_F} \mathrm{d}\epsilon \, \epsilon \, \exp\left[-\frac{3}{2}\beta W^{1/2}\epsilon\right] \\
&= \alpha \exp\left[-\beta W^{3/2}\right] \left(\frac{3}{2}\beta W^{1/2}\right)^{-2} \\
&\quad \times \left[1 - \left(1 + \frac{3}{2}\beta W^{1/2} E_F\right) \exp\left(-\frac{3}{2}\beta W^{1/2} E_F\right)\right] \\
&\approx \alpha \exp\left[-\beta W^{3/2}\right] \left(\frac{3}{2}\beta W^{1/2}\right)^{-2} \\
&= -\frac{em_e}{2\pi^2\hbar^3} \exp\left[-\frac{4\sqrt{2m_e}}{3\hbar e\mathcal{E}} W^{3/2}\right] \frac{4}{9} (\frac{3\hbar e\mathcal{E}}{4\sqrt{2m_e}})^2 W^{-1} \\
&= -\frac{e^3\mathcal{E}^2}{16\pi^2\hbar W} \exp\left[-\frac{4\sqrt{2m_e}}{3\hbar e}\frac{W^{3/2}}{\mathcal{E}}\right].
\end{aligned} \tag{5.34}$$

Equation (5.34) gives the Fowler-Nordheim equation for cold emission current density [7].

5.3.3 *Discussion of the Fowler-Nordheim Equation: Carbon Nanotube (CNT) Electron Emitters*

- As shown in Eq. (5.34), the cold emission current density depends strongly on the work function W of the material, with a smaller W corresponding to a larger emission current. While the power-law dependence, $J \propto W^{-1}$, is relatively weak, the exponential term in Eq. (5.34) dominates, so that J effectively increases exponentially as W decreases, specifically as $\exp(-\text{const} \cdot W^{3/2})$.

  Compared to tungsten ($W \approx 4.5\,\text{eV}$) and carbon ($W \approx 4.3$–$5.0\,\text{eV}$), certain ceramic materials such as lanthanum hexaboride (LaB_6, $W \approx 2.5\,\text{eV}$) have much lower work functions and are therefore ideal for use as cold emitter coatings.
- Equation (5.34) also shows that the cold emission current density depends extremely sensitively on the local electric field strength $\mathcal{E}$: even a small increase in $\mathcal{E}$ leads to an exponentially large increase in the emission current.

  To enhance the local electric field strength near a conductor surface, the most widely used approach is to introduce sharp tips to the conductor surface. To understand the underlying mechanism, consider two planar electrodes separated by a distance d, with the cathode grounded and the anode biased at a voltage V_0. In the absence of any surface protrusions, the electric field between the electrodes is uniform and given by

$$\mathcal{E}_0 = \frac{V_0}{d}. \tag{5.35}$$

  Now, suppose a small conducting sphere with a radius of R is introduced on the cathode surface. Inside the conducting sphere, the electric field vanishes, while immediately outside the sphere the electric field is everywhere normal to the sphere surface. Since there is no free charge outside the conductor ($\rho = 0$), according to the Poisson's equation,

$$\nabla^2 \Phi = 0, \tag{5.36}$$

  where Φ is the electric potential. In spherical coordinates, assuming azimuthal symmetry (i.e., no azimuthal-angle dependence), Eq. (5.37) takes the following form:

$$\nabla^2 \Phi = \frac{1}{r^2}\frac{\partial}{\partial r}\left(r^2 \frac{\partial \Phi}{\partial r}\right) + \frac{1}{r^2 \sin\theta}\frac{\partial}{\partial \theta}\left(\sin\theta \frac{\partial \Phi}{\partial \theta}\right) = 0 \tag{5.37}$$

  Assuming a solution of the form

$$\Phi(r, \theta) = R(r)\Theta(\theta), \tag{5.38}$$

  Equation (5.37) becomes

$$\Phi(r,\theta) = \sum_{l=0}^{\infty} A_l r^l P_l(\cos\theta) + \sum_{l=0}^{\infty} \frac{B_l}{r^{l+1}} P_l(\cos\theta), \tag{5.39}$$

where $P_l(\cos\theta)$ are Legendre polynomials and (A_l, B_l) are constants determined by boundary conditions.

The first boundary condition is that, far from the sphere, Φ approaches that of the uniform field:

$$\lim_{r\to\infty} \Phi(r,\theta) = -r\cos\theta\,\mathcal{E}_0 \tag{5.40}$$

As $P_1(\cos\theta) = \cos\theta$, this boundary means $A_1 = -\mathcal{E}_0$ and $A_l = 0$ for $l \neq 1$.

The second boundary condition is, on the sphere surface, Φ is constant, namely,

$$\Phi(r = R, \theta) \equiv 0. \tag{5.41}$$

Based on the formula of $\Phi(r,\theta)$ in Eq. (5.39),

$$-\mathcal{E}_0 R\cos\theta + \frac{B_0}{R} + \frac{B_1\cos\theta}{R^2} + \sum_{l=2}^{\infty} \frac{B_l}{r^{l+1}} P_l(\cos\theta) \equiv 0 \quad \text{for } \forall\theta. \tag{5.42}$$

This means

$$B_l = 0 \quad \text{for } l \neq 1; \tag{5.43}$$

$$B_1 = \mathcal{E}_0 R^3. \tag{5.44}$$

In summary,

$$\Phi(r,\theta) = -\mathcal{E}_0 r\cos\theta + \frac{\mathcal{E}_0 R^3\cos\theta}{r^2}. \tag{5.45}$$

The electric field is the negative gradient of the potential. In spherical coordinates, its components are

$$\mathcal{E}_r = -\frac{\partial\Phi}{\partial r} = \mathcal{E}_0\cos\theta + 2\frac{\mathcal{E}_0 R^3\cos\theta}{r^3} \tag{5.46}$$

$$\mathcal{E}_\theta = -\frac{1}{r}\frac{\partial\Phi}{\partial\theta} = \mathcal{E}_0\sin\theta\left(1 - \frac{R^3}{r^3}\right). \tag{5.47}$$

At the apex of the sphere ($r = R, \theta = 0$),

$$\mathcal{E}_r = 3\mathcal{E}_0 \tag{5.48}$$

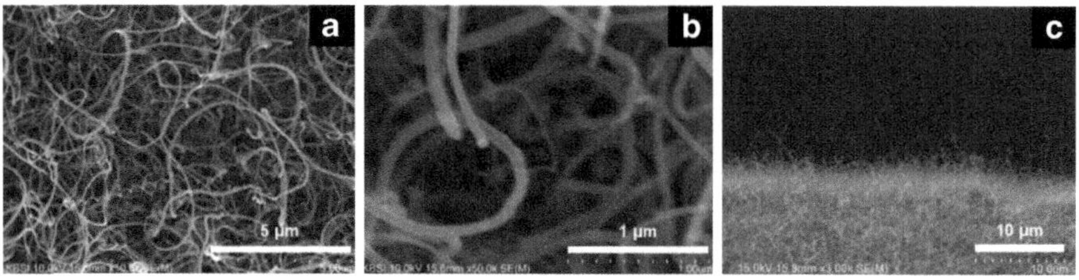

Fig. 5.5 (**a**) (**b**) Scanning electron microscope (SEM) images of CNTs grown on metal substrate synthesized by plasma-enhanced chemical vapor deposition (PECVD). (**c**) Lateral SEM image. Adapted from Reference 19, which is an open-access article distributed under the terms of the Creative Commons Attribution 4.0 International License (CC BY 4.0)

$$\mathcal{E}_\theta = 0. \tag{5.49}$$

As demonstrated by this toy model, a spherical protrusion on the cathode surface enhances the local electric field at its apex by a factor of 3 relative to the macroscopic field. This geometric field enhancement substantially increases the cold emission current density. **If the sphere is further elongated into a sharp, needlelike shape with a major axis much larger than the minor axis, the radius of curvature at the apex becomes much smaller. As a result, the local electric field is enhanced by an even larger factor—often reaching hundreds or even tens of thousands—leading to a dramatic increase in the cold emission current density.**

- This geometric effect is one of the primary reasons **carbon nanotubes (CNTs)** are widely used as cold electron emitters. A typical CNT can be several micrometers in length while having a diameter of only a few nanometers, resulting in an extremely large aspect ratio (Fig. 5.5). Owing to this geometry, CNTs can exhibit field enhancement factors on the order of $\sim 10^2$–10^4, enabling excellent field emission performance at relatively low applied voltages.

  In addition to their strong field enhancement, CNTs possess exceptional mechanical robustness and elasticity. They can sustain large elastic strains without permanent deformation or fracture, even under intense electrostatic forces at the emitter apex. In contrast, metal needle emitters with comparable aspect ratios are susceptible to fracture and fatigue cracking, which degrade the emitter geometry and lead to unstable or short-lived emission. These combined geometric and mechanical advantages make CNTs particularly well suited for high-performance and long-lifetime cold electron emission applications.
- For a given emitter material and geometry, Eq. (5.34) shows that the cold emission current increases monotonically with the applied voltage V_0. In practice, however, when V_0 becomes sufficiently large, catastrophic breakdown of the cold emitter may occur (Fig. 5.6). As the emission current increases, the finite electrical resistivity of the emitter leads to Joule heating, while additional energy exchange associated with electron emission can further heat or cool the emitter tip depending on operating conditions. If the net heat generation exceeds the rate

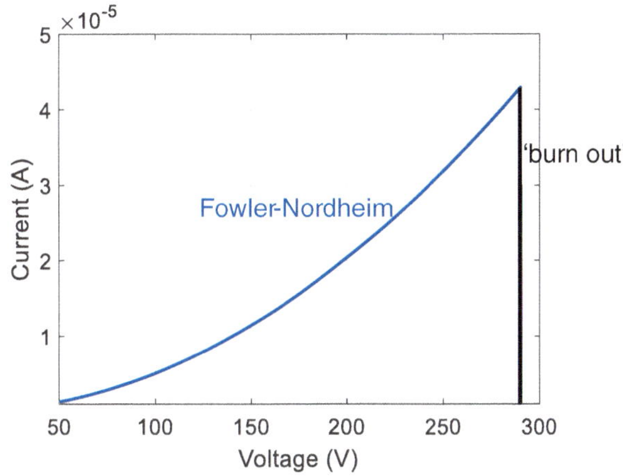

Fig. 5.6 Cold emission current as a function of the voltage applied to the emitter. At sufficiently high voltage, catastrophic breakdown of the emitter may occur

at which heat can be dissipated, the local temperature at the apex can rise rapidly, resulting in melting, evaporation, or field-induced desorption of atoms. These processes "burn out" the emitter tip and may lead to irreversible failure of the cold emitter.

- As described by the Richardson thermionic emission equation in (5.11), a thermionic cathode must be heated to a sufficiently high temperature (~2000–3000 K) to enable electron emission and thus x-ray generation. In applications requiring pulsed x-ray operation—such as cone-beam CT in Linac or C-arm interventional systems—the cathode is typically maintained at a constant high temperature, while x-ray on/off control is achieved electrically.

  One common approach is to switch the cathode-anode high voltage. Alternatively, a control grid positioned near the cathode can be used to gate the electron beam. By applying a suitable bias to the grid, the electric field at the cathode surface is reduced, suppressing electron extraction and preventing electrons from reaching the anode. However, because electrons are already thermally excited in a hot cathode, some residual current persists even in the nominal "off" state. In addition, space-charge effects and the finite capacitance of the tube and associated circuitry limit the achievable switching speed. As a result, the x-ray pulse waveform produced by a hot-cathode tube exhibits finite rise and fall times rather than an ideal rectangular shape.

  In contrast, CNT cathodes do not require heating. Electron emission begins almost instantaneously once the local electric field at the CNT tip exceeds the tunneling threshold, and no space-charge cloud is present when the field is removed. The intrinsic quantum tunneling process is extremely fast, with characteristic response times on the order of picoseconds or shorter. Therefore, the emission mechanism itself does not impose a fundamental limitation on switching speed. Consequently, x-ray pulses generated by CNT-based sources can exhibit very sharp rise and fall times and can closely approximate a rectangular waveform, with the ultimate pulse shape primarily limited by parasitic

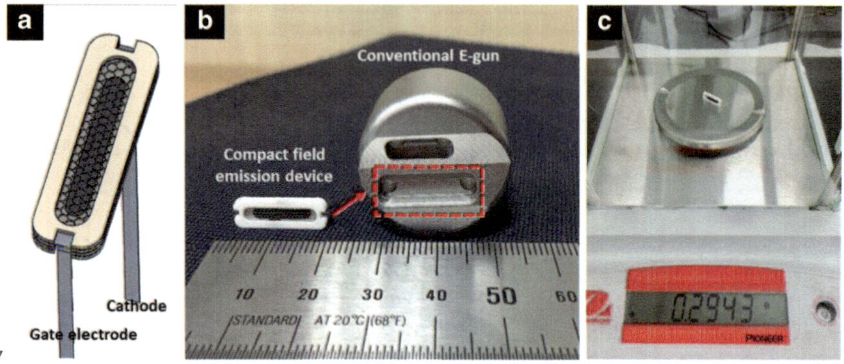

Fig. 5.7 (**a**) 3D design of a compact field emission device with directly grown CNTs on a metal substrate; (**b**) photographic image of the compact field emission device and a conventional dual-type electron gun after removal of the filament emitter; (**c**) photographic image of the 0.2943-gram compact field emission device. Source of this figure and caption: Reference 19, which is an open-access article distributed under the terms of the Creative Commons Attribution 4.0 International License (CC BY 4.0)

capacitances in the driving electronics and tube structure rather than by the electron emission process.

- In conventional thermionic x-ray tubes, the cathode heater assembly occupies significant volume and imposes a minimum practical cathode size. In addition, both the cathode and the anode generate heat during operation. For high-power, heavy-duty tubes, active thermal management is required for both electrodes. A common approach is to house the evacuated glass or metal-ceramic envelope containing the cathode and anode within an oil-filled housing, in which the circulating cooling fluid transfers heat to external heat exchangers, thereby cooling the anode and the cathode.

  CNT-based x-ray sources are generally more compact than conventional thermionic tubes because they do not require a heated cathode or the associated filament and cathode side thermal management structures (Fig. 5.7). Field emission from CNTs occurs at near-ambient temperature and can be gated using integrated microscale electrodes, eliminating bulky heater assemblies and thermionic control grids. The absence of cathode heating also enables CNT emitters to be microfabricated and individually addressed in array form, allowing multisource configurations that are difficult to implement with thermionic cathodes. Together, these factors permit substantial miniaturization of CNT-based x-ray sources.
- It is important to note, however, that CNTs function solely as electron emitters; x-ray generation still occurs via electron-anode interactions. For a given tube power and exposure duration, there is no fundamental difference in the amount of heat deposited at the anode between thermionic and CNT-based x-ray tubes. In high-power applications, system-level thermal management is therefore typically

limited by anode heat dissipation rather than by cathode cooling. Many existing CNT-based x-ray sources are relatively small in size in part because they are designed to operate at lower average power levels.

5.4 Electron Transport from Cathode to Anode

5.4.1 *Space-Charge-Limited Case: The Child-Langmuir Law*

When electrons are thermionically emitted from a hot cathode or tunnel through a cold cathode, the presence of an electric potential V_0 between the cathode and the anode causes the emitted electrons to accumulate in the space between them. This region must be evacuated and maintained under high vacuum (Fig. 5.8); otherwise, collisions between the accelerated electrons and gas molecules would scatter the electrons, reduce their kinetic energy, and prevent them from reaching the anode efficiently. Such collisions may also ionize gas molecules, leading to electrical arcing or damage to the cathode.

As the density of emitted electrons increases, the resulting space charge reduces the electric field at the cathode surface and can eventually suppress further emission. Once this space-charge-limited condition is established, a *steady-state* electric current flows between the cathode and the anode.

Under steady-state conditions, the electric current density $\mathbf{J}$ is independent of spatial position $\mathbf{x}$. This follows from the continuity equation:

$$\frac{\partial \rho(\mathbf{x}, t)}{\partial t} + \nabla \cdot \mathbf{J}(\mathbf{x}) = 0 \tag{5.50}$$

In steady state, the space-charge density does not vary with time, i.e.,

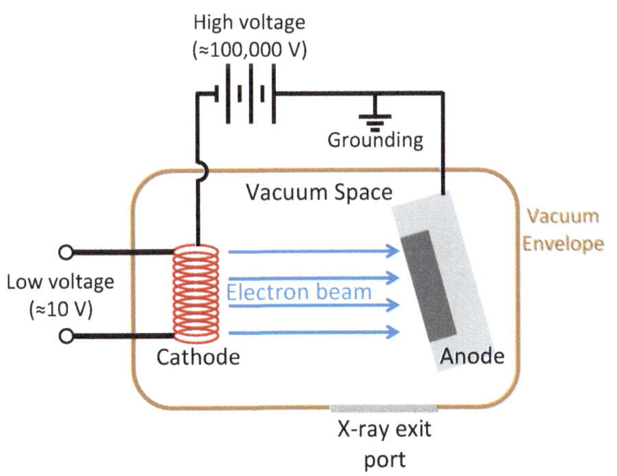

Fig. 5.8 Schematic illustration of the vacuum envelope in a thermionic x-ray tube

$$\frac{\partial \rho(\mathbf{x}, t)}{\partial t} = 0, \tag{5.51}$$

which implies

$$\nabla \cdot \mathbf{J}(\mathbf{x}) = 0. \tag{5.52}$$

Therefore, **J** is spatially invariant.

Our goal is to determine the dependence of the current density J on the applied potential V_0. For simplicity, we consider a 1D geometry along the z-axis, where $z = 0$ corresponds to the cathode and $z = L$ corresponds to the anode.

The following physical laws and boundary conditions apply:

$$\begin{cases} 1. \quad \frac{\mathrm{d}^2\Phi(z)}{\mathrm{d}z^2} = -\frac{\rho(z)}{\epsilon} \\ 2. \quad \Phi(L) - \Phi(0) = V_0 \\ 3. \quad J = \rho(z)\, v(z) \\ 4 \quad \frac{1}{2} m_e v^2(z) = e[\Phi(z) - \Phi(0)]. \end{cases} \tag{5.53}$$

In the equation set (5.53), the first equation is Poisson's equation relating the electric potential $\Phi(z)$ to the space-charge density $\rho(z)$ and permittivity ϵ, the second equation specifies the boundary condition on the potential, the third equation expresses the definition of current density in terms of charge density and electron velocity, and the fourth equation follows from energy conservation in the nonrelativistic regime, relating the electron velocity to the electrostatic potential. These four equations involve four unknowns: J, $\Phi(z)$, $v(z)$, and $\rho(z)$.

To solve for the four unknowns, we first rewrite the last equation in (5.53) as

$$v(z) = \left(\frac{2e}{m_e}\right)^{1/2} \left[\Phi(z) - \Phi(0)\right]^{1/2}. \tag{5.54}$$

Substituting this expression for $v(z)$ into the third equation in (5.53) eliminates the velocity variable and yields

$$\rho(z) = J \left(\frac{2e}{m_e}\right)^{-1/2} \left[\Phi(z) - \Phi(0)\right]^{-1/2}. \tag{5.55}$$

Next, we eliminate $\rho(z)$ by substituting (5.55) into Poisson's equation, the first equation in (5.53), obtaining

$$\frac{\mathrm{d}^2\Phi(z)}{\mathrm{d}z^2} = -\frac{J}{\epsilon} \left(\frac{2e}{m_e}\right)^{-1/2} \left[\Phi(z) - \Phi(0)\right]^{-1/2}. \tag{5.56}$$

Equation (5.56) involves only two unknowns, $\Phi(z)$ and J. Defining

$$\phi(z) = \Phi(z) - \Phi(0), \qquad \alpha = \frac{J}{\epsilon}\left(\frac{2e}{m_e}\right)^{-1/2}, \tag{5.57}$$

Equation (5.56) can be rewritten in the compact form

$$\frac{d^2\phi(z)}{dz^2} = -\alpha\,\phi(z)^{-1/2}. \tag{5.58}$$

There are several ways to solve Eq. (5.58). Here, we adopt a physically motivated *ansatz*. By inspecting the functional dependence on both sides of Eq. (5.58), we assume that the solution for $\phi(z)$ takes the power-law form

$$\phi(z) = \beta\, z^m, \tag{5.59}$$

where the exponent m and prefactor β are constants to be determined.

Substituting Eq. (5.59) into Eq. (5.58) yields

$$\beta\, m(m-1)\, z^{m-2} = -\alpha\, \beta^{-1/2}\, z^{-m/2}. \tag{5.60}$$

For this equation to hold for arbitrary z between the cathode and anode, the powers of z on both sides must be equal, which requires

$$m - 2 = -\frac{m}{2}. \tag{5.61}$$

Solving for m gives

$$m = \frac{4}{3}. \tag{5.62}$$

To determine the coefficient β, we apply the boundary condition given by the second equation in (5.53), namely, $\phi(L) = V_0$. From Eq. (5.59), this condition implies

$$V_0 = \beta\, L^{4/3} \tag{5.63}$$

and therefore

$$\beta = V_0\, L^{-4/3}. \tag{5.64}$$

The electrostatic potential distribution between the cathode and the anode is thus

$$\phi(z) = V_0 \left(\frac{z}{L}\right)^{4/3}. \tag{5.65}$$

We now determine the current density J. Substituting Eq. (5.65) into Eq. (5.58) yields

$$\begin{aligned}\frac{d^2\phi(z)}{dz^2} &= \frac{4}{9}V_0 L^{-4/3} z^{-2/3} \\ &= -\alpha\,\phi(z)^{-1/2} \\ &= -\left[\frac{J}{\epsilon}\left(\frac{2e}{m_e}\right)^{-1/2}\right] V_0^{-1/2} L^{2/3} z^{-2/3}.\end{aligned} \tag{5.66}$$

Equating the coefficients on both sides gives

$$\boxed{J = \frac{4\epsilon}{9L^2}\left(\frac{2e}{m_e}\right)^{1/2} V_0^{3/2}}. \tag{5.67}$$

The characteristic $V_0^{3/2}$ dependence of the anode-cathode current density J is known as the **Child-Langmuir law** [5, 11].

5.4.2 *Discussion of the Child-Langmuir Law*

- Based on the Poisson's equation in (5.53) and the electrostatic potential given in Eq. (5.65), the space-charge density at position z is

$$\begin{aligned}\rho(z) &= -\epsilon\frac{d^2\Phi(z)}{dz^2} = -\epsilon\frac{d^2\phi(z)}{dz^2} \\ &= -\frac{4\epsilon V_0}{9L^{4/3}} z^{-2/3}.\end{aligned} \tag{5.68}$$

  Apparently, $|\rho(z = 0)| \to \infty$, indicating a strong accumulation of negative charge near the cathode surface under space-charge-limited conditions. Physically, this implies that only a fraction of the electrons emitted from the hot cathode are able to reach the anode and contribute to the current, while the remaining electrons are decelerated, trapped near the cathode surface, or returned to the cathode.
- As the cathode temperature T increases, the emission rate increases, and the magnitude of the space-charge density near the cathode grows accordingly. However, under space-charge-limited operation, the anode-cathode current remains unchanged. Only when the applied voltage V_0 is sufficiently large, increasing the cathode temperature leads to an increase in current. This regime is discussed in the following Sect. 5.4.3.

- In deriving the electron velocity $v(z)$ in Eq. (5.54), we implicitly assumed that $v(z = 0) = 0$. This is a crude approximation and is inconsistent with the finite thermal velocity implied by Eq. (5.8), which underlies the Richardson thermionic emission law. When the initial electron velocity v_0 at the cathode is taken into account, the Child-Langmuir law can be generalized to

$$J = -\frac{2\epsilon m_e}{9eL^2}\left[\left(v_0^{3/2} + \frac{2eV_0}{m_e}\right)^{3/4} - v_0^{3/2}\right]^2. \tag{5.69}$$

- When an electron reaches the anode, its velocity is primarily determined by the accelerating voltage V_0. For an x-ray tube operated at $V_0 = 120\,\mathrm{kV}$, relativistic effects become nonnegligible. The electron energy satisfies

$$\frac{m_e c^2}{\sqrt{1-(v_L/c)^2}} = m_e c^2 + eV_0, \tag{5.70}$$

which gives

$$\begin{aligned} 1-(v_L/c)^2 &= \left(\frac{m_e c^2}{m_e c^2 + eV_0}\right)^2 \\ &= \left(\frac{511\,\mathrm{keV}}{511\,\mathrm{keV} + 120\,\mathrm{keV}}\right)^2 = 0.656. \end{aligned} \tag{5.71}$$

Therefore,

$$v_L = \sqrt{1-0.656}\,c \approx 0.59\,c. \tag{5.72}$$

In medical Linacs, the cathode-anode voltage is typically around 40 kV. In this case, electrons reach a speed of approximately $v_L \approx 0.37\,c$ at the anode before entering the microwave accelerator section, whose operating principle will be discussed in Chap. 6.

5.4.3 *Cathode Temperature-Limited Case*

When the cathode-anode current density J is measured experimentally as a function of the applied voltage V_0, the current initially follows the $V_0^{3/2}$ scaling predicted by the Child-Langmuir law. This behavior persists up to a characteristic voltage V_s. In this low-voltage regime ($V_0 < V_s$), the current is space-charge-limited, meaning that the electron flow is restricted by the buildup of negative charge in the interelectrode region rather than by the emission capability of the cathode.

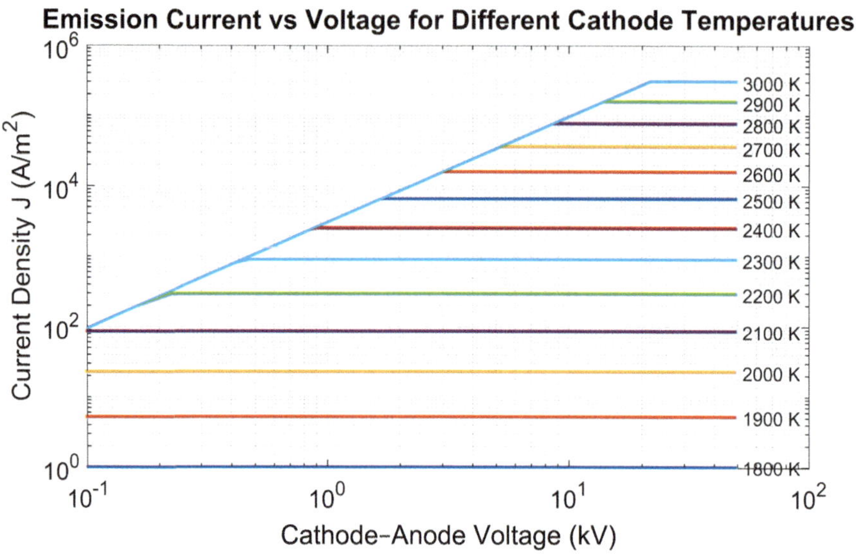

Fig. 5.9 Plots of anode-cathode current versus anode-cathode voltage for a thermionic x-ray tube. Each curve corresponds to a specific cathode filament temperature. At high filament temperatures and/or low voltages, the current is space-charge-limited and follows the Child-Langmuir law, appearing as an upward-sloping straight line on the log-log plot. At sufficiently high voltages, space-charge effects become negligible, and the current saturates at a value determined primarily by the cathode temperature, as described by the Richardson thermionic emission equation

As V_0 increases beyond V_s, the electric field becomes sufficiently strong to sweep away all electrons emitted by the cathode. In this regime, every emitted electron reaches the anode, and the current is no longer limited by space charge. Consequently, further increases in V_0 do not lead to an increase in current, and the J-V_0 characteristic saturates, forming a flat plateau (see Fig. 5.9). The voltage V_s at which this transition occurs is referred to as the **saturation voltage**.

The saturation voltage can be estimated by equating the maximum emission current density from the cathode, given by the Richardson thermionic emission equation, to the space-charge-limited current density given by the Child-Langmuir law. Specifically, equating the right-hand sides of Eqs. (5.11) and (5.67) yields

$$A_0 T^2 \exp\left(-\frac{W}{k_B T}\right) = \frac{4\epsilon}{9L^2}\left(\frac{2e}{m_e}\right)^{1/2} V_s^{3/2}, \tag{5.73}$$

from which the saturation voltage is obtained as

$$V_s = \left[A_0 T^2 \exp\left(-\frac{W}{k_B T}\right)\frac{9L^2}{4\epsilon}\left(\frac{m_e}{2e}\right)^{1/2}\right]^{2/3}. \tag{5.74}$$

This expression shows that the saturation voltage increases with cathode temperature. Physically, a higher temperature results in a larger thermionic emission current, requiring a stronger electric field (i.e., a higher voltage) to remove all emitted electrons and eliminate space-charge effects.

In summary, at low voltages or high emission rates, the anode current is governed by the Child-Langmuir law and is **space-charge-limited**. At sufficiently high voltages or low cathode temperatures, the current becomes **emission-limited** and is accurately described by the Richardson thermionic emission equation.

Medical thermionic x-ray tubes are primarily emission-limited. For a given tube voltage (kV setting), the cathode-anode current (typically measured in milliamperes) determines the x-ray flux. Space-charge effects can appear at very high emission currents. However, conventional medical tubes are usually designed to operate below this regime, so the current is generally limited by cathode emission (mA level) rather than space charge.

Similarly, CNT-based cold x-ray sources are often emission-limited under typical operating conditions, because their maximum emission current is usually lower than that of thermionic tubes. Space-charge effects may occur if the anode-cathode gap is very small or if high current pulses are applied at low voltages.

In Linacs, the cathodes are heated to high temperatures to produce large electron currents. When the cathode-anode voltage is relatively low (e.g., around 40 kV), electrons may temporarily accumulate near the cathode, and the cathode-anode current can be limited by space-charge effects, approximately described by the Child-Langmuir law. This is particularly relevant during the initial pulse formation before the electrons are fully accelerated into the microwave cavity.

5.5 From Electron Deceleration to X-Rays

5.5.1 Electron-Target Interactions

In an x-ray tube, electrons emitted from the cathode (either by thermionic emission or by field emission) are accelerated by a strong electric field toward a target. In x-ray tubes, the target is the anode itself. In LINACs, the target is a separate metallic structure located downstream of the anode at the end of the microwave waveguide (Fig. 5.1b).

Upon striking the target, the high-energy electrons interact with the atomic electrons and nuclei of the target material. This process represents a classic charged-particle-matter interaction problem.

As energetic electrons transport through the target material, they undergo a very large number of inelastic Coulomb interactions. In each individual interaction, only a small fraction of the electron's kinetic energy is transferred. Consequently, the electron energy loss can be treated as a continuous process along its trajectory, rather than as a single catastrophic event. These interactions can be broadly classified into two categories:

1. **Collisional Interactions:** In collisional interactions, incident electrons interact primarily with the orbital electrons of the target atoms. Energy is transferred from the incident electrons to atomic electrons, resulting in excitation or ionization of the atoms.

   An excitation raises an atomic electron to a higher, unstable energy state, which rapidly relaxes back to the ground state. As discussed in Chap. 4, this relaxation may occur through the emission of low-energy optical or infrared photons or via nonradiative thermalization involving phonon emission. In either case, the deposited energy is quickly dissipated within the target material and ultimately converted into heat.

   Ionization results in the ejection of an electron (a δ-ray) from an atomic shell. The ejected electron undergoes further Coulomb interactions as it slows down, producing additional excitations and ionizations. If the ionization creates an inner-shell vacancy, it is subsequently filled by an outer-shell electron. This electronic transition is accompanied by either the emission of a **characteristic (fluorescence) x-ray** or the emission of an Auger electron. Auger electrons deposit their energy locally and contribute directly to heating. Even for characteristic x-rays, only a small fraction escape the target; most are reabsorbed, and their energy is ultimately converted into heat.

   In radiation therapy, collisional interactions between charged particles and tissue are desirable because they contribute directly to absorbed dose. In contrast, for x-ray production, collisional interactions are largely undesirable, as they convert the majority of the electron energy into heat in the anode, with only a small fraction contributing to characteristic x-ray production.
2. **Radiative Interactions:** In radiative interactions, incident electrons are deflected by the strong Coulomb field of atomic nuclei. The resulting deceleration of the electrons leads to the emission of electromagnetic radiation in the form of **bremsstrahlung (braking radiation)**. The emitted x-ray photon carries away a portion of the electron's kinetic energy, producing the continuous x-ray spectrum characteristic of x-ray tubes. Radiative interactions are therefore the desired electron-anode interaction mechanism in x-ray tubes.

The relative importance of radiative interactions can be estimated using the stopping power S, which is defined as the mean kinetic energy loss of a charged particle per unit path length in a given material:

$$S(E) \equiv -\frac{\mathrm{d}E}{\mathrm{d}x} \tag{5.75}$$

The total stopping power can be decomposed into a **collisional stopping power** S_c and a **radiative stopping power** S_r:

$$S(E) = S_\mathrm{c}(E) + S_\mathrm{r}(E) \tag{5.76}$$

Both S_c and S_r depend on the kinetic energy E of the incident particle, its charge, the type of charged particle, and the material properties of the absorber. A rigorous derivation of the expressions for S_c and S_r from first principles would require a detailed treatment of charged-particle-matter interactions and is beyond the scope of this book. Consequently, standard results from the literature are quoted here.

For electrons and neglecting higher-order correction terms, the classical Bethe formula for the **collisional stopping power** is [2]

$$S_c = \frac{4\pi e^4}{m_e c^2 \beta^2} \rho_e \left[\ln\left(\frac{2 m_e c^2 \beta^2}{(1-\beta^2)\, I} \right) - \beta^2 \right], \tag{5.77}$$

where $m_e c^2$ is the electron rest energy, ρ_e is the electron density of the medium (number of electrons per unit volume), and I is the mean excitation energy of the absorber. The relativistic parameter $\beta = v/c$ is related to the kinetic energy E of the charged particle through

$$E = \frac{m_e c^2}{\sqrt{1-\beta^2}} - m_e c^2. \tag{5.78}$$

For electrons with kinetic energy E, an approximate expression for the **radiative stopping power** is [2]

$$S_r = \frac{\alpha e^4}{m_e^2 c^4} Z \rho_e (E + m_e c^2) \bar{B}, \tag{5.79}$$

where $\alpha = 1/137$ is the fine-structure constant and $\bar{B}$ is slowing varying functions of Z and E. The value of $\bar{B}$ is $\frac{16}{3}$ for $E \ll m_e c^2$ and approximately 6 for $E = 2\,\text{MeV}$, 12 for 10 MeV, and 15 for 100 MeV.

The ratio of the radiative stopping power to the collisional stopping power is then given by

$$\begin{aligned} \frac{S_r}{S_c} &= \frac{\dfrac{\alpha \beta^2 \bar{B}}{4\pi m_e c^2}}{\ln\left(\frac{2 m_e c^2 \beta^2}{(1-\beta^2)\, I}\right) - \beta^2} Z (E + m_e c^2) \\ &= \kappa Z (E + m_e c^2). \end{aligned} \tag{5.80}$$

Equation (5.80) shows that the fraction of energy loss due to radiative interactions increases with the atomic number of the target material. This provides an additional advantage to use tungsten ($Z = 74$) as the anode material in x-ray tubes, in addition to its high melting point. The same equation also indicates that the fraction of radiative interactions increases with increasing electron kinetic energy. As illustrated in Fig. 5.10, even for tungsten, the ratio S_r/S_c is only 0.01 at $E = 100\,\text{keV}$

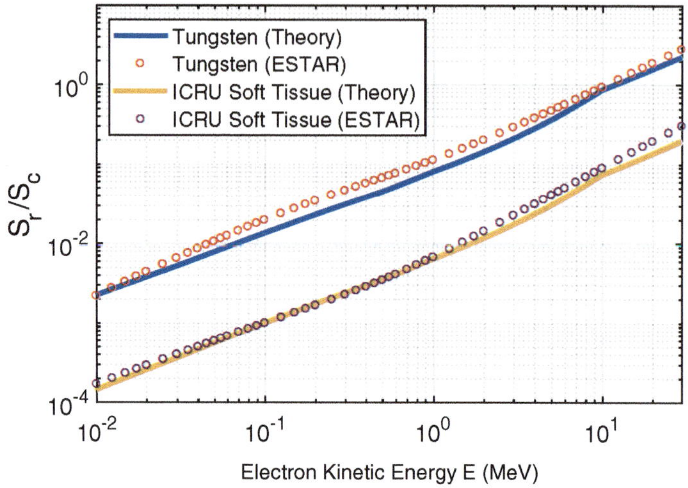

Fig. 5.10 Ratio of radiative stopping power (S_r) to collisional stopping power (S_c) as a function of electron kinetic energy for two materials: tungsten and ICRU 44 soft tissue. Theoretical curves were calculated using Eq. (5.80), while reference values were obtained from the National Institute of Standards and Technology (NIST) ESTAR (Stopping Powers and Ranges for Electrons) Database [13]

but increases to 0.87 at $E = 10\,\text{MeV}$, an electron energy used in radiotherapy linear accelerators.

5.5.2 *Bremsstrahlung X-Rays*

5.5.2.1 Angular Distribution

When an electron reaches the anode, although its initial velocity is approximately along the z axis, the Coulomb field of the target nuclei continuously changes both the direction and the magnitude of the electron velocity. During this deceleration process, bremsstrahlung radiation may be emitted. According to the Liénard-Wiechert radiation formula [9], the instantaneous angular power distribution of radiation from an electron with velocity $\boldsymbol{\beta} = \mathbf{v}/c$ and acceleration $\dot{\boldsymbol{\beta}}$ is given by [9]

$$\frac{\mathrm{d}P}{\mathrm{d}\Omega} = \frac{e^2}{4\pi c} \frac{\left|\mathbf{n} \times \left[(\mathbf{n} - \boldsymbol{\beta}) \times \dot{\boldsymbol{\beta}}\right]\right|^2}{(1 - \mathbf{n} \cdot \boldsymbol{\beta})^5}, \tag{5.81}$$

where $\mathbf{n}$ is the unit vector in the radiation emission direction.

In the nonrelativistic limit ($\beta \ll 1$), Eq. (5.81) can be simplified by approximating $(1 - \mathbf{n} \cdot \boldsymbol{\beta}) \approx 1$ and $(\mathbf{n} - \boldsymbol{\beta}) \approx \mathbf{n}$, yielding

$$
\begin{aligned}
\frac{\mathrm{d}P}{\mathrm{d}\Omega} &\approx \frac{e^2}{4\pi c}\left|\mathbf{n}\times\left(\mathbf{n}\times\dot{\boldsymbol{\beta}}\right)\right|^2 \\
&= \frac{e^2}{4\pi c}\left|\mathbf{n}(\mathbf{n}\cdot\dot{\boldsymbol{\beta}})-\dot{\boldsymbol{\beta}}(\mathbf{n}\cdot\mathbf{n})\right|^2 \\
&= \frac{e^2}{4\pi c}\left[\left|\dot{\boldsymbol{\beta}}\right|^2+\left|\mathbf{n}\cdot\dot{\boldsymbol{\beta}}\right|^2-2\left|\mathbf{n}\cdot\dot{\boldsymbol{\beta}}\right|^2\right] \\
&= \frac{e^2}{4\pi c}\left[\left|\dot{\boldsymbol{\beta}}\right|^2-\left|\dot{\boldsymbol{\beta}}\right|^2\cos^2\theta\right] \\
&= \frac{e^2\dot{\beta}^2}{4\pi c}\sin^2\theta.
\end{aligned}
\tag{5.82}
$$

In Eq. (5.82), θ is the angle between the emission direction $\mathbf{n}$ and the electron **acceleration** vector $\dot{\boldsymbol{\beta}}$, which, for a purely longitudinal slowing down process, points opposite to the electron velocity (often referred to as the **deceleration** direction).

For diagnostic x-ray sources, where the electron velocity is nonrelativistic, the electron acceleration induced by the Coulomb field has both longitudinal and transverse components. As a result, bremsstrahlung photons are emitted over nearly the full solid angle, and the angular distribution is only weakly anisotropic. Consequently, the emission is commonly described as **approximately isotropic** when contrasted with the strongly forward-peaked radiation produced by relativistic electrons in medical LINACs.

For a thick target, although the initially generated bremsstrahlung radiation is approximately isotropic, only photons emitted along directions with a short pathlength through the target material can escape. As a result, the x-rays exiting the target are highly anisotropic, primarily due to differential attenuation within the target itself rather than the intrinsic emission pattern.

5.5.2.2 Energy Distribution

For nonrelativistic electrons interacting with an **infinitesimally thin target**, the energy distribution of bremsstrahlung x-rays is described by **Kramers' law**, first derived theoretically by a Dutch physicist Hendrik A. Kramers in 1923 [10] and experimentally verified shortly thereafter by Warren Nicholas at Cornell University in 1928 (Fig. 5.11, Reference 14). The law states that the bremsstrahlung spectrum exhibits a nearly flat **energy fluence** distribution as a function of photon energy E, up to the cutoff energy $E_{\max} = eU$, where U is the tube potential.

The basic idea behind Kramers' derivation is to relate the probability of a radiative interaction to the distance between the incident electron trajectory and the atomic nucleus, known as the impact parameter. The closer the electron passes to the nucleus, the stronger the Coulomb force and the larger the acceleration, which on average leads to higher-energy emitted photons. However, small impact

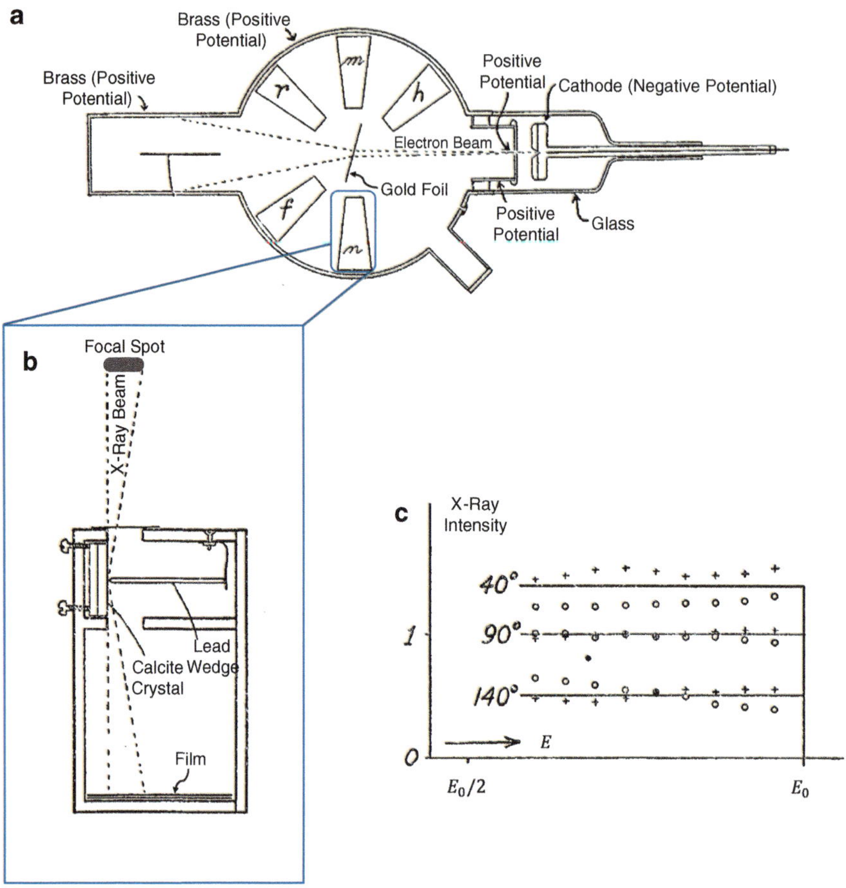

Fig. 5.11 (**a**) Experimental apparatus used by Warren Nicholas in the 1920s to measure the x-ray energy fluence (intensity) spectrum produced from a thin (86 nm) gold foil [14]. The glass section of the x-ray tube was connected to an evacuation system. The main body of the tube was made of brass and designed to minimize the number of backscattered electrons that restrike the foil. Components labeled h, m, r, f, and n are spectrograph systems used to measure the energy distribution of the emitted x-rays. (**b**) Schematic of a spectrograph. A lead wedge was pressed against the face of a calcite crystal so that x-rays exposing the photographic film could only reach it after Bragg diffraction by the crystal. Each x-ray wavelength corresponds to a unique diffraction angle and therefore to a unique position on the film. By analyzing the optical density distribution of the exposed film, the photon energy spectrum of the x-ray intensity can be reconstructed. (**c**) Experimental results obtained by Nicholas. The angles 40°, 90°, and 140° correspond to three observation angles between the detected x-rays and the incident electron beam. The figure is adapted from Ref. 14, published in 1929. The copyright of this public domain report has expired

parameters correspond to small annular areas around the nucleus and therefore smaller differential cross sections. Thus, while close encounters produce "harder" photons, they occur less frequently than distant encounters, which predominantly produce "softer" photons.

Balancing these two competing effects, Kramers showed that, to first order, the average number of photons emitted with energies between E and $E + \mathrm{d}E$ is approximately inversely proportional to the photon energy:

$$\frac{\mathrm{d}N}{\mathrm{d}E} \propto \frac{1}{E} \tag{5.83}$$

Consequently, the corresponding **energy fluence spectrum** $\Psi(E)$ becomes

$$\begin{aligned}\Psi(E) &= \frac{\mathrm{d}(NE)}{\mathrm{d}E}\\ &= E\frac{\mathrm{d}N}{\mathrm{d}E}\\ &= \text{constant.}\end{aligned} \tag{5.84}$$

This explains why the bremsstrahlung energy fluence is nearly flat as a function of photon energy in the thin-target, nonrelativistic limit.

The spectrum terminates at $E_0 = eU$ because of energy conservation: no x-ray photon can be produced with energy exceeding the kinetic energy of the incident electron.

For bremsstrahlung x-rays produced from a **thick target**, the energy fluence spectrum can be understood as the superposition of radiation generated along the entire path of the incident electrons as they continuously lose energy inside the target. Unlike the thin-target case, where each electron undergoes at most a single radiative interaction, electrons in a thick target experience many interactions and are fully stopped within the material.

A useful way to visualize the thick-target process is to decompose the target into a stack of infinitesimally thin foils (Fig. 5.12). An electron enters the first foil with its initial kinetic energy $E_0 = eU$ and produces a thin-target bremsstrahlung spectrum extending from $E = 0$ to $E = E_0$. After traversing the first foil, the electron loses a small amount of energy through collisional and radiative processes, so that in the next foil its kinetic energy is reduced to $E_0 - \mathrm{d}E$. Consequently, the bremsstrahlung spectrum produced in the second foil has the same general shape but a smaller endpoint energy. This process repeats as the electron penetrates deeper into the target, generating a sequence of truncated spectra with progressively decreasing maximum energies.

In the nonrelativistic regime ($E \ll m_e c^2$), according to Eq. (5.79), the radiative stopping power of electrons is proportional to Z and is only weakly dependent on E. As a result, each infinitesimal foil produces an energy fluence spectrum with approximately the same amplitude, differing primarily in the cutoff energy

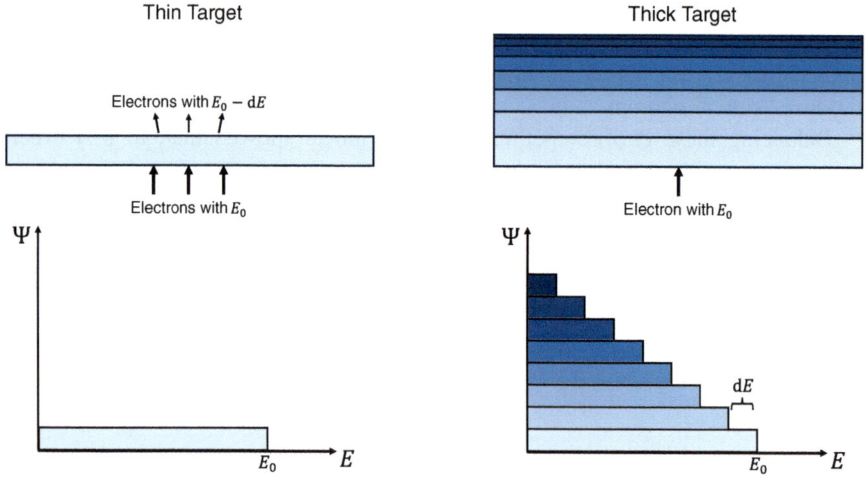

Fig. 5.12 Kramers' energy fluence spectrum models for thin and thick anode targets [14]

determined by the local electron energy. Therefore, when all foil contributions are summed over the entire electron path length, the total energy fluence spectrum becomes linear in photon energy, leading to the well-known **Kramers' thick-target spectrum**,

$$\Psi(E) = C\, N_e\, Z\, (E_0 - E), \tag{5.85}$$

where C is a proportionality constant determined by fundamental constants and target geometry and N_e is the number of incident electrons (proportional to the tube current).

Equation (5.85) describes the **energy fluence spectrum**. The corresponding **photon number spectrum** $\Phi(E)$ is obtained by dividing by E,

$$\begin{aligned}\Phi(E) &= \frac{\Psi(E)}{E} \\ &= C\, N_e\, Z\, \frac{E_0 - E}{E},\end{aligned} \tag{5.86}$$

which diverges toward low photon energies, reflecting the much higher probability of producing soft bremsstrahlung photons.

The total emitted bremsstrahlung energy is obtained by integrating the energy fluence spectrum,

$$\int_0^{E_0} \Psi(E)\, \mathrm{d}E = C N_e Z \int_0^{E_0} (E_0 - E)\, \mathrm{d}E$$

$$= \frac{C}{2} N_e Z E_0^2$$

$$= \frac{Ce^2}{2} N_e Z U^2. \tag{5.87}$$

Equation (5.87) shows that the total bremsstrahlung output scales linearly with the target atomic number and tube current and quadratically with the tube voltage. This strong U^2 dependence explains the rapid increase in x-ray output with increasing x-ray tube potential.

It should be emphasized that the Kramers' spectrum represents the intrinsic bremsstrahlung production inside the target prior to any attenuation effects. In practical x-ray tubes, the observed spectrum is substantially modified by several factors, including self-absorption within the target, inherent filtration, and added filtration. These processes preferentially remove low-energy photons, thereby hardening the spectrum and shifting the effective photon energy upward relative to the ideal Kramers' model, as discussed later in Sect. 5.5.4. In addition, once E_0 exceeds the relevant electron shell binding energies of the target material, discrete characteristic x-ray lines, discussed in the next subsection, are superimposed on the continuous bremsstrahlung background.

5.5.3 Characteristic (Fluorescence) X-Rays

The previous subsection discussed bremsstrahlung x-rays produced by **radiative interactions** of electrons in a target. Another important type of electron-matter interaction is the **collisional interaction**, in which incident electrons eject bound atomic electrons from inner shells of the target atoms. When an atom is left with a vacancy in an electron shell, it is in an excited state and tends to relax to a more stable, lower-energy configuration. This relaxation occurs when outer-shell electrons transition to fill the vacancy, often through a cascade of such transitions.

By energy conservation, each transition is accompanied by the release of energy, which is emitted either in the form of an Auger electron or as a characteristic (fluorescence) x-ray photon. The energy of a characteristic x-ray photon produced in a shell transition is

$$E_c = U_i - U_f, \tag{5.88}$$

where U_i and U_f are the electron energies before and after the transition, respectively. These orbital energies are negative, and their magnitudes correspond to the binding energies of the respective shells. The binding energies are determined primarily by the atomic number Z of the element. Table 5.1 lists these energies for the K, L, and M shells for several elements. For the L and M shells, the listed values are approximate because these outer shells are split into sublevels with slightly

Table 5.1 Electron-binding energies for the K, L, and M shells of several elements (in keV)

Element (Z)	K shell	L shell	M shell
Oxygen (8)	0.5	0.04	–
Calcium (20)	4.0	~0.4	~0.03
Molybdenum (42)	20.0	~2.6	~0.3
Rhodium (45)	23.2	~3.0	~0.3
Silver (47)	25.5	~3.4	~0.4
Tungsten (74)	69.5	~10.2	~1.8

Table 5.2 Characteristic x-ray energies (approximate in keV) for tungsten (W), rhodium (Rh), and molybdenum (Mo)

Line	Shell transition	W	Rh	Mo
K_α	$L \rightarrow K$	~ 59	~ 20	~ 18
K_β	$M \rightarrow K$	~ 67	~ 23	~ 20
K_γ	$N \rightarrow K$	~ 69	~ 23	~ 20
L_α	$M \rightarrow L$	~ 8	~ 3	~ 2
L_β	$N \rightarrow L$	~ 10	~ 3	~ 3
L_γ	$O \rightarrow L$	~ 10	~ 3	~ 3
M_α	$N \rightarrow M$	~ 2	<1	<1
M_β	$O \rightarrow N$	~ 2	<1	<1

different energies due to spin-orbit coupling. In contrast, for the innermost K shell, only a single orbital angular momentum state, zero, is allowed. Consequently, no spin-orbit interaction occurs, so the K shell has no sub-energy levels.

Because the binding energies are unique to each target material, the energies of the emitted photons are also unique. This is why they are called "characteristic" x-rays. This property forms the physical basis of energy-dispersive x-ray spectroscopy (EDX), which is widely used to determine the elemental composition of materials.

As shown in Table 5.2, characteristic x-rays produced by transitions from the L and M shells to the K shell are called the K_α and K_β lines, respectively. The K_α line is further split into $K_{\alpha 1}$ and $K_{\alpha 2}$ due to the presence of sub-levels in the L shell. In most applications, the energy resolution of x-ray detectors is not sufficient to resolve these fine-structure components.

For a given electron-target interaction, the probability of producing a specific characteristic x-ray photon (e.g., K_α) can be written as the product of several factors:

- The probability that the interaction is collisional rather than radiative, which is given by $\frac{S_c}{S}$
- The probability that the collision produces a vacancy in the K shell
- The probability that the vacancy is filled by an electron originating from the L shell rather than from higher shells
- The probability that the relaxation proceeds via a radiative transition rather than an Auger process (aka fluorescence yield)

These probabilities depend primarily on the atomic number of the target material and only weakly on the incident electron energy if it is above the ionization threshold and within the diagnostic energy range.

When the Kramers' bremsstrahlung spectrum and the characteristic x-ray spectrum are combined, the total unfiltered x-ray spectrum may be written as

$$\Psi(E) \approx N_e\, C\, Z\, (E_0 - E) + N_e \sum_j D_j(Z)\, E_j\, \delta(E - E_j), \tag{5.89}$$

$$\Phi(E) \approx N_e\, C\, Z\, \frac{E_0 - E}{E} + N_e \sum_j D_j(Z)\, \delta(E - E_j), \tag{5.90}$$

where index j labels the jth characteristic x-ray line with photon energy E_j and $D_j(Z)$ represents the line production coefficient. The function $\Psi(E)$ is the energy fluence spectrum, $\Phi(E)$ is the photon number spectrum, and $\delta(E)$ denotes the Dirac delta function.

5.5.4 X-Ray Spectrum

5.5.4.1 Filtration by the Anode: Heel Effect

As electrons continuously slow down in the anode, x-rays are produced at a certain depth beneath the anode surface. The expected path length of an electron with incident kinetic energy E_0 in the anode material is given by

$$\begin{aligned} \langle x \rangle &= \int_{E_0}^{0} \left(\frac{\mathrm{d}E}{\mathrm{d}x} \right) \mathrm{d}E \\ &= \int_{0}^{E_0} \left(-\frac{\mathrm{d}E}{\mathrm{d}x} \right) \mathrm{d}E \\ &= \int_{0}^{E_0} S(E)\, \mathrm{d}E, \end{aligned} \tag{5.91}$$

where $S(E)$ is the stopping power of the anode material.

The product of the anode material density ρ and the mean path length $\langle x \rangle$ is known as the **continuous slowing down approximation (CSDA) range**, values of which are tabulated in the NIST ESTAR database, for example, for an electron with $E_0 = 70\,\mathrm{keV}$, the CSDA range in tungsten is $1.67 \times 10^{-2}\,\mathrm{g/cm^2}$. Using the density of tungsten, $\rho = 19.3\,\mathrm{g/cm^3}$, the corresponding mean path length is

$$\langle x \rangle = 8.65 \times 10^{-4}\,\mathrm{cm} = 8.65\,\mu\mathrm{m}.$$

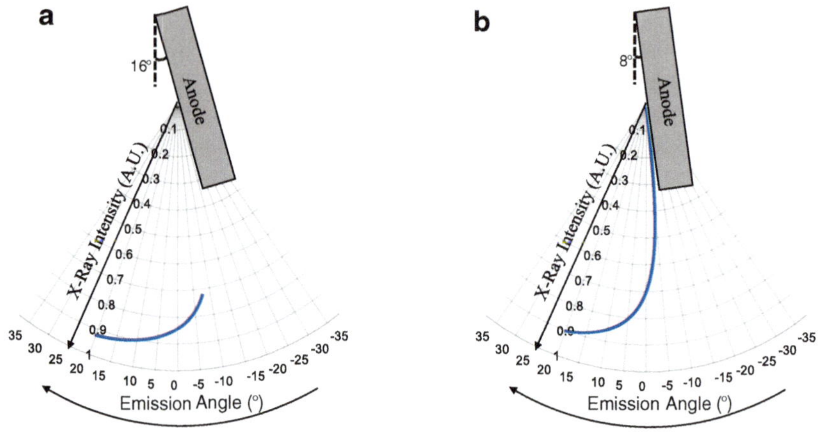

Fig. 5.13 Polar plots of the theoretically calculated x-ray intensity exiting the anode as a function of ray angular position. (**a**) Anode angle (θ_0) = 16°. (**b**) Anode angle (θ_0) = 8°. The x-ray energy is assumed to be 70 keV

Assuming that the average depth of interaction beneath the anode surface, $\langle d \rangle$, is approximately one quarter of this path length, x-rays are generated at a depth of about 2.2 μm below the tungsten anode surface. Although this depth may appear small, tungsten has a very high x-ray attenuation coefficient, and a significant fraction of the generated x-rays can be absorbed within the anode before exiting the tube. For this reason, in x-ray tubes with non-transmissive anodes, the anode surface is tilted by an angle θ_0 with respect to the direction perpendicular to the z-axis. This geometry allows a portion of the produced x-rays to traverse a shorter path within the anode, thereby reducing attenuation and providing sufficient x-ray flux at the designated exit port.

As illustrated in Fig. 5.13, the intensity of x-rays exiting the anode is a strong function of emission angle, with higher intensity on the cathode side and lower intensity on the anode side. This angular variation in x-ray intensity is known as the **heel effect**. Furthermore, as also shown in Fig. 5.13, **decreasing the anode angle θ_0 reduces the angular span over which sufficient x-ray intensity is available, thereby limiting the imaging field of view**.

For a given emission angle θ, lower-energy x-rays are preferentially attenuated by the anode material because the attenuation coefficient of tungsten increases rapidly at low photon energies. Consequently, starting from the triangular Kramers' bremsstrahlung spectrum, the x-ray spectrum exiting the anode can be written as

$$\begin{aligned}\Omega(E) &= \Omega_0(E)\, \mathrm{e}^{-\mu_W(E) L_W} \\ &= a\,(E_{\max} - E)\, \mathrm{e}^{-\mu_W(E) L_W}, \end{aligned} \tag{5.92}$$

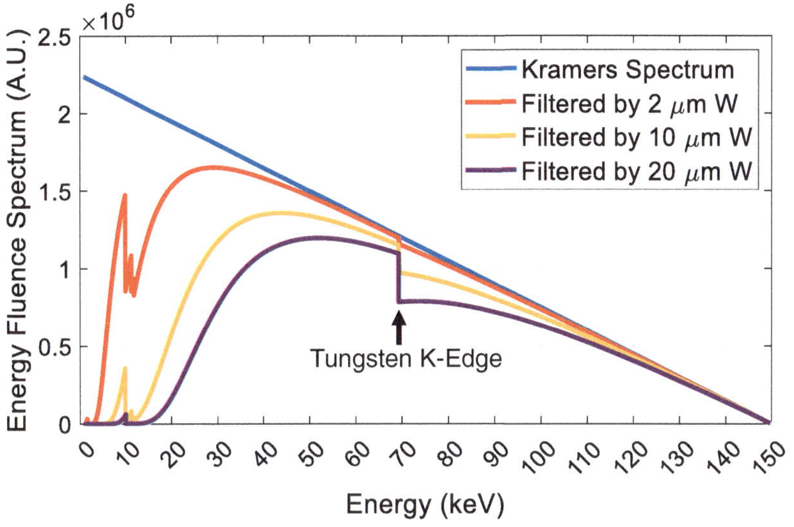

Fig. 5.14 Effect of tungsten (W) anode filtration on the initial Kramers' x-ray spectrum. Low-energy x-rays are preferentially attenuated by the anode. This figure does not consider characteristic x-rays

where $\mu_W(E)$ is the energy-dependent linear attenuation coefficient of tungsten and L_W is the effective x-ray path length within the anode.

Using $\mu_W(E)$ values provided by NIST, the resulting energy fluence spectra for different effective tungsten filtration thicknesses were calculated and are shown in Fig. 5.14. After filtration by the tungsten anode, the original triangular Kramers' spectrum loses its low-energy portion and is no longer monotonically increasing with energy. Instead, the spectrum initially rises with increasing photon energy and then decreases as the bremsstrahlung production probability approaches zero near $E_{\max}$.

As shown in Fig. 5.14, a pronounced drop appears at the K-edge of tungsten ($E_K = 69.5\,\text{keV}$), where the probability of photoelectric absorption increases abruptly. In experimentally measured x-ray spectra, however, this sharp dip is generally not observed. This discrepancy arises for several reasons: First, when a photon with energy exceeding the K-edge undergoes photoelectric absorption in tungsten, the resulting K-shell vacancy is very likely to be filled via radiative transitions, producing secondary K-fluorescence (characteristic) x-rays. Because tungsten has a high atomic number ($Z = 74$), its fluorescence yield is large (approximately 95%). As a result, a significant fraction of the absorbed photons are effectively re-emitted at characteristic energies, partially compensating for the increased absorption at the K-edge. Second, characteristic K_α x-rays generated directly by electron-anode interactions have energies close to the tungsten K-edge. These strong spectral lines add intensity in the vicinity of the edge and further reduce the visibility of any sharp discontinuity in the measured spectrum. Third, the finite energy resolution of practical x-ray spectrometers and detectors leads to spectral broadening, which smooths sharp features such as the K-edge and smears out any abrupt changes in intensity.

5.5.4.2 Inherent Filtration

Besides self-absorption in the anode, x-rays produced in an x-ray tube must penetrate additional materials before exiting the tube. These materials constitute the **inherent filtration**—permanent components in the beam path required for structural support, vacuum containment, insulation, and cooling (Fig. 5.15). Although not

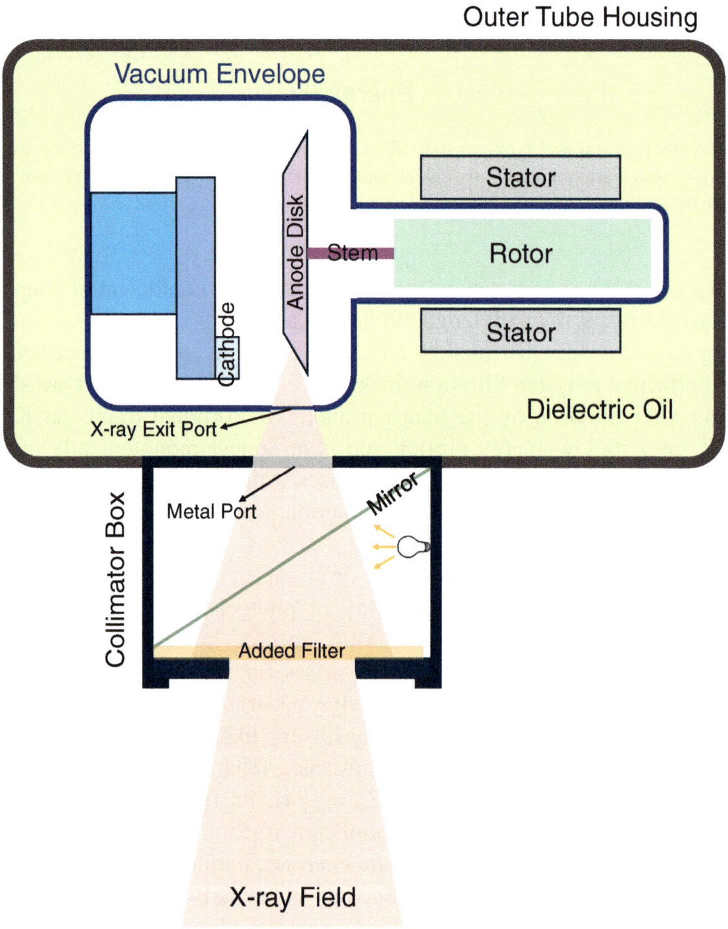

Fig. 5.15 Schematic illustration of the inherent and added filtration components in the beam path of x-rays produced by a rotating-anode x-ray tube

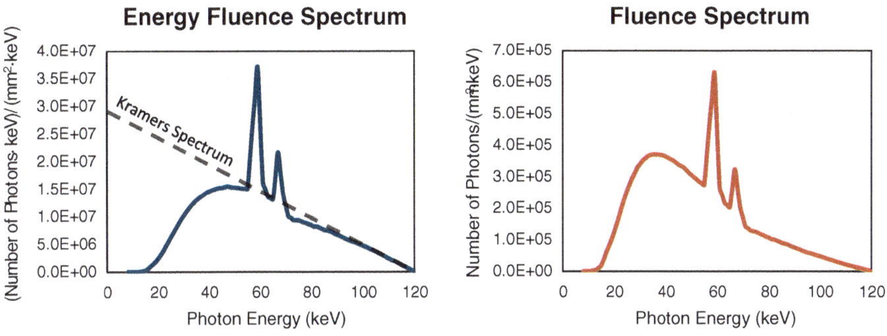

Fig. 5.16 Energy fluence spectrum ($\Psi(E)$, left) and photon fluence spectrum ($\Phi(E) = \Psi(E)/E$, right) for a tungsten anode x-ray tube operated at 120 kV, with a 1.1-mm Al-equivalent inherent filtration. The spectrum was produced using the TASMIP model [3]. The Kramers' spectrum is shown for comparison

designed for spectral shaping, inherent filtration significantly modifies the spectrum by preferentially attenuating low-energy x-rays.

The first source of inherent filtration is the wall of the vacuum envelope. Older tubes commonly use glass envelopes, whereas modern tubes often employ metal or ceramic designs.

At the designed beam exit port, the vacuum envelope includes a window region where the envelope thickness is reduced or a low atomic number beryllium ($Z = 4$) window is used when transmission of low-energy x-rays is desired, as in mammography.

The second inherent filter is the dielectric oil. The vacuum envelope is immersed in oil to provide both electrical insulation and heat removal. Along the beam direction, the oil thickness can be several centimeters and therefore contributes substantial attenuation of low-energy photons.

The third inherent filter is the outer protective housing of the x-ray tube. Most of the housing is designed to absorb stray radiation, but at the beam exit there is a thin metal port window intended to minimize attenuation while maintaining mechanical integrity.

Some x-ray systems also include additional permanent components in the beam path, such as light-field mirrors. Because these components are not removable, their attenuation is also classified as inherent filtration.

With these inherent beam filters, the fraction of low-energy photons in the spectra of Fig. 5.14 is further reduced, making the contribution below 10 keV essentially negligible, as shown in Fig. 5.16.

5.5.4.3 Added Filtration

Added filtration refers to intentionally inserted, removable materials placed in the x-ray beam to shape the spectrum and spatial fluence distribution (Fig. 5.15,

Table 5.3 Typical added filters in diagnostic x-ray imaging systems. Values may vary by manufacturer and models

Modality	Filter material	Typical thickness (mm)
General radiography	Al	0.5–2.0
	Cu	0.1–0.3
Fluoroscopy	Al	0.5–2.0
	Cu	0.1–0.9
Mammography	Mo	0.030
	Rh	0.025
	Ag	0.050

Table 5.3). The purpose of inserting a filter between the x-ray tube and the imaged object is not limited to removing low-energy x-rays. Rather, added filtration is designed to optimize image quality, radiation dose efficiency, and, in some cases, preferentially suppress high-energy x-rays. Based on its intended function, added filtration can be classified into the following categories:

- **High-Energy-Pass Filters:** For chest radiography and other body or extremity imaging applications, low-energy x-rays (e.g., below 30 keV) have negligible probability of penetrating the imaged anatomy and therefore contribute little to image formation. Because these photons are almost completely absorbed in the patient, they increase radiation dose, particularly to the entrance skin, without improving image contrast. High-energy-pass filters are designed to preferentially remove low-energy x-rays, thereby improving dose efficiency. Aluminum (Al) and copper (Cu) are commonly used for this purpose. X-ray CT systems may also employ materials such as tin (Sn) or silver (Ag) to preferentially attenuate low-energy photons, which not only reduces skin dose but also enhances spectral separation for dual-energy CT applications.
- **Low-Energy-Pass Filters:** In applications such as mammography, added filtration is used to preferentially attenuate photons above the filter material's K-edge, thereby producing a spectrum with relatively low mean energy (Fig. 5.17). Because differences in x-ray attenuation coefficients between tissues (e.g., between ductal carcinoma and normal glandular breast tissue) generally increase with decreasing photon energy, low-energy-pass filters can effectively enhance image contrast. Typical materials include Mo, Rh, and Ag (Table 5.3).
- **Fluence-Field-Shaping Filters:** In multi-detector-row CT (MDCT) and cone-beam CT (CBCT), a uniform post-patient x-ray intensity distribution improves image noise uniformity and dose efficiency. Consequently, modern MDCT and CBCT systems employ interchangeable filters to shape the spatial fluence field. In MDCT, these filters are commonly fabricated from combinations of Al, Teflon, graphite, or other materials and are shaped as **bow tie filters**, with less attenuation near the center corresponding to the thickest part of the patient. Multiple bow tie designs are available to accommodate different patient sizes and anatomical regions.

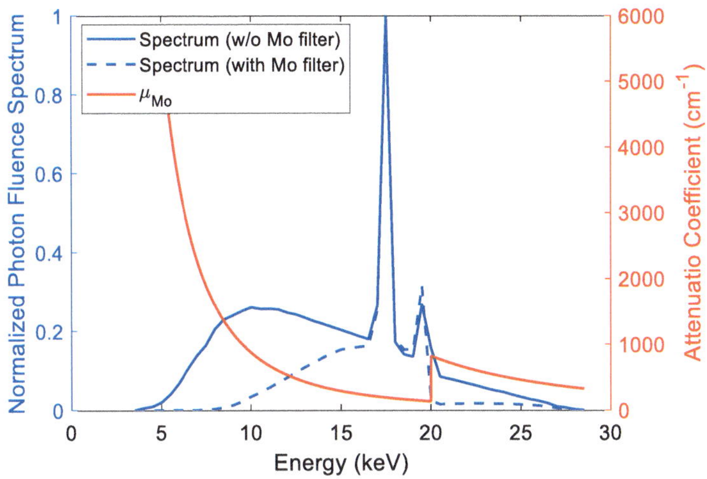

Fig. 5.17 Simulated photon fluence spectra of a Mo-anode x-ray tube for mammography operated at 28 kV are shown with and without 0.030-mm Mo filtration. The Mo attenuation coefficient (μ_{Mo}) is plotted versus x-ray energy to indicate the K-edge. With filtration, x-rays above the Mo K-edge are strongly suppressed

In CBCT systems used in radiotherapy, two primary filters are often provided: a full-fan filter for centered patients and a half-fan filter for laterally offset patients to increase the imaging field of view. Some planar x-ray imaging systems also employ fluence-field-shaping filters to reduce the dynamic range of detected signals, thereby avoiding saturation in thin regions and underexposure in thick regions.

- **Spectral Encoding and Modulation Filters:** In some dual-energy CT systems, two different filter materials are placed in a spatially separated configuration to generate distinct spectra within a single exposure. Such split filters allow simultaneous acquisition of spectrally separated projections for material decomposition.

  In addition, certain systems employ filters with periodic or spatially varying attenuation profiles to encode scatter or other x-ray optics information in the detected signal. These modulation filters can facilitate improved scatter estimation and enable advanced imaging techniques such as x-ray phase-contrast and dark-field imaging by exploiting energy- or angle-dependent interactions.

5.5.4.4 Exposure and Al Equivalence

The filtration effects of inherent and added filters on the x-ray beam are often quantified using **aluminum (Al) equivalence**. To define this term, we first review an important measure of x-ray intensity-**x-ray exposure**, which is defined as

$$X = \frac{Q}{m}. \tag{5.93}$$

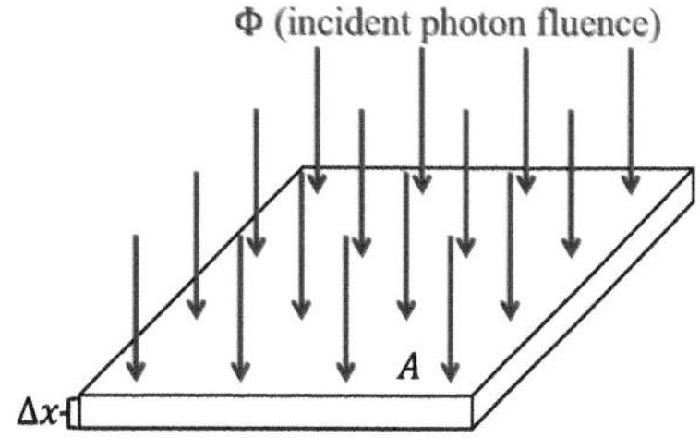

Fig. 5.18 A thin slab of air with cross-sectional area A perpendicular to the incident x-ray beam and infinitesimal thickness Δx. The photon fluence of the incident x-ray is Φ

In Eq. (5.93), Q is the absolute value of the total charge of ions of one sign produced in a given volume of air, and m is the mass of air in that volume. Exposure is the oldest measure of x-ray intensity, since the measurement of charge using an ionization chamber is more straightforward than the direct measurement of deposited energy required for radiation dose. The SI unit for exposure is coulombs per kilogram (C/kg), although the traditional unit, the roentgen (R), is still widely used:

$$1\ \mathrm{R} = 2.58 \times 10^{-4}\ \mathrm{C/kg} \tag{5.94}$$

For dry air, the average energy expended per electron-ion pair (I.P.) created is 33.97 eV. Therefore, the average x-ray energy expended per coulomb of charge created in air is

$$\begin{aligned} \epsilon_{\mathrm{air}} &= \frac{33.97\ \mathrm{eV/I.P.}}{1.602 \times 10^{-19}\ \mathrm{C/I.P.}} \times 1.602 \times 10^{-19}\ \mathrm{J/eV} \\ &= 33.97\ \mathrm{J/C}. \end{aligned} \tag{5.95}$$

As shown in Fig. 5.18, consider a thin slab of air with cross-sectional area A perpendicular to the incident x-ray beam and infinitesimal thickness Δx. Let the incident photon fluence (number of photons per unit area) be Φ, and assume the x-rays are monoenergetic with photon energy E. According to the differential form of the Beer-Lambert law, the expected number of photons that interact within this air slab is

$$\Delta N = (\Phi A)\, \mu_{\mathrm{air}}\, \Delta x. \tag{5.96}$$

Of these ΔN interactions, only a fraction $1 - g$ deposits its energy locally within the volume of interest, while the remaining fraction g escapes as secondary radiation. Because each photon carries energy E and the energy required to produce one coulomb of charge in air is ϵ_{air}, the total charge generated in the slab is

$$\begin{aligned} Q &= \frac{E \Phi A\, [\mu_{\mathrm{air}}(1-g)]\Delta x}{\epsilon_{\mathrm{air}}} \\ &= E \Phi A\, \mu_{\mathrm{en,air}}\, \Delta x\, \epsilon_{\mathrm{air}}^{-1}, \end{aligned} \tag{5.97}$$

where $\mu_{en,air} = \mu_{air}(1 - g)$ is the linear energy-absorption coefficient.

Given that the air volume is $A\Delta x$ with density ρ_{air}, the mass of air is

$$m = A\Delta x\rho_{air}. \tag{5.98}$$

According to Eq. (5.93), the exposure is given by

$$\begin{aligned} X = \frac{Q}{m} &= \frac{E\Phi A\,\mu_{en,air}\,\Delta x\,\epsilon_{air}^{-1}}{A\Delta x\rho_{air}} \\ &= \Psi\left(\frac{\mu_{en}}{\rho}\right)_{air}\epsilon_{air}^{-1}, \end{aligned} \tag{5.99}$$

where $\Psi = E\Phi$ is the energy fluence and $\left(\frac{\mu_{en}}{\rho}\right)_{air} = \mu_{en,air}/\rho_{air}$ is the mass energy-absorption coefficient of air.

Using a similar analysis, for polychromatic x-rays with energy fluence spectrum $\Psi(E)$, the exposure is

$$X = \int \mathrm{d}E\,\Psi(E)\left(\frac{\mu_{en}}{\rho}\right)_{air}(E)\,\epsilon_{air}^{-1}. \tag{5.100}$$

After defining exposure, we can introduce the concept of the Al equivalence of an x-ray filter. It is defined as the thickness of aluminum that produces the same reduction in x-ray exposure as the filter material of interest under specified beam conditions.

Formally, a material of thickness t_m is said to have an Al-equivalent thickness t_{Al} if

$$\frac{X_m}{X_0} = \frac{X_{Al}}{X_0} \tag{5.101}$$

or equivalently

$$\int \mathrm{d}E\,\Psi_0(E)\,\mathrm{e}^{-\mu_m(E)t_m}\left(\frac{\mu_{en}}{\rho}\right)_{air}(E) = \int \mathrm{d}E\,\Psi_0(E)\,\mathrm{e}^{-\mu_{Al}(E)t_{Al}}\left(\frac{\mu_{en}}{\rho}\right)_{air}(E), \tag{5.102}$$

where $\Psi_0(E)$ is the incident x-ray energy fluence spectrum and $\mu_m(E)$ and $\mu_{Al}(E)$ are the energy-dependent linear attenuation coefficients of the material and Al, respectively. Because $\mu(E)$ is strongly energy-dependent, the Al equivalence is determined not only by the filter material but also by the beam spectrum.

Note that a material with Al-equivalent thickness t_{Al} does not, in general, produce the same spectrum as aluminum of the same physical thickness. As shown in Eq. (5.102), the equivalence is defined in terms of exposure, rather than by point-by-point matching of the spectrum at each photon energy.

5.5.4.5 X-Ray Beam Quality and Half-Value Layer

In diagnostic x-ray imaging, the spectral shape of an x-ray beam is commonly referred to as the beam "quality." When an x-ray beam contains a smaller fraction of low-energy photons and a larger fraction of more penetrating high-energy photons, it is said to have higher quality. This terminology originates from early radiological practice, in which low-energy x-rays were recognized to be readily absorbed by the patient, contributing little to image formation while increasing radiation dose, whereas higher-energy x-rays were regarded as more useful and therefore of better "quality" for imaging.

X-ray beam quality is most commonly quantified using the half-value layer (HVL). The HVL is defined as the thickness of a specified material, typically aluminum, that reduces the measured x-ray exposure to one-half of its original value. Mathematically, if X_0 denotes the unattenuated exposure and $X(t)$ the exposure after inserting a thickness t of aluminum, the HVL satisfies

$$X(t_{\mathrm{HVL}}) = \frac{1}{2} X_0. \tag{5.103}$$

Here, X_0 and $X(t_{\mathrm{HVL}})$ are related to the x-ray spectrum by

$$X_0 = \int \mathrm{d}E\, \Psi(E) \left(\frac{\mu_{\mathrm{en}}}{\rho}\right)_{\mathrm{air}} (E), \tag{5.104}$$

$$X(t_{\mathrm{HVL}}) = \int \mathrm{d}E\, \Psi(E)\, \mathrm{e}^{-\mu_{\mathrm{Al}}(E)\, t_{\mathrm{HVL}}} \left(\frac{\mu_{\mathrm{en}}}{\rho}\right)_{\mathrm{air}} (E). \tag{5.105}$$

A larger HVL corresponds to a harder beam with a higher average photon energy and therefore higher beam quality. The HVL incorporates the combined effects of tube potential, inherent filtration, and added filtration on the spectral shape.

It should be noted, however, that in modern x-ray imaging a more penetrating spectrum does not necessarily yield better image quality, for example, in mammography, an overly penetrating beam reduces image contrast and contrast-to-noise ratio for a given exposure, even though the beam quality as measured by HVL is higher.

5.5.4.6 X-Ray Spectrum Models

The x-ray photon fluence spectrum and energy fluence spectrum strongly influence radiation dose efficiency and image quality. Consequently, accurate knowledge of the x-ray spectrum is valuable in diagnostic imaging. However, direct measurement of the spectrum is often impractical without sophisticated energy-discriminating spectrometers. As shown in Eq. (5.100), commonly available metrics such as exposure are determined by integrals over the spectrum and therefore do not uniquely specify the spectral shape.

Rather than directly measuring the spectrum, validated spectral models are commonly used. Once calibrated, these models can be adapted to specific x-ray beams by adjusting a small number of free parameters using measurable quantities such as exposure, half-value layer, tube potential (V), and known filtration conditions.

One of the earliest spectrum models is the analytic Kramers' model introduced in Sect. 5.5.2.2. While instructive, it is highly idealized and not representative of practical x-ray spectra, for example, the Kramers' model in Eq. (5.89) does not include self-filtration in the anode or other inherent filtrations in the tube assembly. As a result, it serves mainly as a pedagogical model rather than a realistic description of clinical beams.

To improve realism, T. R. Fewell developed in the 1970s a semiempirical analytic model for diagnostic x-ray tube spectra that incorporates radiation physics and experimental fitting to represent bremsstrahlung production, characteristic radiation, anode self-absorption, inherent filtration, and added filtration [6]. The **Fewell model** may be viewed as a corrected Kramers' spectrum fitted to measured tubes. For the bremsstrahlung component, the spectrum can be written as

$$\Psi(E) = CZ(E_0 - E)\, f(E, E_0, \theta)\, \mathrm{e}^{-\mu(E)t}, \tag{5.106}$$

where the empirical correction factor $f(E, E_0, \theta)$ accounts for production efficiency, angular effects, and target self-absorption and the term $\mathrm{e}^{-\mu(E)t}$ accounts for attenuation by added filtration. The final Fewell spectrum is given by adding discrete characteristic lines to the bremsstrahlung spectrum in Eq. (5.106).

In the 1990s, Boone and Seibert developed another approach to model x-ray spectra produced by diagnostic x-ray tubes in the range of 20–140 kV with W, Mo, or Rh anodes [3, 4]. For tungsten anodes, the approach is referred to as the **tungsten anode spectral model using interpolating polynomials (TASMIP)**. Using either Monte Carlo simulations or the Fewell model as a starting point, the authors generated x-ray spectra at discrete tube potentials V (e.g., 30, 40, ..., 150 kV). For a given photon energy bin centered at E, the photon fluence is modeled as a polynomial function of V:

$$\Phi(E, V) \approx \sum_{j=0}^{n} a_j(E)\, V^j \tag{5.107}$$

That is, for each energy bin, the method fits a polynomial in V to the fluence values obtained at discrete tube potentials, yielding a set of coefficients $\{a_j(E)\}$. The spectrum at an arbitrary tube potential is then obtained by evaluating these calibrated polynomials. One example of TASMIP spectrum is provided in Fig. 5.16.

Compared with the Fewell and Kramers' models, TASMIP provides a more realistic representation of clinical x-ray spectra because the polynomial coefficients were calibrated via a data-driven approach. The output of TASMIP is a tabulated spectrum with a typical energy interval of 1 keV, rather than a closed-form analytic expression as in the Fewell or Kramers' models.

When adapting the TASMIP model to an x-ray tube that differs from the system used by Boone and Seibert, the spectrum is adjusted by modifying the assumed added filtration so that the HVL predicted by the model matches the measured HVL of the actual beam. For Mo or Rh anode tubes, the corresponding models must be used, namely, the **molybdenum anode spectral model using interpolating polynomials (MASMIP)** and the **Rhodium anode spectral model using interpolating polynomials (RASMIP)**.

Knowledge of an x-ray beam's photon fluence spectrum $\Phi(E)$ or energy fluence spectrum $\Psi(E)$ has numerous applications, for example, it can be used in conjunction with a measured exposure X to estimate the number of photons incident on a given area of the detector. Let X denote the reading of an exposure measurement performed at a location within an x-ray beam. For the same exposure time and at the same location, the expected number of x-ray photons per unit area is

$$\bar{N} = \int \mathrm{d}E \, \frac{\Psi(E)}{E}. \tag{5.108}$$

Let κ denote the ratio between $\bar{N}$ and X. According to Eqs. (5.100) and (5.108),

$$q \equiv \frac{\bar{N}}{X} = \epsilon_{\mathrm{air}} \frac{\int \mathrm{d}E \, \Psi(E) \, E^{-1}}{\int \mathrm{d}E \, \Psi(E) \left(\frac{\mu_{\mathrm{en}}}{\rho}\right)_{\mathrm{air}}(E)}. \tag{5.109}$$

Equation (5.109) shows that q is determined solely by the *shape* of $\Psi(E)$. If the spectral shape is known, q can be calculated, allowing the expected photon fluence to be estimated from an exposure measurement as

$$\bar{N} = q \, X. \tag{5.110}$$

As shown in Fig. 5.19, both $\left(\frac{\mu_{\mathrm{en}}}{\rho}\right)_{\mathrm{air}}(E)$ and E^{-1} decrease with increasing photon energy E. Over the range $E \in [50, 100]$ keV, their ratio is relatively constant. This explains why q exhibits only weak variation across beam filtration conditions at a given tube potential [3].

5.6 Heat Management in X-Ray Tubes

As shown in Fig. 5.9, when the tube potential reaches 100 kV or higher, the tube current is no longer space-charge-limited and is primarily determined by the cathode temperature. Consequently, the **anode input power** of an x-ray tube—defined as the product of the tube potential and the tube current—can be increased relatively easily

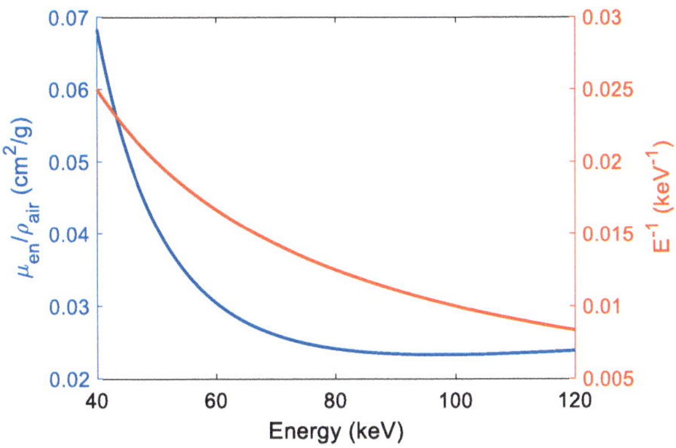

Fig. 5.19 Plots of the air mass energy-absorption coefficient $[\left(\frac{\mu_{en}}{\rho}\right)_{air}(E)]$ and E^{-1} versus photon energy E

by using a more powerful electron emitter. For this reason, power itself is generally not the primary factor distinguishing low-end from high-end medical x-ray tubes.

Instead, the most delicate and costly component of an x-ray tube is its heat storage and dissipation system. As a result, **heat units (HU)** or **heat storage capacity** are more meaningful performance metrics, and the cost of an x-ray tube is typically positively correlated with this capability.

To understand the heat transfer mechanisms in x-ray tubes, it is helpful to consider an analogy with heat dissipation in a car engine. Combustion within the cylinders produces extremely high temperatures, and the heat is first transferred to the surrounding metal components by **thermal conduction**. The heat then conducts to liquid coolant surrounding the engine block.

If the heat were not removed rapidly, the engine components could overheat or melt. Therefore, the heat removal rate must exceed the heat generation rate even when the engine operates at full throttle. This requirement is met by circulating the coolant from the engine to a radiator located behind the vehicle's front grille. Air flowing through the radiator removes heat from the coolant, representing a typical **convention process**. In addition, you can feel heat radiating from the hood after driving, even without direct contact, illustrating **radiative heat transfer**.

5.6.1 Conduction

To analyze thermal conduction, we introduce the concept of **thermal conductivity** κ, which quantifies a material's ability to transport heat. Materials with larger thermal conductivity transfer heat more efficiently than those with smaller values. The thermal energy crossing a unit area per unit time (i.e., the heat flux $\mathbf{q}$) is related to the spatial temperature gradient ∇T through Fourier's law:

$$\mathbf{q} = -\kappa \nabla T \tag{5.111}$$

The SI unit of κ is $\mathrm{W\,m^{-1}\,K^{-1}}$. The negative sign indicates that heat flows from regions of higher temperature to lower temperature, i.e., down the temperature gradient.

Consider a material occupying a volume $\mathcal{V}$ with mass density ρ and specific heat capacity c. By conservation of energy, the rate of change of thermal energy stored in the volume equals the rate at which an internal heat source deposits energy minus the net rate at which heat flows out through the boundary surface $\mathcal{S}$:

$$\int_{\mathcal{V}} \frac{\partial(\rho c T)}{\partial t}\,\mathrm{d}v = \int_{\mathcal{V}} p_i\,\mathrm{d}v - \oint_{\mathcal{S}} \mathbf{q}\cdot\mathbf{da} \tag{5.112}$$

Here, p_i denotes the rate at which the internal heat source deposits thermal energy per unit volume within $\mathcal{V}$.

Using the divergence theorem, the surface integral can be converted to a volume integral:

$$\int_{\mathcal{V}} \frac{\partial(\rho c T)}{\partial t}\,\mathrm{d}v = \int_{\mathcal{V}} p_i\,\mathrm{d}v - \int_{\mathcal{V}} (\nabla\cdot\mathbf{q})\,\mathrm{d}v \tag{5.113}$$

Because this relation holds for any arbitrary volume, the integrands must be equal:

$$\begin{aligned}\rho c\frac{\partial T}{\partial t} &= p_i - \nabla\cdot\mathbf{q}\\ &= p_i + \kappa\,\nabla\cdot(\nabla T)\\ &= p_i + \kappa\nabla^2 T\end{aligned} \tag{5.114}$$

In summary, the governing heat transfer equation is

$$\frac{\partial T}{\partial t} = \frac{p_i}{\rho c} + \frac{\kappa}{\rho c}\nabla^2 T. \tag{5.115}$$

Let the volume of interest be the anode. We assume that anode does not have any internal heater or cooler; therefore, $q_i = 0$, and Eq. (5.115) becomes

$$\boxed{\frac{\partial T}{\partial t} - \frac{\kappa}{\rho c}\nabla^2 T = 0.} \tag{5.116}$$

In the 1940s, W. J. Oosterkamp discussed the solutions of Eq. (5.116) for the following cases.

5.6.1.1 Short X-Ray Exposure in Stationary Anodes

The relevant initial and boundary conditions are

$$\begin{cases} T(z, t=0) = T_0, \\ -\kappa \left.\dfrac{\partial T}{\partial z}\right|_{z=0} = \dfrac{P}{A}, & \text{(within the focal spot)}, \\ -\kappa \left.\dfrac{\partial T}{\partial z}\right|_{z=0} = 0, & \text{(outside the focal spot)}. \end{cases} \tag{5.117}$$

Here, the z direction is along the cathode-anode axis, P denotes the rate of thermal energy deposition by the electron beam on the anode, and A is the effective focal spot area, and P/A represents the incident surface heat flux.

Given these conditions and treating the anode as a semi-infinite solid for short-exposure times, Oosterkamp showed that the solution of the heat equation in Eq. (5.116), evaluated at the surface $z = 0$ and at the center of the focal spot, is

$$T(z=0, t) - T_0 = \frac{2P}{A}\sqrt{\frac{t}{\pi\kappa\rho c}}. \tag{5.118}$$

Equation (5.118) indicates that, for short x-ray exposures, the surface temperature rise is proportional to the square root of the exposure time. Moreover, for a given anode material, the maximum allowable temperature $T_{\max}$, which is typically limited by the material's melting point, constrains the instantaneous tube power. Imposing $T(z = 0, t) \leq T_{\max}$ yields the short-exposure tube power limit [16]

$$P_{\max} = \frac{1}{2}(T_{\max} - T_0)\, A\sqrt{\frac{\pi\kappa\rho c}{t}}. \tag{5.119}$$

In summary, the Oosterkamp stationary tube model in Eq. (5.119) shows the following:

- $P_{\max}$ increases with the target material melting point.
- $P_{\max} \propto \sqrt{\rho c}$, the heat capacity per unit volume of the target material.
- $P_{\max} \propto \sqrt{\kappa}$, the thermal conductivity of the target.
- $P_{\max}$ is inversely proportional to the square root of the exposure time.
- $P_{\max} \propto \sqrt{\rho c}$, the focal spot area.

As indicated by Table 5.4, tungsten (W) has a very high melting point and high density, which are key advantages for use as an anode material. However, its thermal conductivity and heat capacity are relatively low. Consequently, practical x-ray tube anodes are implemented as **composite structures**: a thin tungsten or tungsten-rhenium alloy layer provides efficient x-ray production, while an underlying Mo, Cu, and/or graphite backing body is used to enhance heat spreading or storage. This composite design effectively decouples x-ray generation from thermal management, enabling substantially higher tube loading than would be possible with bulk tungsten alone.

Table 5.4 Thermal and physical properties of several metals at room temperature. κ thermal conductivity, c specific heat, ρ mass density

Material	Melting point (K)	κ ($\mathrm{W\,m^{-1}\,K^{-1}}$)	c ($\mathrm{J\,kg^{-1}\,K^{-1}}$)	ρ ($\mathrm{g\,cm^{-3}}$)
W	3695	174	134	19.30
Mo	2896	138	251	10.22
Rh	2237	150	243	12.41
Cu	1357	401	385	8.96
Ag	1235	429	235	10.50

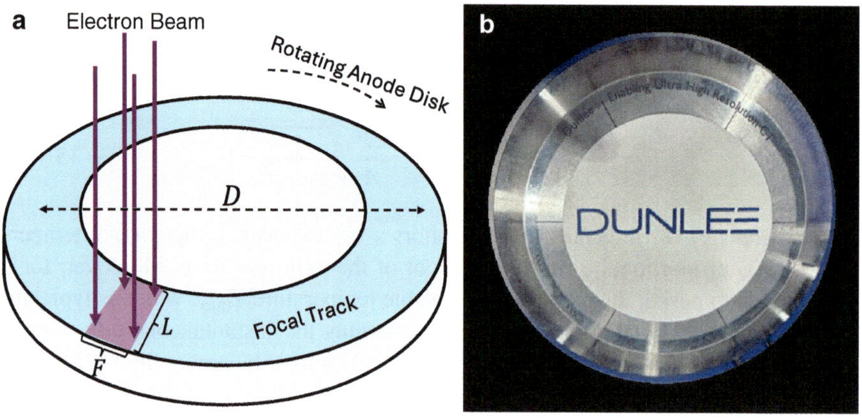

Fig. 5.20 (**a**) Illustration of the focal track and focal spot area on a rotating anode disk. (**b**) A Dunlee™ anode disk designed and manufactured by Koninklijke Philips N.V. for ultrahigh-resolution CT applications

5.6.1.2 Short X-Ray Exposure in Rotating Anodes

Figure 5.20 illustrates a rotating-anode disk in which the focal track has diameter D and rotates at ν revolutions per second. The physical dimensions of the focal spot are $F \times L$. For a given point on the focal track, the exposure time during one revolution is

$$t = \frac{F}{\pi D} \cdot \frac{1}{\nu} = \frac{F}{\pi D \nu}. \tag{5.120}$$

If t is sufficiently short, the heat generated during this interval can be assumed to diffuse away before the same point returns to the focal spot, so that the local temperature relaxes back to T_0 after one revolution. Under this assumption, the rotating anode problem can be treated using the stationary anode model in Eq. (5.119), giving

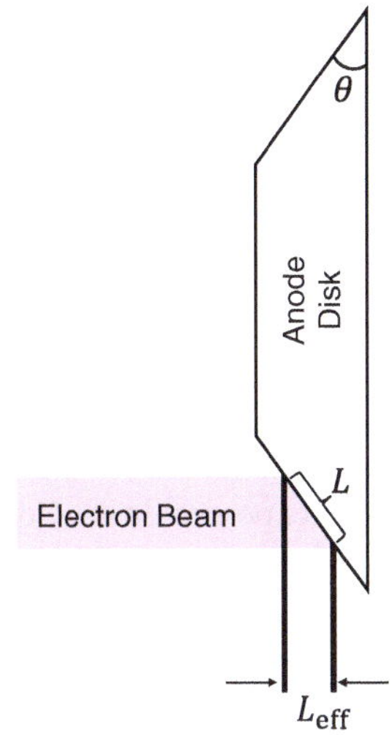

Fig. 5.21 Geometric relationship among the physical focal spot size (L), effective focal spot size (L_{eff}), and the anode angle (θ)

$$P_{\max} = \frac{1}{2}\,(T_{\max} - T_0)\,(FL)\sqrt{\frac{\pi\kappa\rho c}{\frac{F}{\pi D\nu}}}$$
$$= \frac{\pi}{2}\,(T_{\max} - T_0)\,L\left(F\kappa\rho c D\nu\right)^{1/2}. \tag{5.121}$$

As shown in Fig. 5.21, the projected (effective) focal length is related to the anode angle θ by

$$L_{\text{eff}} = L\sin\theta, \tag{5.122}$$

while the projected width remains F. In imaging system design, it is common to choose the geometry such that $L_{\text{eff}} \approx F$, producing approximately isotropic focal spot blur. Under this approximation,

$$F \approx L\sin\theta, \quad \text{or} \quad L \approx \frac{F}{\sin\theta}. \tag{5.123}$$

Combining Eqs. (5.121) and (5.123) yields [17]

$$\boxed{P_{\max} \approx \frac{\pi}{2\sin\theta}\,(T_{\max} - T_0)\,(\kappa\rho c D\nu)^{1/2}\,F^{3/2}.} \tag{5.124}$$

Equation (5.124) has several important implications:

- $P_{\max} \propto \sqrt{D}$: Increasing the anode track diameter allows higher instantaneous power. It also provides a larger heat sink, reducing temperature rise.
- $P_{\max} \propto \sqrt{\nu}$: Higher rotation speed increases the allowable instantaneous tube power.

- $P_{\max} \propto F^{3/2}$: A larger focal spot permits higher input power and tube current.
- $P_{\max} \propto 1/\sin\theta$: Smaller anode angles increase allowable power. However, as shown earlier in Fig. 5.13, smaller θ also increase the heel effect and reduce usable field of view. Thus, the anode angle must balance power handling and imaging coverage. Applications such as cardiac angiography favor smaller angles for high instantaneous power and limited field of view, whereas chest radiography favors larger angles for wide field coverage at lower instantaneous power.

It should be noted the model in Eq. (5.124) represents an idealized upper bound. In practice, heat is not completely removed from a given focal track location between successive passes, especially at high repetition rates, large duty cycles, or long exposure times. Consequently, residual heat accumulates in the anode, and the local temperature does not fully return to T_0 before the next irradiation. The present treatment therefore applies most accurately to short, isolated exposures and provides an optimistic estimate of the maximum allowable instantaneous tube power.

5.6.1.3 Continuous X-Ray Exposure

For the short-exposure problem discussed in Sects. 5.6.1.1–5.6.1.2, an important assumption is that the heat generated within the focal spot can be sufficiently "digested" by the bulk of the anode.

In some imaging procedures, such as CT, x-rays must be produced continuously by the x-ray tube for periods ranging from tens of seconds to several minutes. Under these continuous-load conditions, the short-exposure tube loading models may no longer be applicable, because thermal conduction within the anode will eventually saturate the heat storage capacity of the anode disk [18]. To maintain the temperature of the anode disk and focal track below their melting points, heat must ultimately be removed from the tube assembly. In practice, convection through the cooling fluid surrounding the tube housing is typically used to carry heat away.

Heat can also be transferred away from the anode through the stem and rotor assembly, which, unlike the anode disk, can exchange heat with surrounding components that are thermally coupled to the cooling fluid. In conventional tube designs, this thermal pathway is intentionally minimized by fabricating the stem that connects the anode disk to the rotor from a material with low thermal conductivity. This design limits heat transfer to the rotor, because excessive heating of the rotor can damage the bearing balls between the rotor and stator, potentially leading to tube failure.

However, in x-ray tubes that employ liquid-metal bearings between the rotor and stator, the risk of bearing damage from elevated temperatures is significantly reduced. This design allows greater heat transfer through the rotor assembly, thereby providing an additional thermal pathway for removing heat from the anode disk.

New learners of x-ray tubes may have a misconception that CT tubes have a higher **peak input power** than radiography or angiography tubes. This misconception likely arises from the observation that CT tubes are typically **larger and**

heavier. In everyday experience, the peak power of many electrical devices (e.g., air conditioners, motors, and engines) often increases with device size or weight.

Recall that the input power of an x-ray tube is determined by the product of the tube current and the electron acceleration voltage. In CT imaging, the instantaneous power is not necessarily higher than that used in other imaging systems. In fact, some angiographic imaging sequences in interventional x-ray systems may operate at even higher instantaneous power. Each of these high-power exposures, however, occurs only for a **very short duration**.

The key requirement for CT is the ability to **produce x-rays continuously for extended periods** up to several minutes. Consequently, CT tubes are designed to provide the highest heat storage capacity and the most efficient heat dissipation among medical x-ray tubes, which typically makes them bulkier. For example, heat storage capacity is often increased by enlarging the anode disk diameter and mass, while the cooling rate is improved through more efficient thermal conduction and convection within the tube assembly and housing—features that typically increase the overall tube size and weight.

5.6.2 Thermal Radiation

As shown in Figs. 5.15 and 5.22, the anode—where most of the heat is initially generated—is largely isolated from the surrounding environment by a highly

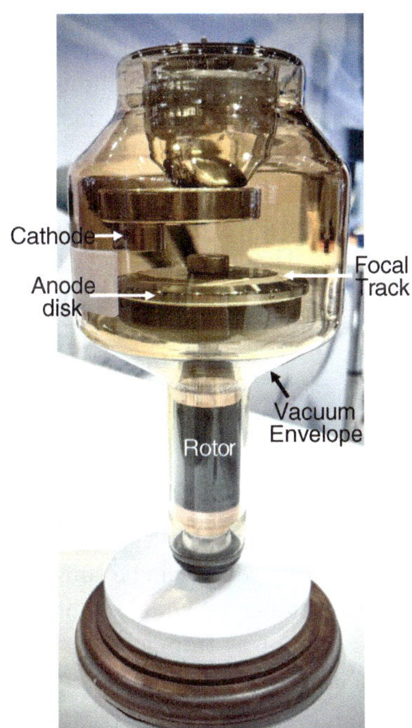

Fig. 5.22 Photo of an RTC700 rotating anode x-ray tube insert manufactured by I.A.E. Industria Applicazioni Elettroniche S.P.A. The anode disk has a diameter of 10.2 cm and an anode angle of 13°. The focal track is made of a tungsten-rhenium alloy, and the anode heat storage capacity is 800 kHU. The anode rotating speed ranges from 3000 to 10,000 rpm

evacuated glass or ceramic envelope (i.e., the tube insert). Because the envelope contains essentially no gas molecules, heat transfer by conduction through gas or convection cannot occur. As a result, **thermal radiation becomes the dominant mechanism for transferring heat away from the hot anode surface** within the tube insert.

Unlike thermal conduction, thermal radiation does not require the presence of a material medium. The radiative heat transfer power emitted by a hot surface depends on the fourth power of its absolute temperature T [12],

$$P = \alpha A T^4, \tag{5.125}$$

where A is the surface area of the radiator and α is a material-dependent factor.

Because both the anode and the surrounding tube envelope have finite temperatures, each surface emits thermal radiation. The net radiative heat transfer from the anode to the envelope can therefore be expressed as

$$P = \alpha_a A_a T_a^4 - \alpha_e A_e T_e^4, \tag{5.126}$$

where the subscripts a and e denote the *anode* and the *envelope*, respectively.

Due to the strong T^4 dependence, radiation-based heat transfer becomes very effective when the anode temperature reaches high values, typically in the range of 1000–2500° C during x-ray production. At these temperatures, the electromagnetic radiation emitted by the hot anode lies predominantly in the infrared region of the spectrum. This radiative energy travels across the vacuum and is absorbed by the inner wall of the tube envelope.

Equation (5.125) also shows that increasing the surface area of the anode increases the radiative heat transfer rate. This provides another advantage of using a larger anode disk, which increases the radiating surface area available for heat removal.

Equation (5.126) further indicates that the net radiative heat transfer from the anode depends on the temperature difference between the anode and the tube envelope. Therefore, the heat deposited in the envelope wall must be efficiently removed to keep T_e relatively low. This is typically achieved through thermal conduction and convection in the cooling fluid surrounding the vacuum envelope, as discussed in the following subsection.

5.6.3 Convection

Thermal convection can be considered a special form of heat transfer that combines thermal conduction and fluid motion. It occurs when a **moving fluid (liquid or gas)** carries heat away from a hot surface. Initially, heat is conducted from the surface to a thin layer of fluid immediately adjacent to the hot surface. This heated fluid layer

Fig. 5.23 Photo of a MCS-6074 CT tube (Varex Imaging Corp.), with the oil pump and heat exchanger in close proximity of the tube housing

is then swept away and mixed with the cooler bulk fluid downstream, enhancing the overall heat transfer rate.

In the case of x-ray tubes, the fluid is typically a dielectric insulating oil, which is not only a heat transfer medium but also provides electrical insulation for the high-voltage components inside the tube. The tube housing has an inlet and outlet for the oil, and a pump drives the oil to circulate between an external cooler called the heat exchanger and the tube. Once a thin layer of oil near the tube insert wall absorbs heat from the wall, it is replaced by cooler oil from the upstream. As shown by the Fourier's law for thermal conduction (Eq. (5.111)), a large spatial gradient in temperature enhances heat flow from the wall to the oil.

Downstream, the heated oil flow downstream and get cooled down, especially when reaching the external heat exchanger, where fans are typically used to introduce air or water flow as a secondary layer of convention. Because the air flow rate and the air-oil contact surface area can be much larger than those at the oil-insert interface, the convective heat flow ensures that heat can be sufficiently removed and the insert wall is maintained at low temperatures.

Convective cooling is particularly important for CT tubes, which often need to produce x-rays continuously over extended periods. To enhance convective heat removal, several design considerations are critical: First, the **oil volume must be sufficiently large** so that, when the heated layer of oil near the tube insert mixes with the bulk oil downstream, the overall temperature rise of the oil remains small. Second, the **oil pump and external heat exchangers must be powerful** enough to remove heat rapidly and maintain the oil at a low temperature. Third, the **heat exchanger should be located close to the x-ray tube** (Fig. 5.23), within the rotating CT gantry, to minimize cooldown delays between consecutive scans.

To better understand the importance of proximity, consider an extreme case where the oil immediately downstream of the tube insert is sufficiently cooled. This maximizes the temperature gradient between the insert wall and the surrounding fluid, enhancing heat transfer and helping prevent overheating of both the tube insert and the external housing.

To improve the anode cooling rate of CT tubes and enable advanced CT applications such as dual-source CT, a so-called **Straton** tube (Siemens Healthineers) was introduced in the early 2000s [21]. As shown in Fig. 5.24, the anode disk is attached

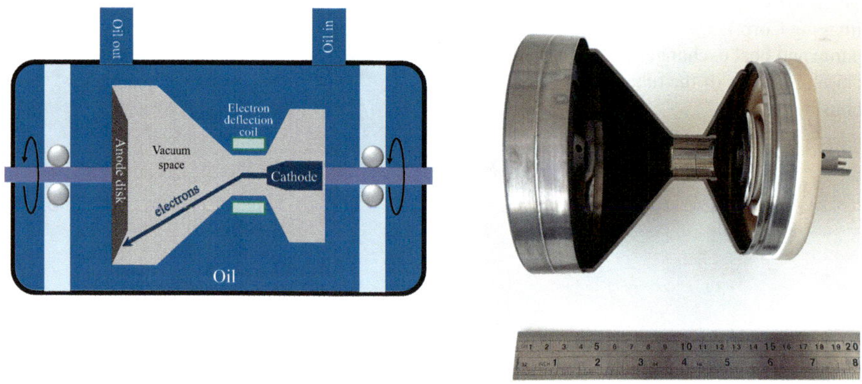

Fig. 5.24 (Left) Schematic illustration of the Straton x-ray tube. (Right) Photo of the Straton tube insert, with part of the insert wall removed to reveal the internal structure. The external tube housing is not shown in this photo

directly to the wall of the vacuum envelope, and **the entire envelope rotates** during operation. This differs from conventional rotating anode tubes, in which the anode rotates inside a stationary vacuum envelope.

Because the anode disk is in direct contact with the metallic wall of the vacuum envelope, heat generated in the anode can be conducted directly to the envelope wall and then removed by the surrounding cooling oil through convection. Since heat transfer by **thermal conduction** is significantly more efficient than **thermal radiation**, this design allows the mass of the anode body to be substantially reduced, making the overall x-ray tube more compact.

The Straton tube achieves a cooling rate of approximately 5 MHM/min, which is several times higher than that of conventional CT tubes, while also reducing the overall tube weight (Table 5.5).

5.7 Summary of Chap. 5

1. The basic building blocks of an x-ray tube are an electron emitter, an electron accelerator, and a target where electrons rapidly decelerate and emit x-rays.
2. Most electron emitters are based on thermionic emission, which is the liberation of electrons from a heated material. The emission current density is related to the temperature and the work function of the material by the Richardson equation:

$$J = A\,T^2 \exp\left(-\frac{W}{k_B T}\right)$$

Table 5.5 Typical parameters of x-ray tubes used in different imaging modalities. Values are representative ranges and may vary by manufacturer and model. *MHU* mega heat units. 1 MHU= 0.71×10^6 J

Model application	Varex A-195 radiography	Varex G-1582BI angiography	Varex MCS-6074 CT	Siemens Straton CT
Nominal focal spot size (mm)	0.6/1.25	0.3/0.6/1.0	0.9/1.2	0.6/0.8
Anode diameter (cm)	10.2	13.3	20	9.6
Anode angle (°)	12	10	7	7
Maximum tube voltage (kV)	150	125	140	150
Nominal anode input power (kW)	40/100	18/60/112	24/53.2	120
Maximum anode heat load (MHU)	0.35	1.5	6.3	N/A
Maximum cooling rate (MHU/min)	0.08	0.3	0.84	5

3. Some x-ray tubes employ cold electron emitters based on the quantum tunneling effect. In this case, the emission current density depends on the surface electric field strength and the work function of the material, as described by the Fowler-Nordheim equation:

$$J = -\frac{e^3 \mathcal{E}^2}{16\pi^2 \hbar W} \exp\left[-\frac{4\sqrt{2m_e}}{3\hbar e}\frac{W^{3/2}}{\mathcal{E}}\right]$$

4. When the electron emission rate exceeds the rate at which electrons leave the cathode surface, space charge builds up near the emitter. The emitted electrons can be repelled by this space charge and return to the emitter. When an external voltage V_0 is applied between the cathode and the anode, more electrons are drawn toward the anode. In this space-charge-limited regime, the current density between the anode and cathode is no longer determined by temperature but instead depends strongly on V_0 according to the Child-Langmuir law:

$$J = \left[\frac{4\epsilon_0}{9L^2}\left(\frac{2e}{m_e}\right)^{1/2}\right] V_0^{3/2}$$

5. When accelerated electrons interact with the anode material, the majority of the interactions are collisional and produce heat; only a small fraction are radiative and produce x-rays. The initial bremsstrahlung spectrum can be approximated by Kramers' model:

$$\Psi(E) = \begin{cases} K_1\, N_e\, Z & \text{thin target} \\ K_2\, N_e\, Z\,(E_0 - E) & \text{thick target} \end{cases}$$

The initial spectrum is further modified by attenuation in the anode, tube window, and other inherent or external filters. As a result, the low-energy portion of the Kramers' spectrum is significantly reduced. Models such as TASMIP have been developed to provide more realistic descriptions of the exiting x-ray spectrum.

6. Heat storage capacity is a key performance metric of x-ray tubes. Note that this metric does not directly depend on the electron emitter. Therefore, replacing a thermionic emitter with a carbon nanotube emitter does not fundamentally address the heat transfer challenge in x-ray tubes, since the vast majority of heat is generated at the anode. Among the three heat transfer mechanisms, thermal conduction largely determines the maximum peak input power of an x-ray tube, while thermal convection determines the cooling rate. Increasing the anode diameter and rotating the anode at higher speeds help increase the anode heat storage capacity. In addition, replacing radiative heat transfer with conductive heat transfer, as implemented in Straton tubes, can greatly improve the cooling rate.

References

1. The Official Site of Thomas Alva Edison: https://www.thomasedison.org/.
2. Attix FH. Introduction to radiological physics and radiation dosimetry. Berlin: Wiley-VCH; 2004.
3. Boone JM, Seibert JA. An accurate method for computer-generating tungsten anode x-ray spectra from 30 to 140 kV. Med Phys 1997; 24(11):1661–1670. https://doi.org/10.1118/1.597953.
4. Boone JM, Fewell TR, Jennings RJ. Molybdenum, rhodium, and tungsten anode spectral models using interpolating polynomials with application to mammography. Med Phys 1997; 24(12):1863–1874. https://doi.org/10.1118/1.598100.
5. Child CD. Discharge from hot CaO. Phys Rev (Series I) 1911; 32:492–511. https://doi.org/10.1103/PhysRevSeriesI.32.492.
6. Fewell TR, Shuping RE. Handbook of mammographic x-ray spectra. Washington: HEW Publication (FDA); 1978.
7. Fowler RH, Nordheim LW. Electron emission in intense electric fields. Proc R Soc A 1928; 119(781):173–181. https://doi.org/10.1098/rspa.1928.0091.
8. Guarnieri M. The age of vacuum tubes: Early devices and the rise of radio communications [historical]. IEEE Ind Electron Mag 2012; 6(1):41–43. https://doi.org/10.1109/MIE.2012.2182822.
9. Jackson JD. Classical electrodynamics, 3rd ed. Hoboken: Wiley; 1998.
10. Kramers HA. XCIII. On the theory of X-ray absorption and of the continuous X-ray spectrum. Lond Edinb Dublin Philos Mag J Sci 1923; 46(275):836–871. https://doi.org/10.1080/14786442308565244.
11. Langmuir I. The effect of space charge and residual gases on thermionic currents in high vacuum. Phys Rev 1913; 2:450–486. https://doi.org/10.1103/PhysRev.2.450.
12. Lienhard IV JH, Lienhard V JH. A heat transfer textbook, 6th ed. Cambridge MA: Phlogiston Press; 2024.
13. National Institute of Standards and Technology (NIST). Estar (stopping powers and ranges for electrons) database. Technical Report, 2017. https://physics.nist.gov/PhysRefData/Star/Text/ESTAR.html.

14. Nicholas WW. Continuous spectrum x rays from thin targets. Bureau Standards J Res 1929; 2:837–870.
15. NobelPrize.org. Nobel prize in physics 1928; 1928.
16. Oosterkamp WJ. The heat dissipation in the anode of an x-ray tube. i. introduction; loads of short duration applied to stationary anodes. Philips Res Rep 1948; 3:49–59.
17. Oosterkamp WJ. The heat dissipation in the anode of an x-ray tube. ii. loads of short duration applied to rotating anodes. Philips Res Rep 1948; 3:161–173.
18. Oosterkamp WJ. The heat dissipation in the anode of an x-ray tube. iii. continuous loads. Philips Res Rep 1948; 3:303–317.
19. Park S, Gupta AP, Yeo SJ, Jung J, Paik SH, Mativenga M, Kim SH, Shin JH, Ahn JS, Ryu J. Carbon nanotube field emitters synthesized on metal alloy substrate by PECVD for customized compact field emission devices to be used in x-ray source applications. Nanomaterials 2018; 8(6). ISSN 2079-4991. https://doi.org/10.3390/nano8060378.
20. Richardson OW. Thermionic phenomena and the laws which govern them. Nobel Lecture 1929. https://www.nobelprize.org/prizes/physics/1928/richardson/lecture/.
21. Schardt P, Deuringer J, Freudenberger J, Hell E, Knupfer W, Mattern D, Schild M. New x-ray tube performance in computed tomography by introducing the rotating envelope tube technology. Med Phys 2004; 31(9):2699–2706. https://doi.org/10.1118/1.1783552.
22. Schwabl F. Quantum mechanics, 4th ed. Berlin: Springer; 2007.

Chapter 6
Waveguide Physics in Medical Linear Accelerators (Linacs)

Contents

K. Li, *Essential Physics of Medical X-ray Sources and Detectors*,
https://doi.org/10.1007/978-3-032-27800-5_6

6.1 Introduction

A linear accelerator (Linac) is a device that accelerates charged particles such as electrons along a straight line using electromagnetic (EM) fields (Fig. 6.1). When the accelerated electrons interact with a target, their rapid deceleration produces x-rays. As shown in Fig. 5.1 in Chap. 5, a Linac shares some similar building blocks with an x-ray tube: an electron emitter, an electron accelerator, and a target where electrons interact to emit x-ray radiation.

However, there is a major difference between x-ray tubes and Linacs: **the kinetic energy of the electrons**. For medical Linacs, the desired x-ray energies are on the **MeV** scale ("megavoltage x-rays"). To accelerate electrons to this energy level using a DC voltage alone would require an ultrahigh-voltage power supply on the order of one million volts. Even if a constant MV voltage source were available, applying such a high electric potential between two electrodes would often lead to severe electrical arcing. Therefore, electron acceleration in a Linac does not rely solely on a DC voltage potential.

As shown in Fig. 5.1b, in a Linac, electrons emitted from the cathode first gain some initial kinetic energy from a DC voltage (e.g., 40 kV) applied between the cathode and an anode. Unlike the anode in an x-ray tube, the anode in a Linac contains a circular aperture through which the accelerated electrons pass to enter a microwave acceleration waveguide.

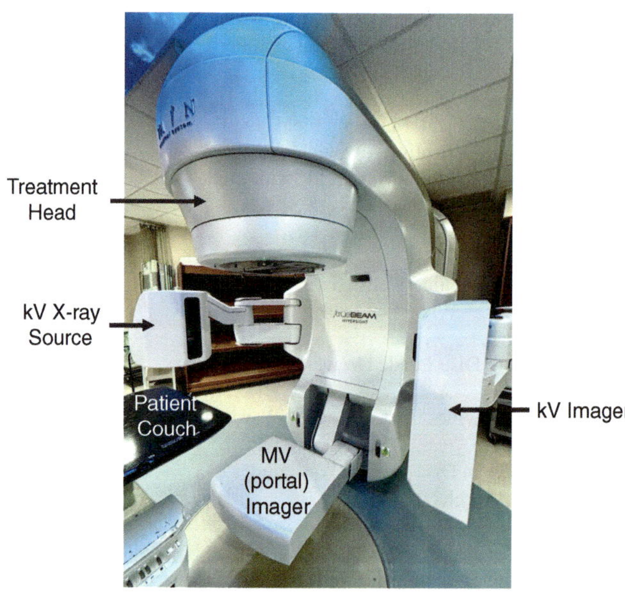

Fig. 6.1 Photograph of a Varian TrueBeam HyperSight medical Linac

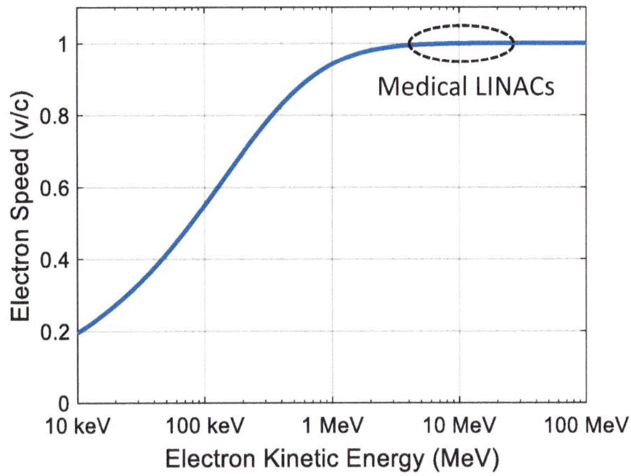

Fig. 6.2 Relationship between electron speed and kinetic energy. In medical Linacs, the final speeds of the accelerated electrons are close to the speed of light in vacuum, c

To calculate the electron speed (v) at the anode, we can use the relativistic energy conservation law,

$$\frac{m_e c^2}{\sqrt{1 - v^2/c^2}} = m_e c^2 + eV_0, \tag{6.1}$$

where $V_0 = 40\,\text{kV}$ is the DC acceleration voltage and $m_e c^2 = 511\,\text{keV}$ is the electron rest mass energy. Based on Eq. (6.1), the electron speed v is

$$1 - \frac{v^2}{c^2} = \frac{(m_e c^2)^2}{(m_e c^2 + eV_0)^2}$$
$$\Rightarrow \quad v = c\sqrt{1 - \frac{(m_e c^2)^2}{(m_e c^2 + eV_0)^2}} = c\sqrt{1 - \frac{511^2}{(511 + 40)^2}} = 0.4c. \tag{6.2}$$

For a 6-MeV Linac, according to Eq. (6.2), the final electron speed before reaching target is

$$v = c\sqrt{1 - \frac{(m_e c^2)^2}{(m_e c^2 + eV_0)^2}} = c\sqrt{1 - \frac{511^2}{(511 + 6000)^2}} = 0.997c, \tag{6.3}$$

which is very close to the speed of light. The electron speeds corresponding to other kinetic energies commonly used in radiotherapy are plotted in Fig. 6.2. As shown, these speeds are all close to c. **The primary objective of this chapter is to explain how electrons can be accelerated to relativistic speeds using a microwave accelerator.**

The term **microwave** refers to EM waves in the frequency range of approximately 300 MHz–300 GHz. For example, household microwave ovens typically operate at about 2.45 GHz ($\lambda \approx 12$ cm), while the microwave frequency in medical Linacs is typically 2.856 GHz ($\lambda \approx 10$ cm).

The frequency range of radio-frequency (RF) waves spans roughly 3 kHz–300 GHz. For example AM (amplitude modulation) radio broadcasting in the United States covers frequencies from 540 kHz to 1700 kHz, while FM (frequency modulation) broadcasting ranges from 88.0 MHz to 108.0 MHz. In medical MRI systems, RF signals typically range from about 1 MHz to 300 MHz. **Compared with microwaves, the RF spectrum extends to much lower frequencies; however, it also includes the microwave band.** Consequently, the terms "RF accelerators" and "microwave accelerators" are often used interchangeably in the literature [4, 6, 8, 12]. In this chapter, the term "microwave" is used to refer specifically to the EM waves employed in medical Linacs for electron acceleration.

In free space, microwaves propagate at a speed (c) that is essentially the same as the speed of accelerated electrons in a Linac. One might therefore naively imagine that a microwave traveling in free space could accelerate electrons in a manner analogous to ocean waves pushing surfers. Unfortunately, there is a fundamental difficulty with this analogy. As reviewed in Sect. 6.3, for EM waves in free space, the electric field (**E**) and magnetic field (**B**) are perpendicular to each other and both lie in the transverse plane, perpendicular to the direction of wave propagation. For electrons traveling along a direction collinear with the wave, the electric field has no longitudinal component, and thus the electrons experience no accelerating force.

There are ways to overcome this difficulty. One approach is to introduce an angle between the incident wave and the electron beam and use conducting "mirrors" to reflect and redirect the wave toward the electron beam. In practice, however, **the electron accelerator in a Linac is essentially a carefully designed waveguide** that redirects microwaves so that a strong electric field parallel to the electron beam can be established and enhanced through resonant processes. The central objective of this chapter is to explain how this can be achieved. The explanation inevitably involves some electrodynamics and mathematics; however, we will use a simplified toy model of microwave waveguides so that the focus remains on the essential physics rather than on complicated mathematical details.

6.2 A Brief Overview of Linac Structure and Components

While the accelerating waveguide can be regarded as the core of a Linac, it cannot operate in isolation. Its performance depends critically on a range of supporting subsystems, including the microwave source, vacuum system, cooling system, target, collimation, as well as other software and hardware systems (Fig. 6.3). Each of these components is a substantial subject in its own right—particularly the microwave source, which warrants extensive treatment.

This chapter does not aim to provide a comprehensive overview of the entire Linac system, which is available in References [2, 4, 6, 9]. Instead, it focuses on the physics of the accelerating waveguide, often considered the most conceptually

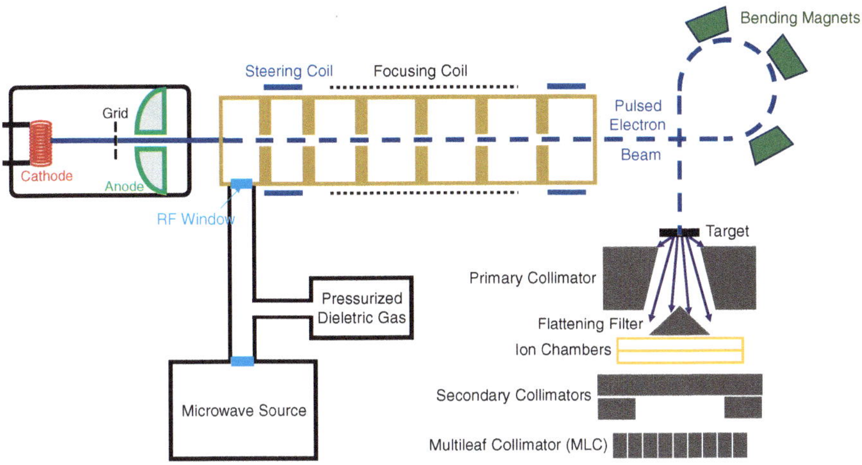

Fig. 6.3 Schematic illustration of Linac hardware components. Not shown: pulsed modulator, circulator, water cooling system, treatment couch, onboard imaging systems, and the power supply and control electronics for the cathode and microwave source

challenging aspect for newcomers to Linac technology. To place the waveguide in context, however, a very brief overview of the major hardware components of a Linac is provided in this section.

6.2.1 *Electron Gun*

The electron gun in a medical Linac operates based on thermionic emission, similar to conventional x-ray tubes. A DC voltage (typically around 40 kV) is applied between the cathode and the anode to provide the initial acceleration of emitted electrons. Unlike in x-ray tubes, however, the anode in a Linac does not produce x-rays. Instead, it functions as an electrode with a central aperture (iris) that allows the electron beam to pass into the microwave waveguide for further acceleration.

In a Linac, microwaves produced by their source (klystron or magnetron) are not continuous; instead, they are generated in short, high-power pulses with a repetition rate typically ranging from 100 to 400 pulses per second. The electron gun is synchronized by pulsed modulator with these RF pulses so that electrons are injected only when the accelerating electric field is present. To achieve this timing control, many Linac electron guns incorporate a grid between the cathode and anode to gate electron emission. This grid ensures that electrons are released only during the RF pulse, when acceleration is effective. An electron gun equipped with such a control grid is referred to as a **triode electron gun** (Fig. 6.4).

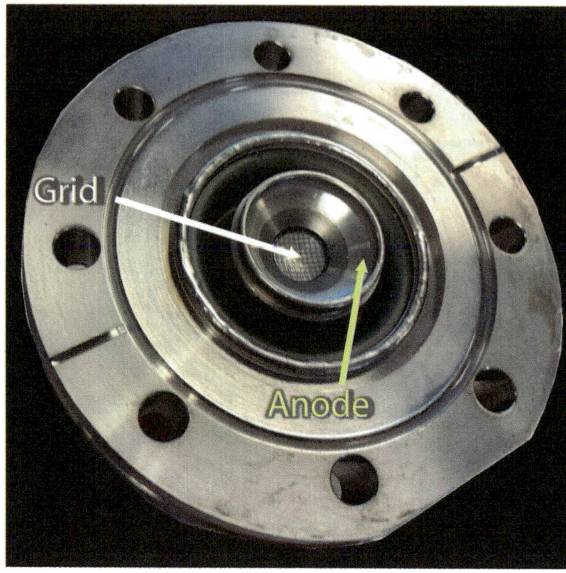

Fig. 6.4 Photograph of the grid and anode assembly from a triode electron gun. (Image courtesy of Dr. Jingwei Duan)

6.2.2 Microwave Production and Waveguide Systems

Each linac is equipped with its own RF power system, using either a magnetron or a klystron amplifier driven by a low-power RF source [3, 7], which generates the microwaves required for electron acceleration.

The 2-5 MW, 3 GHz S-band microwaves are delivered to the accelerating structure through a transmission waveguide, typically with a rectangular cross section. Because no electron beam is present in this section, the transmission waveguide does not need to be evacuated. Instead, it is filled with a pressurized insulating gas (typically SF_6) to increase dielectric strength and prevent arcing.

The **transmission waveguide** is coupled to the accelerating waveguide through an **RF window**. This component allows the two regions to be electromagnetically connected while remaining physically isolated, since the transmission line contains pressurized gas whereas the accelerating structure must be maintained under high vacuum. The RF window is typically made of a high-strength dielectric material and shaped to provide impedance matching between the two waveguides, thereby minimizing reflection and power loss. The dielectric is brazed to metal waveguide flanges, forming a vacuum-tight seal on one side and a pressure boundary on the other.

The **accelerating waveguide** is the primary focus of this chapter. It is designed to support the required time-varying electric field pattern for efficient electron acceleration. The structure is typically fabricated from a high-conductivity copper. Sections 6.4–6.6 provide a detailed discussion of the underlying physical principles of the accelerating waveguide.

Electron-focusing and steering coils are placed around the waveguide to confine the beam transversely and maintain alignment along the central axis. Because the waveguide walls and internal structures have finite conductivity, microwave-induced wall currents produce Joule heating. To ensure stable operation, a water cooling system is integrated around the waveguide to remove heat and maintain a controlled temperature.

6.2.3 Beam Bending Magnet

This is achieved using an "achromatic" bending magnet system, which is designed to cancel out energy-dependent dispersion. Such systems are typically implemented with a 270° bend and carefully shaped, nonuniform magnetic fields generated by contoured pole pieces [6]. Within this arrangement, electrons of different energies naturally follow slightly different trajectories: lower-energy electrons curve more sharply, while higher-energy electrons follow paths with larger radii. The spatial variation of the magnetic field is engineered to compensate for these differences, so that electrons of all nominal energies converge at the same exit point with the same direction and focusing characteristics.

Beyond providing energy-independent beam direction, the bending system also aids in beam focusing and energy selection [6]. Electrons outside the desired energy range are not properly transported through the magnetic optics and are intercepted by apertures or collimators within the treatment head, enhancing the energy uniformity and stability of the delivered beam.

6.2.4 Target

After being bent by the bending magnets, the electrons strike a target, where their deceleration produces bremsstrahlung x-rays and generates heat. As shown in Fig. 5.10 of Chap. 5, the fraction of radiative interactions relative to collisional interactions increases with both the target atomic number (Z) and the electron energy. Even at 6 MeV, collisional interactions remain significant, and a portion of the electron energy is deposited as heat in the target. Consequently, the target is equipped with an active cooling system to maintain its temperature well below the material's melting point.

For electron energies above approximately 15 MeV, copper ($Z = 29$) is often used as the target material. As indicated in Table 5.4, copper has high thermal conductivity, which allows heat to be conducted away more efficiently. Although its atomic number is lower than tungsten, at these higher energies, the increase in bremsstrahlung yield achievable by using tungsten instead of copper is relatively modest, while copper offers superior thermal management and less x-ray attenuation.

Regardless of whether the target is made of tungsten or copper, it is a transmission target in a Linac, meaning that the forward-directed x-rays are extracted for the intended application.

For electron radiotherapy procedures, such as the treatment of keloids, the x-ray target is removed from the beam path, allowing the electron beam to be delivered to the patient without conversion to x-rays. In this mode, the beam is typically broadened using scattering foils and collimated with the treatment jaws in combination with electron applicators or custom cutouts.

6.2.5 Treatment Beam Shaping and Modulation Systems

The treatment beam is initially shaped into a cone by a **primary tungsten collimator** positioned downstream of the target. This collimator also serves to shield radiation outside the primary beam area, preventing unnecessary exposure to surrounding personnel or the patient.

In photon radiotherapy, the x-ray field typically exhibits higher intensity along the central axis because bremsstrahlung production is forward-peaked and central photons traverse less material than off-axis photons. To achieve a more uniform intensity distribution across the field, a **flattening filter** is placed downstream of the primary collimator.

In electron radiotherapy, the flattening filter is replaced by one or two **scattering foils**, which broaden the initially narrow electron beam to provide adequate field coverage.

The beam size is further defined by movable **secondary tungsten collimators (jaws)**, which set the treatment field to a square or rectangular shape (typically up to $40 \times 40\,\text{cm}^2$ measured at a standard distance (100 cm) from the source). These jaws do not control the spatial variation of intensity within the field; they only define the geometric field edges.

For advanced techniques such as intensity-modulated radiation therapy (IMRT) or volumetric modulated arc therapy (VMAT), where the beam can be shaped irregularly and intensity-modulated to deliver highly nonuniform fluence patterns, a multileaf collimator (MLC) is employed (Fig. 6.5). An MLC consists of multiple parallel tungsten leaves (e.g., 60 or more), each controlled by an individual stepper motor. The leaves move in and out of the x-ray field to modulate fluence spatially. At a constant dose rate, regions that remain open longer receive a higher dose, while regions blocked for longer periods receive a lower dose. By dynamically adjusting leaf positions during irradiation, the MLC can generate complex, patient-specific dose distributions with high precision.

Fig. 6.5 Photograph of High-Definition (HD) multileaf collimator (MLC) leaves in a Varian TrueBeam HyperSight Linac

6.2.6 Guidance and Verification Systems

The electron acceleration and x-ray production components are mounted on a rotating structure known as the gantry, which allows the photon or electron beam to be delivered from multiple angles around the patient. This rotation enables flexible beam arrangements and improves dose conformity to the target.

The gantry also incorporates several imaging and safety systems, including alignment lasers, an on-board kV x-ray source with a corresponding kV detector (**kV imager**) for patient positioning and image guidance (Fig. 6.1). In addition, collision sensors are integrated to ensure safe operation. The gantry also includes an MV imager or **portal imager**. The term "portal" originates from "treatment portal," meaning the radiation field used for therapy. The portal imager detects residual treatment beam after it has passed through the patient, enabling verification of treatment field alignment and dose.

As of this writing, both the portal and kV imagers use scintillator-based, indirect-conversion, energy-integrating detectors. Because treatment x-rays are significantly more penetrating, portal imagers typically employ thicker CsI layers or additional photon-to-electron converters to improve detection efficiency.

Within the treatment head, typically between the flattening filter (or scattering foil) and the secondary collimators, two transmission ionization chambers are installed to monitor dose rate, integrated dose, and field symmetry. They control the delivered dose independently, ensuring accurate dose delivery and terminating the beam when needed (Fig. 6.6).

Fig. 6.6 Photograph of an ion chamber from a Linac. (Image courtesy of Dr. Jingwei Duan)

6.3 A Brief Review of Electrodynamics

6.3.1 Maxwell's Equations

Maxwell's equations are fundamental to the study of electron acceleration in a Linac because they completely describe how electromagnetic fields behave inside the Linac waveguide. Maxwell's equations determine the spatial distributions of the electric field **E** and magnetic field **B** inside the guide, and they govern important quantities such as the power flow in the waveguide, as well as the phase velocity and group velocity of the propagating microwaves.

In the vacuum space within the Linac waveguide, Maxwell's equations are given by

$$\begin{cases} (1) \quad \nabla \cdot \mathbf{E} = \dfrac{\rho}{\epsilon_0} & \text{Gauss's law} \\ (2) \quad \nabla \cdot \mathbf{B} = 0 & \text{Absence of magnetic monopoles} \\ (3) \quad \nabla \times \mathbf{E} = -\dfrac{\partial \mathbf{B}}{\partial t} & \text{Faraday's law of induction} \\ (4) \quad \nabla \times \mathbf{B} = \mu_0 \mathbf{J} + \mu_0 \epsilon_0 \dfrac{\partial \mathbf{E}}{\partial t} & \text{Ampère's law with Maxwell's addition} \end{cases} \tag{6.4}$$

where ϵ_0 denotes the vacuum permittivity and μ_0 denotes the vacuum permeability. These constants are related to the speed of light in vacuum (c) by

$$c = \frac{1}{\sqrt{\epsilon_0 \mu_0}}. \tag{6.5}$$

In the vacuum region of the waveguide where there are no free charges or currents, the space charge density (ρ) and current density ($\mathbf{J}$) are zero. Maxwell's equations therefore reduce to

$$\begin{cases} (1) & \nabla \cdot \mathbf{E} = 0 \\ (2) & \nabla \cdot \mathbf{B} = 0 \\ (3) & \nabla \times \mathbf{E} = -\dfrac{\partial \mathbf{B}}{\partial t} \\ (4) & \nabla \times \mathbf{B} = \dfrac{1}{c^2}\dfrac{\partial \mathbf{E}}{\partial t}. \end{cases} \tag{6.6}$$

Taking the curl ($\nabla\times$) of both sides of the third equation in Eq. (6.6) gives

$$\begin{aligned} \nabla \times (\nabla \times \mathbf{E}) &= -\frac{\partial}{\partial t}(\nabla \times \mathbf{B}) \\ &= -\frac{1}{c^2}\frac{\partial}{\partial t}\left(\frac{\partial \mathbf{E}}{\partial t}\right) \\ &= -\frac{1}{c^2}\frac{\partial^2 \mathbf{E}}{\partial t^2}. \end{aligned} \tag{6.7}$$

For the left-hand side of Eq. (6.7), we use the vector identity known as the "curl of the curl" identity:

$$\nabla \times (\nabla \times \mathbf{E}) = \nabla(\nabla \cdot \mathbf{E}) - \nabla^2 \mathbf{E} \tag{6.8}$$

From the first equation in Eq. (6.6), we have $\nabla \cdot \mathbf{E} = 0$. Therefore, the left-hand side of Eq. (6.7) reduces to $-\nabla^2\mathbf{E}$, and Eq. (6.7) becomes

$$\nabla^2 \mathbf{E} - \frac{1}{c^2}\frac{\partial^2 \mathbf{E}}{\partial t^2} = 0. \tag{6.9}$$

Similarly, it can be shown that the magnetic field $\mathbf{B}$ in vacuum without space charge or current satisfies

$$\nabla^2 \mathbf{B} - \frac{1}{c^2}\frac{\partial^2 \mathbf{B}}{\partial t^2} = 0. \tag{6.10}$$

6.3.2 *Time-Harmonic Electromagnetic Waves*

Excitation sources of EM waves often oscillate sinusoidally at a given temporal frequency f. The resulting EM fields are referred to as **time-harmonic** electromagnetic waves.[1] For such waves, the electric and magnetic fields can be expressed in the form

$$\begin{cases} \mathbf{E}(\mathbf{x}, t) = \mathbf{E}(\mathbf{x})\mathrm{e}^{-\mathrm{i}\omega t} \\ \mathbf{B}(\mathbf{x}, t) = \mathbf{B}(\mathbf{x})\mathrm{e}^{-\mathrm{i}\omega t} \end{cases}, \tag{6.11}$$

where $\omega = 2\pi f$ is the angular frequency. The spatial functions $\mathbf{E}(\mathbf{x})$ and $\mathbf{B}(\mathbf{x})$ represent the complex amplitudes of the fields.

Substituting Eq. (6.11) into the wave equations (6.9)–(6.10) yields

$$\begin{cases} \nabla^2\mathbf{E}(\mathbf{x}) + \dfrac{\omega^2}{c^2}\mathbf{E}(\mathbf{x}) = 0 \\ \nabla^2\mathbf{B}(\mathbf{x}) + \dfrac{\omega^2}{c^2}\mathbf{B}(\mathbf{x}) = 0. \end{cases} \tag{6.12}$$

Equation (6.12) is known as the **Helmholtz equation**.

6.3.3 *Time-Harmonic Plane Waves*

One solution of the Helmholtz equation in Eq. (6.12) is

$$\begin{cases} \mathbf{E}(\mathbf{x}) = \mathbf{E}_0\,\mathrm{e}^{\mathrm{i}\mathbf{k}\cdot\mathbf{x}} \\ \mathbf{B}(\mathbf{x}) = \mathbf{B}_0\,\mathrm{e}^{\mathrm{i}\mathbf{k}\cdot\mathbf{x}} \end{cases}, \tag{6.13}$$

where the vectors $\mathbf{E}_0$ and $\mathbf{B}_0$ are constants independent of the spatial location $\mathbf{x}$. The vector $\mathbf{k}$ is the **wave vector** whose direction specifies the direction of wave propagation. According to the Helmholtz equation, the magnitude of $\mathbf{k}$ satisfies

$$k = \frac{\omega}{c}. \tag{6.14}$$

Substituting Eq. (6.13) into the time-harmonic expressions in Eq. (6.11), we obtain the full space-time dependence of the fields:

$$\begin{cases} \mathbf{E}(\mathbf{x}, t) = \mathbf{E}_0\,\mathrm{e}^{\mathrm{i}(\mathbf{k}\cdot\mathbf{x}-\omega t)} \\ \mathbf{B}(\mathbf{x}, t) = \mathbf{B}_0\,\mathrm{e}^{\mathrm{i}(\mathbf{k}\cdot\mathbf{x}-\omega t)} \end{cases} \tag{6.15}$$

[1] Even if an EM wave does not oscillate sinusoidally, it can always be decomposed into a superposition of sinusoidal components using Fourier analysis.

If the fields of an EM wave follow the form given in Eq. (6.15), the wave is called a **plane wave**. This is because surfaces of constant phase,

$$\mathbf{k} \cdot \mathbf{x} - \omega t = \text{constant},$$

are planes perpendicular to the wave vector **k**. On each such plane, the magnitudes of the electric and magnetic fields, **E** and **B**, remain constant.

If we establish a Cartesian coordinate system such that the z axis is along the direction of **k**, Eq. (6.15) becomes

$$\begin{cases} \mathbf{E}(\mathbf{x}, t) = \mathbf{E}_0 \, \mathrm{e}^{\mathrm{i}(kz - \omega t)} \\ \mathbf{B}(\mathbf{x}, t) = \mathbf{B}_0 \, \mathrm{e}^{\mathrm{i}(kz - \omega t)} \end{cases} . \tag{6.16}$$

Recall that Maxwell's equations in Eq. (6.6) require

$$\nabla \cdot \mathbf{E} = 0.$$

Since the fields in Eq. (6.16) depend only on z, the divergence becomes

$$\nabla \cdot \mathbf{E} = \frac{\partial E_z}{\partial z} = \mathrm{i}k(E_0)_z \, \mathrm{e}^{\mathrm{i}(kz - \omega t)}. \tag{6.17}$$

For this expression to vanish everywhere, we must have

$$(E_0)_z = 0.$$

Similarly, applying $\nabla \cdot \mathbf{B} = 0$ leads to

$$(B_0)_z = 0.$$

Therefore, **electromagnetic waves in free space are transverse waves**: both electric and magnetic fields are perpendicular to the direction of wave propagation (see Fig. 6.7). This transverse nature presents a challenge for electron acceleration, since a purely transverse wave cannot continuously accelerate electrons along the direction of propagation. In contrast, for applications such as microwave heating or cooking, the transverse character of microwaves does not pose a limitation (see Fig. 6.8).

In Sect. 6.4, we will discuss how a longitudinal electric field can be generated from plane EM waves using a waveguide.

Maxwell's equations in Eq. (6.6) also require

$$\nabla \times \mathbf{E} = -\frac{\partial \mathbf{B}}{\partial t}.$$

Using the plane-wave expressions for **E** and **B** in Eq. (6.16), we obtain

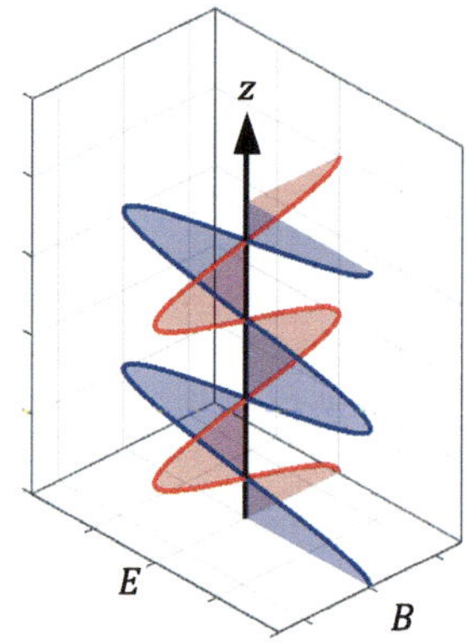

Fig. 6.7 A plane EM wave propagating along the z-direction in free space. The electric field (**E**) and magnetic field are mutually perpendicular

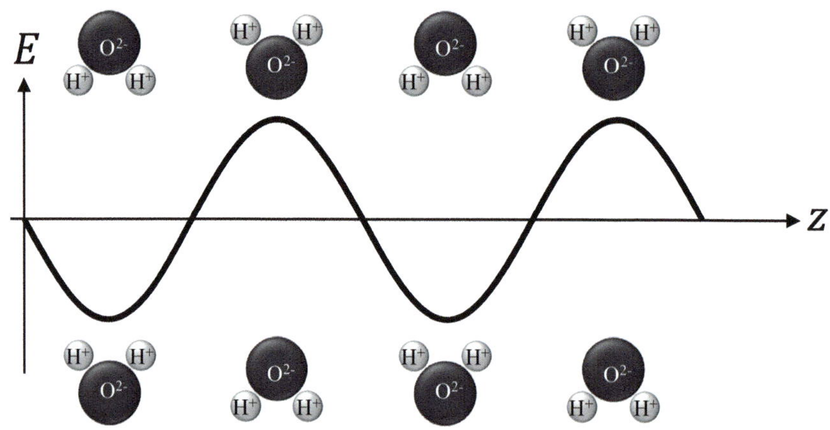

Fig. 6.8 The transverse oscillating electric field of a microwave interacts with water molecules that possess an electric dipole moment, causing them to rotate or oscillate. The resulting molecular motion, along with intermolecular friction, converts this energy into heat, as occurs in a microwave oven or during microwave ablation procedures

$$\nabla \times \left(\mathbf{E}_0\, \mathrm{e}^{\mathrm{i}(kz-\omega t)}\right) = -\frac{\partial}{\partial t}\left(\mathbf{B}_0\, \mathrm{e}^{\mathrm{i}(kz-\omega t)}\right). \tag{6.18}$$

Evaluating the spatial and temporal derivatives gives

$$\mathrm{i}k\,\hat{\mathbf{z}} \times \mathbf{E}_0 = \mathrm{i}\omega\,\mathbf{B}_0, \tag{6.19}$$

which leads to

$$\mathbf{B}_0 = \frac{k}{\omega}\,\hat{\mathbf{z}} \times \mathbf{E}_0$$

$$\text{or equivalently} \qquad \mathbf{B}_0 = \frac{1}{c}\,\hat{\mathbf{z}} \times \mathbf{E}_0. \tag{6.20}$$

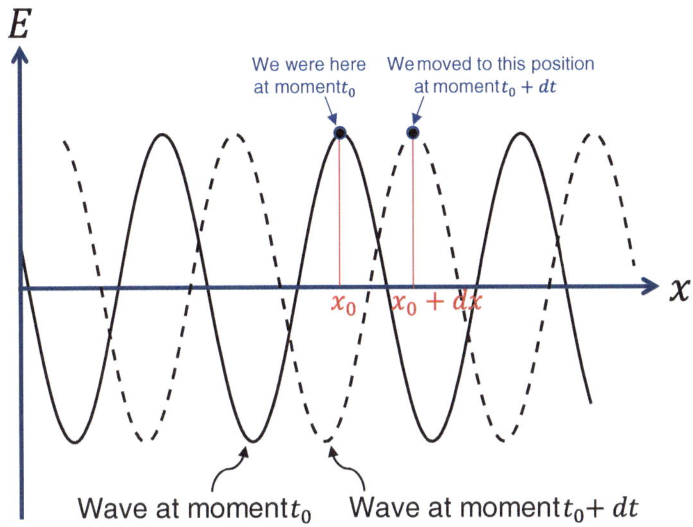

Fig. 6.9 Motion of a wave crest over time

Therefore, **E and B are in phase and mutually perpendicular.** As illustrated in Fig. 6.7, if **E** is along the $\pm x$ direction, then **B** must be along the $\pm y$ direction (Fig. 6.9).

6.3.4 *Phase Velocity of EM Waves*

Imagine a sinusoidal wave

$$E(x, t) = \cos(kx - \omega t), \tag{6.21}$$

and suppose we land on a crest of the wave at location x_0 at time t_0. We then attach ourselves to that crest (i.e., we remain at a fixed phase of the wave) and move together with it. After a short time interval $\mathrm{d}t$, the crest has moved to $x_0 + \mathrm{d}x$. Because our **phase location** on the wave remains unchanged, the following relation must hold:

$$k(x_0 + \mathrm{d}x) - \omega(t_0 + \mathrm{d}t) = kx_0 - \omega t_0 \tag{6.22}$$

By definition, the velocity of the crest is

$$v = \frac{\mathrm{d}x}{\mathrm{d}t}. \tag{6.23}$$

Using Eq. (6.22), we obtain

$$v = \frac{\mathrm{d}x}{\mathrm{d}t} = \frac{\omega}{k}. \tag{6.24}$$

This example shows that the **phase velocity** of a wave is given by ω/k, where k is the magnitude of the wave vector **k**.

For a plane wave in free vacuum space, the k-ω relationship in Eq. (6.14) gives

$$v = \frac{\omega}{k} = c. \tag{6.25}$$

This is the fundamental origin of the relation $c = \frac{1}{\sqrt{\epsilon_0 \mu_0}}$ as shown in Eq. (6.5).

6.3.5 Boundary Conditions at Ideal Conducting Waveguide Walls

This chapter frequently uses the following boundary conditions for microwaves at the waveguide walls:

1. Inside an ideal conductor, $\mathbf{E} = \mathbf{0}$. If $\mathbf{E} \neq \mathbf{0}$, it would exert a force on free electrons, causing them to move. The electrons would rapidly rearrange themselves on the conductor surface until the internal electric field is completely canceled.
2. Inside an ideal conductor, there can be no time-varying magnetic field. This follows from Faraday's law,

   $$\nabla \times \mathbf{E} = -\frac{\partial \mathbf{B}}{\partial t}.$$

   Since $\mathbf{E} = \mathbf{0}$ everywhere inside an ideal conductor, we must have

   $$\frac{\partial \mathbf{B}}{\partial t} = 0.$$

   Thus, any magnetic field inside an ideal conductor must be time-independent.
3. Just outside an ideal conductor, there can be no component of a time-varying magnetic field that is normal to the surface. Because $\nabla \cdot \mathbf{B} = 0$, the normal component of **B** must be continuous across a boundary. Since a time-varying magnetic field cannot exist inside the ideal conductor, the normal component of a time-varying **B** field must also vanish at the surface.
4. Just outside an ideal conductor, **E** must be perpendicular to the surface. This can be shown using Faraday's law (the third equation in Maxwell's equations in Eq. (6.4)). Consider a very small rectangular loop that crosses the boundary

between the conductor and the vacuum. Two sides of the loop are parallel to the surface, and two sides are perpendicular to it. Let the loop height be h, with half inside the conductor and half outside. Applying Faraday's law in integral form,

$$\oint \mathbf{E} \cdot \mathrm{d}\mathbf{l} = -\frac{\mathrm{d}}{\mathrm{d}t} \int \mathbf{B} \cdot \mathrm{d}\mathbf{a}. \tag{6.26}$$

As h shrinks and the loop area approaches zero, the magnetic flux term on the right-hand side of Eq. (6.26) approaches zero. The line integral on the left-hand side then receives contributions only from the two sides parallel to the surface, giving

$$E_{\parallel}^{\text{outside}}\, l = E_{\parallel}^{\text{inside}}\, l. \tag{6.27}$$

From the first boundary condition, $E_{\parallel}^{\text{inside}} = 0$. Therefore,

$$E_{\parallel}^{\text{outside}} = 0. \tag{6.28}$$

Hence, for an ideal conductor, the electric field just outside the conductor has no tangential component and must be perpendicular to the conductor surface.

- For a real conductor with finite conductivity, the electric field inside the conductor is generally small but nonzero. Consequently,

$$E_{\parallel}^{\text{outside}} \neq 0. \tag{6.29}$$

However, the following boundary condition still holds:

$$E_{\parallel}^{\text{outside}} = E_{\parallel}^{\text{inside}} \tag{6.30}$$

A more detailed discussion of electromagnetic boundary conditions for arbitrary materials can be found in References [5, 11].

6.4 Microwave Dynamics in a Hollow Waveguide with Perfectly Conducting Walls

6.4.1 Toy Model

In real medical Linacs, the accelerating waveguides typically have circular cross sections rather than rectangular or parallel-plate geometries. However, the electromagnetic field analysis of circular waveguides requires the use of Bessel functions. To explain the underlying physics while minimizing mathematical complexity, we instead consider a simplified parallel-plate (toy-model) waveguide. In this geometry, the theoretical description involves only basic sinusoidal functions.

Consider a time-harmonic plane wave whose electric field **E** is polarized to be within the x-z plane and whose magnetic field **B** is along the $\pm y$ direction. The wave propagates in the x-z plane at an angle with respect to the z axis. Now, suppose two perfectly conducting planes are placed at $x = 0$ and $x = a$. The wave is "reflected" back and forth between these conducting boundaries, and the superposition of the incident and reflected waves forms a guided mode within the region between the plates.

The reflections at the conducting planes can be understood in terms of the fourth boundary conditions reviewed in Sect. 6.3.5: the tangential component of the electric field must vanish at the surface of a perfect conductor. In this specific case,

$$E_z(0, z, t) = 0 \tag{6.31}$$

$$E_z(a, z, t) = 0. \tag{6.32}$$

A single obliquely propagating plane wave does not generally satisfy these boundary conditions. Physically, the oscillating electric field of the incident wave drives free electrons in the metal surface to oscillate, producing surface currents that radiate an additional electromagnetic field. This radiated field combines with the incident wave to cancel the tangential electric field at the conductor surface, which appears macroscopically as a reflected wave.

The reflection of EM waves from conducting surfaces underlies the operation of a mirror. Visible light, a sub-type of EM waves, is reflected by a thin silver film deposited on a glass substrate. The glass primarily provides mechanical support and optical flatness, with minimal attenuation of light. The actual reflective (mirroring) effect is produced by the conducting silver layer.

This strong reflection of EM waves from conducting surfaces also explains why metal containers are generally unsuitable for use in microwave ovens. Instead of allowing microwave energy to penetrate into the food, the metal reflects most of the incident radiation, thereby reducing the efficiency of energy deposition within the food inside the metal containers.

6.4.2 Longitudinal Electric Field

When the incident wave is reflected by the conducting surfaces, the superposition of the incident and reflected waves produces a field distribution that satisfies the boundary conditions at both conducting planes. The interference of these waves forms a standing-wave pattern in the transverse (x) direction, while the wave continues to propagate along the longitudinal (z) direction. As a result, the field amplitudes become sinusoidally distributed between the plates, which is the characteristic behavior of guided modes in a parallel-plate waveguide.

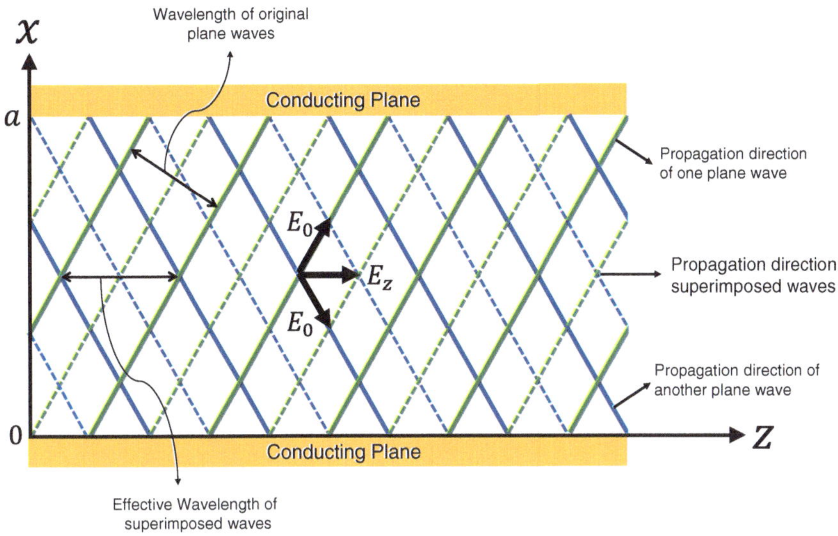

Fig. 6.10 Plane waves between two perfectly conducting planar walls. The y axis is perpendicular to the page. The two conducting walls are parallel to the y-z plane and intersect the x-axis at $x = 0$ and $x = a$. The reflection and superposition of the waves produce a standing-wave pattern in the transverse (x) direction while forming a propagating wave along z, which can therefore be regarded as the overall direction of wave propagation

As illustrated in Fig. 6.10, the resulting field configuration may contain a nonzero electric field component along z. In Cartesian coordinates, the total electric field of the superimposed waves can therefore be written as

$$\mathbf{E} = E_x e^{-i\omega t}\,\hat{\mathbf{x}} + E_z e^{-i\omega t}\,\hat{\mathbf{z}}. \tag{6.33}$$

Here, **the subscripts $_x$ and $_z$ indicate the directions of the corresponding electric field components.** The magnitudes of E_x and E_z generally depend on the spatial position (x, y, z). This spatial dependence can be written explicitly, for example, as $E_z(x, y, z)$.

In a vacuum region, the electric field satisfies the Helmholtz equation in Eq. (6.12). Substituting the above form shows that the equation separates into two scalar equations governing the $\hat{\mathbf{x}}$ and $\hat{\mathbf{z}}$ components of $\mathbf{E}$:

$$\nabla^2 E_x + \frac{\omega^2}{c^2} E_x = 0, \tag{6.34}$$

$$\nabla^2 E_z + \frac{\omega^2}{c^2} E_z = 0 \tag{6.35}$$

We expect E_z in this planar waveguide to propagate along the z direction. As a result, E_z can be written as the product of a propagation factor $e^{ik_z z}$ and a function describing the spatial variation along the x direction:

$$E_z(x, z) = u(x)e^{ik_z z} \tag{6.36}$$

k_z is an important physical quantity in waveguide analysis: **it represents the longitudinal component of the wavevector and describes how the phase of the electromagnetic wave evolves along the propagation direction** z.

Substituting this expression into Eq. (6.35) and dividing out the common factor $e^{ik_z z}$ yield

$$\frac{d^2}{dx^2}u(x) + \left(\frac{\omega^2}{c^2} - k_z^2\right)u(x) = 0. \tag{6.37}$$

The general solution of this ordinary differential equation is

$$u(x) = A\sin(\beta x) + B\cos(\beta x), \tag{6.38}$$

where

$$\boxed{\beta^2 = \frac{\omega^2}{c^2} - k_z^2.} \tag{6.39}$$

At the surfaces of the conducting walls shown in Fig. 6.10, the tangential component of the electric field must vanish. Therefore, $u(x)$ must satisfy the boundary conditions

$$u(0) = 0, \tag{6.40}$$

$$u(a) = 0. \tag{6.41}$$

Applying these boundary conditions to Eq. (6.38) gives

$$\begin{aligned} B &= 0, \\ \beta &= \frac{m\pi}{a}, \qquad m = 1, 2, 3, \ldots. \end{aligned} \tag{6.42}$$

Each integer value of m corresponds to a valid solution for E_z, known as a **mode**. The case $m = 0$ is excluded because it leads to the trivial solution $u(x) = 0$.

For a given mode, using Eqs. (6.36) and (6.38),

$$\begin{aligned} E_z^{(m)}(x, z) &= u^{(m)}(x)\, e^{ik_z^{(m)} z} \\ &= A\sin\left(\frac{m\pi x}{a}\right) e^{ik_z^{(m)} z} \end{aligned} \tag{6.43}$$

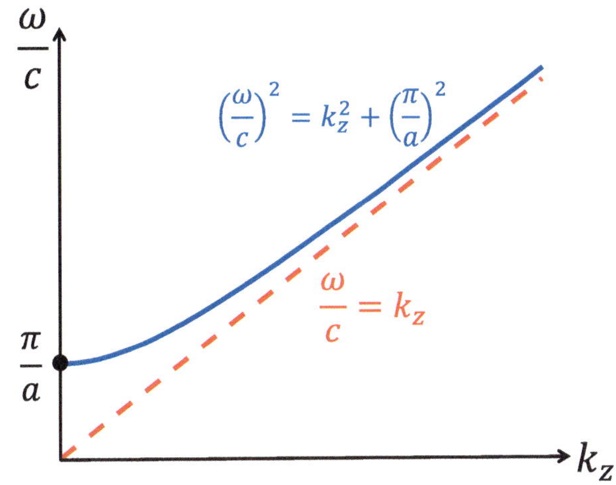

Fig. 6.11 Dispersion diagram for microwaves in a hollow waveguide with mode $m = 1$

Including the time dependence gives

$$E_z^{(m)}(x, z, t) = A \sin\left(\frac{m\pi x}{a}\right) e^{i(k_z^{(m)} z - \omega t)}. \tag{6.44}$$

For a given mode index m, k_z can be obtained from Eq. (6.39):

$$k_z^{(m)} = \sqrt{\left(\frac{\omega}{c}\right)^2 - \left(\frac{m\pi}{a}\right)^2}. \tag{6.45}$$

Equivalently,

$$\left(\frac{\omega}{c}\right)^2 = \left(k_z^{(m)}\right)^2 + \left(\frac{m\pi}{a}\right)^2. \tag{6.46}$$

Equation (6.46) is known as the dispersion relation between k_z and ω. It shows that the allowable longitudinal wavenumber k_z depends on the microwave frequency ω (see Fig. 6.11). The term "dispersion" refers to the fact that k_z, and therefore the phase velocity of the wave in the waveguide varies with frequency. This behavior contrasts with propagation in free space, where the phase velocity is equal to c and independent of frequency.

According to Eq. (6.46), if $\frac{\omega}{c} = \frac{m\pi}{a}$, then $k_z = 0$, meaning the wave cannot propagate along the waveguide. The minimum ω required for wave propagation along the z direction of the waveguide is referred to as the **cutoff frequency**. According to Eq. (6.46),

$$\omega_c = \frac{m\pi c}{a}. \tag{6.47}$$

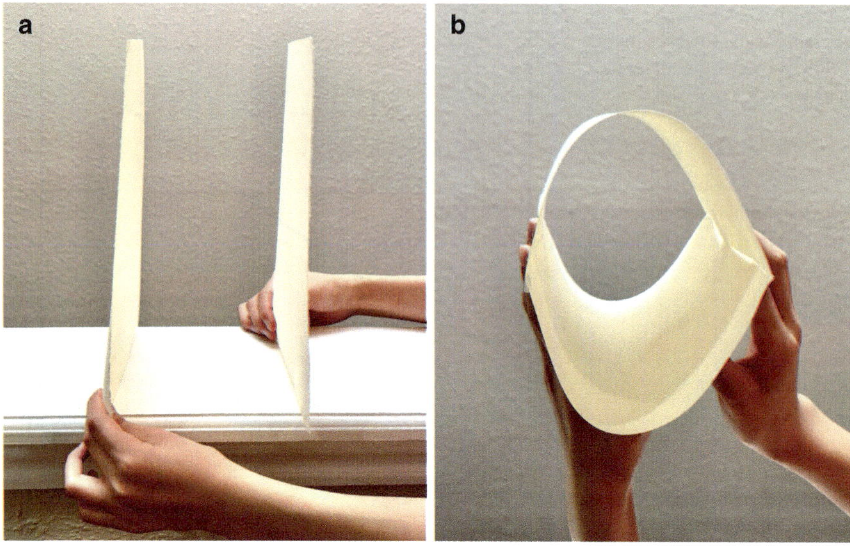

Fig. 6.12 When two planes shown in (**a**) are joined at their ends and smoothly bent, the structure forms a cylinder, as illustrated in (**b**)

For an EM wave with $\omega < \omega_C$, k_z becomes imaginary, and the field decays exponentially along z, so its propagation along the long axis of waveguide is not possible.

A moment to temporarily forget about the math

By properly designing a waveguide with conducting walls to reflect microwaves, a nonzero electric field along the wave propagation and electron beam direction can be created. The magnitude of the acceleration electric field depends on the input microwave power, the frequency, and the conductivity of the waveguide walls, as shown in the next subsection.

In practice, the conducting walls are typically formed into cylindrical cavities, for which the electromagnetic field solutions involve Bessel functions [8]—an added mathematical complexity that we have intentionally avoided here. The toy model shown in Fig. 6.10 is designed to illustrate the essential mechanism by which a longitudinal electric field can be generated for electron acceleration. If the two planar conducting walls in this model are joined at their ends and smoothly curved (Fig. 6.12), the structure becomes a cylindrical waveguide. Although the explicit expressions for the electric and magnetic fields differ between these geometries, the underlying principles and essential physics remain the same.

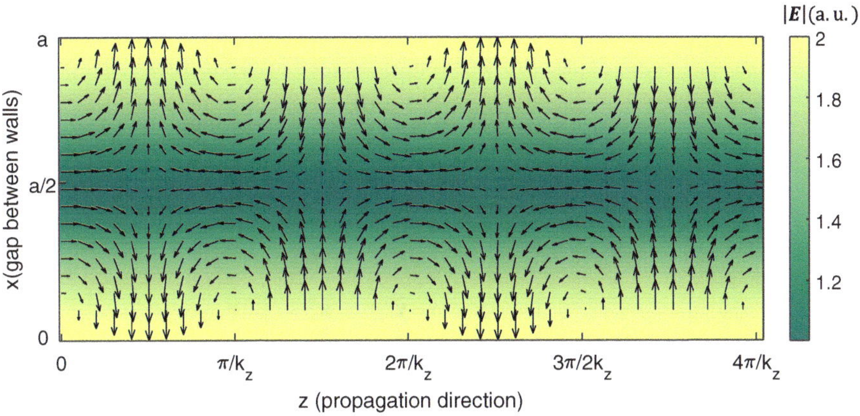

Fig. 6.13 Electric field magnitude (colormap) and direction (vector field) in the toy-model waveguide for the $m = 1$ mode

6.4.3 Transverse Electric Field, Magnetic Field, and Poynting Vector

As illustrated in Fig. 6.10, the toy-model waveguide problem discussed in this section assumes that the electric field and the wave vector lie within the x-z plane, while the magnetic field is along the y direction. Equation (6.44) provides the longitudinal component of the electric field, E_z. Meanwhile, the transverse component E_x is not necessarily zero. To solve for E_x, we use Gauss's law (the first equation in (6.6)):

$$\begin{aligned} 0 &= \nabla \cdot \mathbf{E} \\ &= \frac{\partial E_x}{\partial x} + \frac{\partial E_y}{\partial y} + \frac{\partial E_z}{\partial z} \\ &= \frac{\partial E_x}{\partial x} + 0 + \mathrm{i}k_z^{(m)} A \sin\left(\frac{m\pi x}{a}\right) \mathrm{e}^{\mathrm{i}(k_z^{(m)} z - \omega t)}, \end{aligned} \tag{6.48}$$

which yields

$$E_x(x, z, t) = \mathrm{i}\frac{k_z^{(m)} a}{m\pi} A \cos\left(\frac{m\pi x}{a}\right) \mathrm{e}^{\mathrm{i}(k_z^{(m)} z - \omega t)}. \tag{6.49}$$

Here, we have assumed the integration constant is zero, consistent with the absence of a uniform, static electric field in the waveguide.

Figure 6.13 shows a snapshot of the **E** field strength and direction calculated based on Eqs. (6.44) and (6.49) with mode $m = 1$.

Given **E**, the magnetic field can be obtained from Faraday's law (the third equation in (6.6)),

$$\nabla \times \mathbf{E} = -\frac{\partial \mathbf{B}}{\partial t} = \mathrm{i}\omega \mathbf{B}, \tag{6.50}$$

which leads to

$$\mathbf{B} = \frac{1}{\mathrm{i}\omega} \begin{vmatrix} \hat{\mathbf{x}} & \hat{\mathbf{y}} & \hat{\mathbf{z}} \\ \frac{\partial}{\partial x} & \frac{\partial}{\partial y} & \frac{\partial}{\partial z} \\ E_x(x,z,t) & 0 & E_z(x,z,t) \end{vmatrix} = 0\,\hat{\mathbf{x}} + \frac{1}{\mathrm{i}\omega}\left(\frac{\partial E_x}{\partial z} - \frac{\partial E_z}{\partial x}\right)\hat{\mathbf{y}} + 0\,\hat{\mathbf{z}}. \tag{6.51}$$

This confirms that magnetic field has no components in the x-z plane and its y-component is

$$\begin{aligned} B_y(x,z,t) &= \frac{1}{\mathrm{i}\omega}\left(\frac{\partial E_x}{\partial z} - \frac{\partial E_z}{\partial x}\right) \\ &= -\frac{a}{\mathrm{i}\omega m\pi}\left[(k_z^{(m)})^2 + \left(\frac{m\pi}{a}\right)^2\right] A\cos\left(\frac{m\pi x}{a}\right) \mathrm{e}^{\mathrm{i}(k_z^{(m)} z - \omega t)} \\ &= \mathrm{i}\frac{\omega a}{m\pi c^2} A\cos\left(\frac{m\pi x}{a}\right) \mathrm{e}^{\mathrm{i}(k_z^{(m)} z - \omega t)}. \end{aligned} \tag{6.52}$$

Since $B_z = 0$ everywhere in the waveguide, the wave is called a **transverse magnetic (TM)** wave.

The instantaneous Poynting vector, which represents the rate of electromagnetic energy flow per unit area, is given by [5]

$$\begin{aligned} \mathbf{S} &= \frac{1}{\mu_0}\mathbf{E} \times \mathbf{B} \\ &= \frac{1}{\mu_0}\begin{vmatrix} \hat{\mathbf{x}} & \hat{\mathbf{y}} & \hat{\mathbf{z}} \\ E_x & 0 & E_z \\ 0 & B_y & 0 \end{vmatrix} \\ &= -\frac{1}{\mu_0}E_z B_y\,\hat{\mathbf{x}} + \frac{1}{\mu_0}E_x B_y\,\hat{\mathbf{z}}. \end{aligned} \tag{6.53}$$

Thus, there is no power flow along the y-direction, as expected for this waveguide mode.

For time-harmonic fields, the **time-averaged Poynting vector** is

$$\langle \mathbf{S} \rangle = \frac{1}{2\mu_0} \Re \left(\mathbf{E} \times \mathbf{B}^* \right), \tag{6.54}$$

where the complex conjugate ensures the correct averaging of real fields over time.

The x-component of $\langle \mathbf{S} \rangle$ is related to the product of $E_z B_y^*$. Because E_z is real and B_y is purely imaginary, $\Re(S_x) = 0$, indicating $\langle S_x \rangle = 0$ and **no net energy flow toward the perfectly conducting walls.**

The z-component of $\langle \mathbf{S} \rangle$ is given by

$$\begin{aligned} \langle S_z \rangle &= \frac{1}{2\mu_0} \Re \left(E_x B_y^* \right) \\ &= \frac{1}{2\mu_0} \Re \left(\mathrm{i} \frac{k_z^{(m)} a}{m\pi} A \cos\left(\frac{m\pi x}{a}\right) \mathrm{e}^{\mathrm{i}(k_z^{(m)} z - \omega t)} \cdot (-\mathrm{i}) \frac{\omega a}{m\pi c^2} A \cos\left(\frac{m\pi x}{a}\right) \right. \\ &\qquad \left. \mathrm{e}^{-\mathrm{i}(k_z^{(m)} z - \omega t)} \right) \\ &= \frac{k_z^{(m)} \omega}{2\mu_0} \left(\frac{aA}{m\pi c} \right)^2 \cos^2\left(\frac{m\pi x}{a}\right). \end{aligned} \tag{6.55}$$

Equation (6.55) shows that although the microwaves reflect between the conducting planes, their interference produces a **net forward energy flow along the z-direction**.

Equation (6.55) also provides a connection between the input microwave power and the electric field amplitude A. The total microwave power carried by the waveguide is obtained by integrating $\langle S_z \rangle$ over the waveguide cross section $a \times b$ where $b \to +\infty$:

$$\begin{aligned} P_{\text{in}} &= \int_0^b \mathrm{d}y \int_0^a \mathrm{d}x \, \langle S_z \rangle \\ &= \frac{k_z^{(m)} \omega}{2\mu_0} \left(\frac{aA}{m\pi c} \right)^2 b \int_0^a \mathrm{d}x \cos^2\left(\frac{m\pi x}{a}\right) \\ &= \frac{ab k_z^{(m)} \omega}{4\mu_0} \left(\frac{aA}{m\pi c} \right)^2 \end{aligned} \tag{6.56}$$

Equation (6.56) shows how the electric field amplitude A depends on the input microwave power in the hollow waveguide with perfectly conducting walls.

Next, we discuss the electromagnetic energy stored per unit volume in the waveguide (energy density) [5]:

$$u = \frac{1}{2} \left(\epsilon_0 |\mathbf{E}|^2 + \frac{1}{\mu_0} |\mathbf{B}|^2 \right) \tag{6.57}$$

For time-harmonic fields, the cycle-averaged energy density is

$$\begin{aligned}\langle u\rangle &= \frac{1}{4}\left(\epsilon_0|\mathbf{E}|^2 + \frac{1}{\mu_0}|\mathbf{B}|^2\right)\\ &= \frac{1}{4}\left(\epsilon_0|E_z(x,z)|^2 + \epsilon_0|E_x(x,z)|^2 + \frac{1}{\mu_0}|B_y(x,z)|^2\right).\end{aligned} \tag{6.58}$$

According to Eqs. (6.44), (6.49), and (6.52),

$$\begin{cases} |E_z(x,z)|^2 = A^2\sin^2\left(\frac{m\pi x}{a}\right)\\ |E_x(x,z)|^2 = \left(\frac{k_z^{(m)}a}{m\pi}\right)^2 A^2\cos^2\left(\frac{m\pi x}{a}\right)\\ |B_y(x,z)|^2 = \left(\frac{\omega a}{m\pi c^2}\right)^2 A^2\cos^2\left(\frac{m\pi x}{a}\right).\end{cases}$$

Substituting these expressions into the energy density gives

$$\begin{aligned}\langle u\rangle = \frac{A^2}{4}\Bigg[&\epsilon_0\sin^2\left(\frac{m\pi x}{a}\right) + \epsilon_0\left(\frac{k_z^{(m)}a}{m\pi}\right)^2\cos^2\left(\frac{m\pi x}{a}\right)\\ &+ \frac{1}{\mu_0}\left(\frac{\omega a}{m\pi c^2}\right)^2\cos^2\left(\frac{m\pi x}{a}\right)\Bigg].\end{aligned} \tag{6.59}$$

Using the relation

$$c^2 = \frac{1}{\mu_0\epsilon_0},$$

the magnetic term can be rewritten as

$$\frac{1}{\mu_0}\left(\frac{\omega a}{m\pi c^2}\right)^2 = \epsilon_0\left(\frac{\omega a}{m\pi c}\right)^2.$$

Therefore,

$$\langle u\rangle = \frac{\epsilon_0 A^2}{4}\left[\sin^2\left(\frac{m\pi x}{a}\right) + \left((k_z^{(m)})^2 + (\omega/c)^2\right)\left(\frac{a}{m\pi}\right)^2\cos^2\left(\frac{m\pi x}{a}\right)\right]. \tag{6.60}$$

Using the waveguide dispersion relation

$$\frac{\omega^2}{c^2} = (k_z^{(m)})^2 + \left(\frac{m\pi}{a}\right)^2, \tag{6.61}$$

we obtain

$$(k_z^{(m)})^2 + (\omega/c)^2 = 2(k_z^{(m)})^2 + \left(\frac{m\pi}{a}\right)^2 .$$

Substituting this result gives

$$\begin{aligned}\langle u\rangle &= \frac{\epsilon_0 A^2}{4}\left[\sin^2\left(\frac{m\pi x}{a}\right) + \left(2\left(\frac{k_z^{(m)}a}{m\pi}\right)^2 + 1\right)\cos^2\left(\frac{m\pi x}{a}\right)\right] \\ &= \frac{\epsilon_0 A^2}{4}\left[1 + 2\left(\frac{k_z^{(m)}a}{m\pi}\right)^2 \cos^2\left(\frac{m\pi x}{a}\right)\right] . \end{aligned} \tag{6.62}$$

The total electromagnetic energy stored per unit length of the waveguide is therefore

$$\begin{aligned}\langle U\rangle &= b\int_0^a \langle u\rangle\, \mathrm{d}x \\ &= \frac{b\epsilon_0 A^2}{4}\int_0^a \left[1 + 2\left(\frac{k_z^{(m)}a}{m\pi}\right)^2 \cos^2\left(\frac{m\pi x}{a}\right)\right] \mathrm{d}x \\ &= \frac{ab\epsilon_0 A^2}{4}\left[1 + \left(\frac{k_z^{(m)}a}{m\pi}\right)^2\right] . \end{aligned} \tag{6.63}$$

Using again the dispersion relation, one finds

$$1 + \left(\frac{k_z^{(m)}a}{m\pi}\right)^2 = \left(\frac{a\omega}{m\pi c}\right)^2 ,$$

so that

$$\langle U\rangle = \frac{ab\epsilon_0 A^2}{4}\left(\frac{a\omega}{m\pi c}\right)^2 . \tag{6.64}$$

According to Eq. (6.52),

$$A^2\left(\frac{\omega a}{m\pi c^2}\right)^2 = |B_y(x=0,z)|^2 , \tag{6.65}$$

where $|B_y(x=0,z)|$ denotes the magnitude of the magnetic field at the surface of the waveguide wall. Therefore, the stored energy can also be written as

$$
\begin{aligned}
\langle U \rangle &= \frac{ab\epsilon_0 c^2}{4} |B_y(x=0,z)|^2 \\
&= \frac{ab}{4\mu_0} |B_y(x=0,z)|^2 \\
&= \frac{ab\mu_0}{4} |H_y(x=0,z)|^2,
\end{aligned}
\tag{6.66}
$$

where $H_y = B_y/\mu_0$ is the magnetizing field.

A moment to temporarily forget about the math
This subsection contains quite a few formulas, and I understand that not everyone enjoys equations. For those readers, here is the key idea without mathematics: in a waveguide of known geometry, knowledge of the power flow through its cross section allows us to determine the magnitudes of the electric and magnetic fields in all directions. This, in turn, connects the input microwave power or stored energy density to the accelerating electric field strength. For waveguide walls with finite conductivity, as discussed in Sect. 6.5, we can also calculate the induced currents and power losses in the walls, providing a direct link between the power cost of sustaining a given accelerating field.

6.4.4 Phase Velocity and Effective Wavelength

Although the hollow waveguide can establish an electric field and energy flow along the z direction, it does not guarantee electron acceleration. As shown in Eq. (6.43), E_z oscillates in both time and space. An electron traveling along $+z$ may encounter an E_z pointing in the $+z$ direction, which would decelerate this negatively charged particle. To achieve cumulative acceleration, the electron should remain in phase with the accelerating field. **The simplest approach is to synchronize the electron velocity with the phase velocity of the microwave so that the electron consistently experiences an accelerating E_z.**

Using the expression for k_z from Eq. (6.45) for mode $m = 1$, the phase velocity of the microwave is

$$
\begin{aligned}
v_p^{(m=1)} &= \frac{\omega}{k_z^{(m=1)}} \\
&= \frac{\omega}{\sqrt{\left(\frac{\omega}{c}\right)^2 - \left(\frac{\pi}{a}\right)^2}}
\end{aligned}
$$

$$= \frac{c}{\sqrt{1 - \left(\frac{\omega_c}{\omega}\right)^2}} > c. \tag{6.67}$$

In Eq. (6.67), $\omega_c = \pi c/a$ is the cutoff frequency. Clearly, the phase velocity exceeds the speed of light, c. For higher-order modes, the phase velocity is even larger. **This superluminal phase velocity is not suitable for sustained electron acceleration**, so a method described in Sect. 6.6 is required to slow down the phase velocity of the microwave.

The phase velocity of electromagnetic waves in a hollow waveguide exceeds c because the superposition of obliquely reflected waves produces a longer effective wavelength along the propagation direction z, as illustrated in Fig. 6.10. For mode $m = 1$, the effective wavelength along z is related to the corresponding phase velocity and angular frequency by

$$\begin{aligned} \tilde{\lambda}^{(m)} &= \frac{v_p^{(m)}}{f} = \frac{v_p^{(m)}}{\omega/(2\pi)} \\ &= \frac{2\pi c}{\sqrt{\omega^2 - \omega_c^2}}, \end{aligned} \tag{6.68}$$

which is indeed longer than the free-space wavelength of the microwave, $\lambda_0 = 2\pi c/\omega$.

6.5 Microwave Dynamics in a Hollow Waveguide with Finite Conductivity Walls

In Sect. 6.4, we demonstrated that by confining a plane microwave between two perfectly conducting walls, reflections from the walls can generate an electric field component along the longitudinal direction, which in principle could accelerate electrons traveling in the same direction. However, because the phase velocity of microwaves in the hollow waveguide is faster than the velocity of electrons, it does not provide cumulative acceleration to the electrons.

The waveguides used in practical Linacs are modified versions of this hollow waveguide configuration. In particular, metallic discs are inserted into the waveguide to reduce the phase velocity of the microwave so that it better matches the velocity of the accelerating electrons.

Another important difference between realistic waveguides and the idealized model discussed in Sect. 6.4 lies in the electrical conductivity of the walls. In practice, the wall material (typically copper) has a finite electrical conductivity. As a result, electromagnetic fields induce surface currents that experience resistive losses, generating heat. Quantities such as the Q factor and shunt impedance are commonly used metrics to characterize the power loss properties of a given waveguide. For a waveguide with perfectly conducting walls, there are no resistive losses, so the power dissipated in the walls is zero. Consequently, both the Q factor and the shunt impedance approach infinity in this idealized limit.

6.5.1 Skin Effect

In Sect. 6.3.5, we noted that inside a perfectly conducting metal the electric field $\mathbf{E}$ must vanish. In a realistic waveguide wall with finite electrical conductivity σ, however, the electric field inside the conductor is not necessarily zero, especially near the metal surface. The presence of this field drives conduction currents, which lead to power dissipation in the metal walls.

For microwaves with a frequency of about 3 GHz, any charge imbalance inside a good conductor disappears extremely rapidly. Free charges redistribute almost instantaneously, so that the bulk charge density remains essentially zero—any charge accumulation is confined to an extremely thin surface layer. Therefore, for the analysis of microwaves inside the waveguide walls, we can still assume $\nabla \cdot \mathbf{E} = 0$ [5].

According to Ohm's law, the electric current density is proportional to the electric field:

$$\mathbf{J} = \sigma \mathbf{E}. \tag{6.69}$$

When $\mathbf{J} \neq 0$, Maxwell's equations in (6.6) must include the conduction current $\mathbf{J} = \sigma \mathbf{E}$:

$$\begin{cases} (1) \quad \nabla \cdot \mathbf{E} = 0 \\ (2) \quad \nabla \cdot \mathbf{B} = 0 \\ (3) \quad \nabla \times \mathbf{E} = -\dfrac{\partial \mathbf{B}}{\partial t} \\ (4) \quad \nabla \times \mathbf{B} = \mu\epsilon \dfrac{\partial \mathbf{E}}{\partial t} + \mu\sigma\mathbf{E} \end{cases} \tag{6.70}$$

For time-harmonic electromagnetic waves with time dependence $\mathrm{e}^{-\mathrm{i}\omega t}$, the equations become

$$\begin{cases} (1) \quad \nabla \cdot \mathbf{E} = 0 \\ (2) \quad \nabla \cdot \mathbf{B} = 0 \\ (3) \quad \nabla \times \mathbf{E} = \mathrm{i}\omega\mathbf{B} \\ (4) \quad \nabla \times \mathbf{B} = -\mathrm{i}\omega\mu\epsilon\mathbf{E} + \mu\sigma\mathbf{E}. \end{cases} \tag{6.71}$$

It is convenient to define a **complex permittivity**

$$\epsilon' = \epsilon + \mathrm{i}\frac{\sigma}{\omega}. \tag{6.72}$$

With this definition, the fourth equation in (6.71) can be written as

$$\nabla \times \mathbf{B} = -\mathrm{i}\omega\mu\epsilon'\mathbf{E}. \tag{6.73}$$

This equation has the same form as Maxwell's equation in vacuum, except that ϵ' replaces ϵ and μ replaces μ_0. Consequently, the electric field satisfies a Helmholtz equation similar to Eq. (6.12),

$$\nabla^2\mathbf{E}(\mathbf{x}) + k^2\mathbf{E}(\mathbf{x}) = 0, \tag{6.74}$$

where

$$\begin{aligned} k^2 &= \omega^2\mu\epsilon' \\ &= \omega^2\mu\epsilon + \mathrm{i}\omega\mu\sigma. \end{aligned} \tag{6.75}$$

A plane-wave solution of Eq. (6.74) can be written as

$$\mathbf{E}(\mathbf{x}) = \mathbf{E}_0\, \mathrm{e}^{\mathrm{i}\mathbf{k}\cdot\mathbf{x}}. \tag{6.76}$$

Since k^2 is complex, the wave vector $\mathbf{k}$ is also complex and can be expressed as

$$\mathbf{k} = \boldsymbol{\beta} + \mathrm{i}\boldsymbol{\alpha}, \tag{6.77}$$

where $\boldsymbol{\beta}$ is the phase propagation vector and $\boldsymbol{\alpha}$ is the attenuation vector. Substituting this form into the plane-wave solution gives

$$\mathbf{E}(\mathbf{x}) = \mathbf{E}_0\, \mathrm{e}^{-\boldsymbol{\alpha}\cdot\mathbf{x}}\, \mathrm{e}^{\mathrm{i}\boldsymbol{\beta}\cdot\mathbf{x}} \tag{6.78}$$

or

$$\mathbf{E}(\mathbf{x}, t) = \mathbf{E}_0\, \mathrm{e}^{-\boldsymbol{\alpha}\cdot\mathbf{x}}\, \mathrm{e}^{\mathrm{i}(\boldsymbol{\beta}\cdot\mathbf{x}-\omega t)}. \tag{6.79}$$

The second exponential term represents the usual propagation phase, while the first exponential term describes the exponential decay of the field amplitude inside the conductor due to its finite conductivity.

For a good conductor such as copper,

$$\sigma \gg \omega\epsilon. \tag{6.80}$$

Under this condition,

$$k^2 \approx i\omega\mu\sigma. \tag{6.81}$$

Taking the square root yields

$$\begin{aligned} k &= \sqrt{\mathrm{i}\omega\mu\sigma} \\ &= \sqrt{\mathrm{e}^{\mathrm{i}\frac{\pi}{2}}}\sqrt{\omega\mu\sigma} = \mathrm{e}^{\mathrm{i}\frac{\pi}{4}}\sqrt{\omega\mu\sigma} \\ &= (1+\mathrm{i})\sqrt{\frac{\omega\mu\sigma}{2}}. \end{aligned} \tag{6.82}$$

Comparing this expression with Eq. (6.77) gives

$$\boxed{\alpha \approx \beta \approx \sqrt{\frac{\omega\mu\sigma}{2}}}. \tag{6.83}$$

According to the boundary condition discussed in Eq. (6.30), $E_{\parallel}^{\text{outside}} = E_{\parallel}^{\text{inside}}$, which requires the tangential electric field to be continuous across the interface. For a wave propagating along a uniform surface, the tangential electric field immediately outside the conductor does not exhibit exponential decay along the surface. Because the tangential fields must match at the boundary, the field inside the conductor cannot decay along the surface either. Consequently, the exponential attenuation of the field can occur only in the direction **normal to the surface**, implying that **the attenuation vector $\boldsymbol{\alpha}$ is perpendicular to the surface and points into the conductor.**

The exponential decay of the EM field inward from the conductor surface produces the so-called **skin effect**. Let δ denotes the distance δ over which the EM field decays to $1/\mathrm{e}$ of its surface value, namely,

$$\mathrm{e}^{-\alpha\delta} = \mathrm{e}^{-1}. \tag{6.84}$$

According to Eq. (6.83),

$$\boxed{\delta = \sqrt{\frac{2}{\omega\mu\sigma}}}. \tag{6.85}$$

Example: Skin depth in copper

The electrical conductivity of copper (Cu) is

$$\sigma = 5.8 \times 10^7 \text{ S/m}.$$

The magnetic permeability of copper is approximately

$$\mu_{\text{Cu}} \approx \mu_0 = 1.26 \times 10^{-6} \text{ H/m}.$$

Therefore, for a 3-GHz microwave propagating in copper, the skin depth is

$$\begin{aligned}\delta &= \sqrt{\frac{2}{\omega\mu\sigma}}\\ &= \sqrt{\frac{2}{(2\pi \times 3 \times 10^9)(1.26 \times 10^{-6})(5.8 \times 10^7)}}\\ &\approx 1.2 \times 10^{-6} \text{ m}.\end{aligned} \tag{6.86}$$

As shown in the example above, the EM field penetrates only about one micron into copper, which is significantly shorter than the microwave wavelength. This explains why the phenomenon is called the skin effect and δ is often referred to as the **skin depth**.

6.5.2 EM Field in the Waveguide Wall

The skin effect discussed in Sect. 6.5.1 was derived for conductors of arbitrary geometry. In this subsection, we apply the same idea to a specific geometry—the toy-model waveguide shown in Fig. 6.10—in order to analyze the electromagnetic fields inside the waveguide wall. For this configuration, the magnetic field is along the y direction, while the electric field lies in the x-z plane.

In general, **E** and **B** depend on both x and z. However, inside a good conductor, the skin depth δ is typically much smaller than the microwave wavelength. Consequently, **the fields vary much more rapidly in the direction normal to the wall (x) than along the tangential directions (y or z)**. To a good approximation, the fields inside the conductor can therefore be treated as functions of x only.

Starting from the Ampère-Maxwell equation (fourth equation in (6.71)),

$$\nabla \times \mathbf{B} = \mu\mathbf{J} - i\omega\mu\epsilon\mathbf{E}, \tag{6.87}$$

or equivalently

$$\nabla \times \mathbf{H} = \mathbf{J} - \mathrm{i}\frac{\omega\epsilon}{\sigma}\mathbf{J}, \tag{6.88}$$

where $\mathbf{H} = \mathbf{B}/\mu$ is the magnetizing field. As shown earlier in Eq. (6.80), inside a good conductor, $\frac{\omega\epsilon}{\sigma} \ll 1$. Therefore, we obtain the approximate relation

$$\nabla \times \mathbf{H} \approx \mathbf{J}. \tag{6.89}$$

For the geometry considered in Fig. 6.10, the magnetic or magnetizing field has only a y-component, i.e., $\mathbf{H} = (0, H_y, 0)$. Therefore,

$$\begin{aligned} \mathbf{J} &\approx \begin{vmatrix} \hat{\mathbf{x}} & \hat{\mathbf{y}} & \hat{\mathbf{z}} \\ \frac{\partial}{\partial x} & \frac{\partial}{\partial y} & \frac{\partial}{\partial z} \\ 0 & H_y(x,t) & 0 \end{vmatrix} \\ &= 0\,\hat{\mathbf{x}} + 0\,\hat{\mathbf{y}} + \frac{\partial H_y(x,t)}{\partial x}\,\hat{\mathbf{z}}. \end{aligned} \tag{6.90}$$

Thus, the induced current density inside the conductor is primarily along the z direction:

$$J_z(x,t) = \frac{\partial H_y(x,t)}{\partial x}. \tag{6.91}$$

Integrating $J_z(x,t)$ from the surface into the conductor gives

$$\int_0^\infty J_z(x,t)\,dx = \int_0^\infty \frac{\partial H_y(x,t)}{\partial x}\,\mathrm{d}x. \tag{6.92}$$

For the right side,

$$\int_0^\infty \frac{\partial H_y(x,t)}{\partial x}\,\mathrm{d}x = H_y(\infty,t) - H_y(0,t) = -\,H_y(0,t), \tag{6.93}$$

where we used the fact that inside a good conductor, $H_y(\infty,t) \approx 0$.

For the left side of Eq. (6.92),

$$\int_0^\infty J_z(x,t)\,\mathrm{d}x = \int_0^\infty J_z(0,t)\mathrm{e}^{-x/\delta}\,\mathrm{d}x = \delta\,J_z(0,t). \tag{6.94}$$

Equating the two results yields the relation between the tangential magnetic field and the surface current density,

$$\boxed{J_z(0,t) = -\frac{H_y(0,t)}{\delta}}, \tag{6.95}$$

which shows that the induced current flows in the z direction and decays exponentially inside the conductor:

$$\boxed{J_z(x,t) = -\frac{H_y(0,t)}{\delta}\mathrm{e}^{-x/\delta}} \tag{6.96}$$

Finally, using Ohm's law $\mathbf{J} = \sigma\mathbf{E}$ gives the electric field inside the wall:

$$\begin{aligned} E_z(x,t) &= \frac{J_z(x,t)}{\sigma} \\ &= -\frac{H_y(0,t)}{\sigma\delta}\mathrm{e}^{-x/\delta} \end{aligned} \tag{6.97}$$

A moment to temporarily forget about the math
In a waveguide with finite-conductivity walls, electromagnetic fields penetrate only a thin layer of the wall due to the skin effect. Because the skin depth is much smaller than the microwave wavelength, the fields vary mainly in the direction perpendicular to the wall.

The tangential magnetic field at the surface induces currents along the direction of wave propagation, which decay exponentially with depth. The induced electric field inside the wall is proportional to these currents via the material's conductivity.

Knowing the surface field, the wall conductivity, and the microwave frequency allows us to determine the induced currents and the resulting Joule power loss. This links the accelerating field to the power dissipated in the walls and provides the foundation for defining the shunt impedance—a topic of Sect. 6.5.6.

6.5.3 *Surface Resistance*

Without loss of generality, we use the geometry shown in Fig. 6.10 to define the surface resistance. Let the tangential electric field at the conductor surface be $\mathbf{E}(x = 0)$. According to Ohm's law, the current density at the surface is

$$\mathbf{J}(x = 0) = \sigma\mathbf{E}(x = 0). \tag{6.98}$$

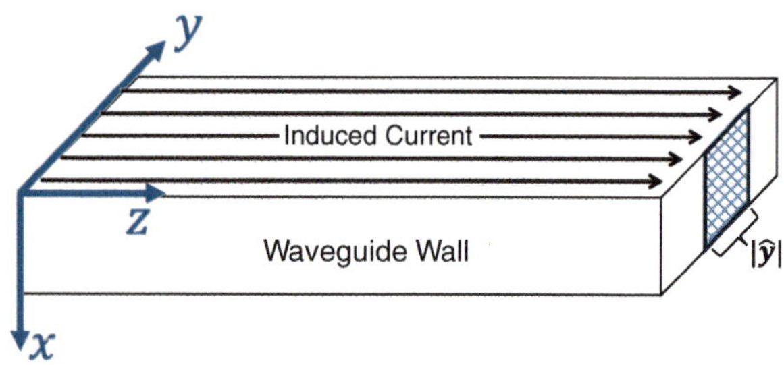

Fig. 6.14 Schematic illustration of microwave-induced current near the surface of a conducting waveguide wall

Due to the skin effect, the current density inside the conductor decays exponentially with depth. For a good conductor, the magnitude of the current density can be approximated as

$$\mathbf{J}(x) = \mathbf{J}(x = 0)\mathrm{e}^{-x/\delta}. \tag{6.99}$$

Consider a unit width in the y direction. The total current flowing through the plane indicated in Fig. 6.14 is

$$\begin{aligned} I &= |\hat{\mathbf{y}}| \int_0^{+\infty} J(x)\,\mathrm{d}x \\ &= |\hat{\mathbf{y}}| J(x = 0) \int_0^{+\infty} e^{-x/\delta} \mathrm{d}x \\ &= |\hat{\mathbf{y}}|\, \delta\, J(x = 0). \end{aligned} \tag{6.100}$$

For the conductor volume corresponding to a unit surface area ($|\hat{\mathbf{y}}| \times |\hat{\mathbf{z}}|$), the resistance is defined as

$$\begin{aligned} R_s &= \frac{E(x = 0)|\hat{\mathbf{z}}|}{I} \\ &= \frac{E(x = 0)}{\delta J(x = 0)} \\ &= \frac{1}{\delta \sigma} \\ &= \sqrt{\frac{\omega \mu}{2\sigma}}. \end{aligned} \tag{6.101}$$

The quantity R_s is known as the **surface resistance** of the waveguide wall. It characterizes the **effective resistance experienced by the microwave surface current flowing within the skin depth of the conductor.**

A moment to temporarily forget about the math

In Linacs, the currents induced by microwaves do not flow through the bulk of the waveguide wall. Instead, due to the skin effect, they are confined to an extremely thin surface layer of thickness δ. As a result, the effective resistance experienced by these currents depends strongly on the material conductivity: a lower conductivity leads to a higher surface resistance, as it increases Ohmic losses.

Because the induced currents are restricted to the surface, they are also highly sensitive to the microscopic geometry of that surface. Any surface roughness forces the current to follow a longer, more tortuous path and can lead to local current crowding. Both effects increase the dissipated power and therefore increase the effective surface resistance.

Finally, the surface resistance increases with increasing microwave frequency. At higher frequencies, the skin depth δ becomes smaller, which confines the current to an even thinner region near the surface. This reduces the effective cross-sectional area available for current flow, thereby increasing the resistance.

6.5.4 Power Loss in Waveguide Walls

The microwave power is dissipated in the conductor because the electric field does work on the free charges. The time-averaged power loss per unit conductor volume at depth x is

$$
\begin{aligned}
\langle p_{\mathrm{L}} \rangle &= \frac{1}{2} \Re\left(\mathbf{J} \cdot \mathbf{E}^*\right) \\
&= \frac{1}{2} \frac{|J(x)|^2}{\sigma} \\
&= \frac{1}{2\sigma} |J(x=0)|^2 \, \mathrm{e}^{-2x/\delta}. \qquad (6.102)
\end{aligned}
$$

Integrating over the conductor depth gives the time-averaged power loss **per unit surface area of the wall**:

$$
\begin{aligned}
\langle P_{\mathrm{L}} \rangle &= \int_0^\infty \langle p_{\mathrm{L}} \rangle \, \mathrm{d}x \\
&= \frac{1}{2\sigma} |J(x=0)|^2 \int_0^\infty \mathrm{e}^{-2x/\delta} \, \mathrm{d}x
\end{aligned}
$$

$$
\begin{aligned}
&= \frac{|J(x=0)|^2\delta}{4\sigma} \\
&= \frac{|H_y(x=0)|^2}{4\sigma\delta} \\
&= \frac{R_s|H_y(x=0)|^2}{4}
\end{aligned}
\tag{6.103}
$$

When both walls shown in Fig. 6.10 are considered,

$$
\langle P_{\mathrm{L}} \rangle = \frac{1}{2} R_s |H_y(x=0)|^2. \tag{6.104}
$$

6.5.5 *Q* Factor

The **quality factor (*Q* factor)** is a dimensionless parameter that measures **how efficiently a waveguide system stores electromagnetic energy relative to how quickly it loses energy**. It is defined as

$$
\boxed{Q = \omega \frac{\text{electromagnetic energy stored per unit length}}{\text{average power loss per unit length}}}. \tag{6.105}
$$

A larger Q corresponds to a smaller energy loss rate: energy remains in the waveguide for many oscillation cycles. A smaller Q means that the energy decays more rapidly.

According to Eq. (6.66), the electromagnetic energy stored per unit length in the waveguide is

$$
\langle U \rangle = \frac{ab\mu_0}{4} |H_y(x=0)|^2. \tag{6.106}
$$

According to Eq. (6.104), the power loss per unit length due to the finite conductivity of the two waveguide walls is

$$
\langle P_{\mathrm{L}} \rangle = \frac{b}{2} R_s |H_y(x=0)|^2. \tag{6.107}
$$

Therefore,

$$
\begin{aligned}
Q &= \omega \frac{\frac{ab\mu_0}{4}|H_y(x=0)|^2}{\frac{b}{2}R_s|H_y(x=0)|^2} \\
&= \frac{a\omega\mu_0}{2R_s}.
\end{aligned}
\tag{6.108}
$$

For a good conductor, the surface resistance is

$$R_s = \sqrt{\frac{\omega\mu}{2\sigma}}. \tag{6.109}$$

Assuming a nonmagnetic conductor ($\mu \approx \mu_0$), the quality factor becomes

$$\boxed{Q = a\sqrt{\frac{\omega\sigma\mu_0}{2}}}. \tag{6.110}$$

A moment to temporarily forget about the math

The Q factor characterizes the balance between energy storage and energy loss in the waveguide. It determines the cavity fill time and how effectively electromagnetic energy is retained over time.

As shown by Eq. (6.110), the Q factor increases with the spacing between the conducting walls. A larger separation allows more electromagnetic energy to be stored within the waveguide volume, while the resistive loss continues to occur only at the surface of the waveguide wall.

The Q factor also increases with the square root of the electrical conductivity of the wall material. This is straightforward to understand: higher conductivity reduces the surface resistance, thereby decreasing the Ohmic power loss.

Although the surface resistance increases with frequency due to the skin effect, the total electromagnetic energy stored in the waveguide grows more rapidly than the corresponding increase in wall loss. As a result, **the overall Q factor increases with frequency**.

It is important to note that while the Q factor quantifies the efficiency of energy storage, it does not directly measure the efficiency of electron acceleration. The stored energy does not necessarily produce the required electric field orientation or phase velocity for sustained acceleration. For that purpose, another metric—the shunt impedance—is used.

6.5.6 Shunt Impedance

Shunt impedance quantifies how efficiently microwave power is converted into charge particle accelerating voltage. It is defined as

$$\mathcal{Z} = \frac{V_{\text{acc}}^2}{2 \times \text{average power loss}}, \tag{6.111}$$

where V_{acc} is the *peak accelerating voltage* along the z direction. The factor of two arises because the RF power is time-averaged, while V_{acc} is a peak quantity.

It is often convenient to define the *length-normalized shunt impedance*:

$$\boxed{Z = \frac{\mathcal{Z}}{L_z} = \frac{(V_{\text{acc}}/L_z)^2}{2 \times \text{average power loss}/L_z}} \tag{6.112}$$

The term "shunt impedance" originates from the equivalent resistor-inductor-capacitor (RLC) circuit model of an microwave cavity [6]. In this model, the wall loss is represented by a resistor $\mathcal{Z}$ in parallel with the accelerating voltage, giving

$$P = \frac{V^2}{2\mathcal{Z}}. \tag{6.113}$$

In a Linac, the waveguide is designed such that the phase velocity of the microwave matches the velocity of the electrons being accelerated. As discussed in Sect. 6.4.4, a hollow waveguide does not satisfy this requirement. Therefore, conducting discs are introduced into the waveguide to reduce the phase velocity of the microwave. This approach will be discussed in detail in Sect. 6.6. For the purpose of analyzing the shunt impedance, we assume that practical Linac structures provide a mechanism to ensure that the electron velocity remains synchronized with the microwave phase velocity. Under this condition, the electrons continuously experience the same accelerating field. Consequently, the ratio V_{acc}/L_z in Eq. (6.112) is simply equal to the amplitude of E_z. Therefore,

$$\frac{V_{\text{acc}}}{L_z} = |E_z|. \tag{6.114}$$

Using

$$E_z^{(m)}(x, z) = A \sin\left(\frac{m\pi x}{a}\right) e^{ik_z z}, \tag{6.115}$$

for $m = 1$ and $x = a/2$, we obtain

$$\frac{V_{\text{acc}}}{L_z} = A. \tag{6.116}$$

According to Eq. (6.104), the average wall loss per unit length of the waveguide is

$$\langle P_{\text{L}} \rangle = \frac{b}{2} R_s |H_y(x = 0)|^2. \tag{6.117}$$

Thus,

$$Z = \frac{A^2}{bR_s|H_y(x=0)|^2}. \tag{6.118}$$

According to Eqs. (6.44) and (6.52),

$$|H_y(x=0)| \propto A. \tag{6.119}$$

Therefore,

$$\begin{aligned} Z &\propto \frac{1}{R_s} \\ &\propto \sqrt{\sigma}. \end{aligned} \tag{6.120}$$

A moment to temporarily forget about the math
A higher shunt impedance implies that a larger accelerating field can be achieved for a given input microwave power. Therefore, the shunt impedance is an important figure of merit for evaluating the efficiency of electron acceleration in microwave Linacs.

Improved electrical conductivity of the waveguide walls reduces Ohmic power loss, which in turn increases the shunt impedance.

In medical Linacs, the shunt impedance is on the order of 10–100 MΩ/m [1].

6.6 Microwave Dynamics in Disc-Loaded Waveguides

As discussed in Sect. 6.4.4, the phase velocity of a microwave in a hollow waveguide exceeds c. Since electrons to be accelerated travel at velocities below c, the microwave in a simple hollow waveguide cannot provide cumulative acceleration. **The main focus of this section is to explore methods for reducing the phase velocity of the microwave so that it matches the electron velocity.**

To achieve this, periodic metallic structures such as irises (or discs) are introduced to divide the waveguide into a series of coupled "cells" along the z direction (Fig. 6.15). These periodic perturbations modify the propagation of the microwave in a way analogous to how a periodic potential in a crystal modifies electron wavefunctions, as discussed in Sect. 1.3.1 of Chap. 1. Each iris contains a central aperture, allowing partial transmission of the field while reflecting the remainder, thereby coupling adjacent cells.

In contrast to a simple hollow waveguide, such a periodic system supports slow-wave modes whose dispersion relation is fundamentally altered. While an isolated

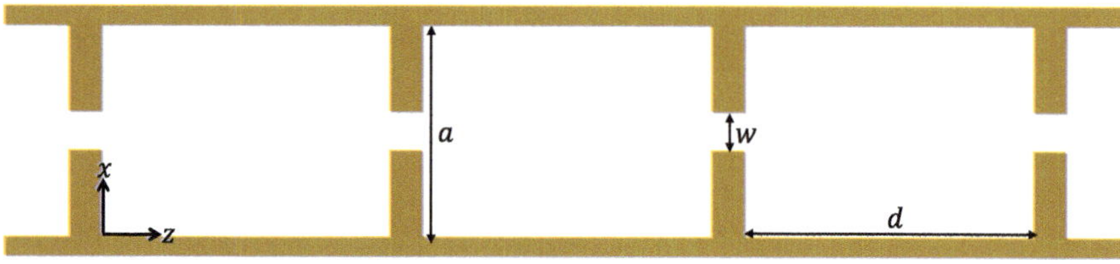

Fig. 6.15 Schematic illustration of a disc-loaded toy-model waveguide, where each disc has a central aperture of width w. The magnetic field is polarized along the y-direction, while the electric field lies in the x-z plane. The spacing between adjacent discs is d

resonant cavity supports standing waves (with no net phase advance along z), the coupling between adjacent cavities enables the formation of traveling waves with a finite but reduced phase velocity matched to the electron velocity.

6.6.1 Resonant Cavities

To understand how the discs reduce the phase velocity, let us first consider the limiting case in which the iris aperture is completely closed ($w = 0$). In this limit, each cell becomes an isolated microwave resonant cavity, in which the fields form standing waves and there is no net propagation. Correspondingly, the **phase advance between cells vanishes**.

Compared to a hollow waveguide, additional boundary conditions are imposed at the disc surfaces: the tangential components of the electric field must vanish on all perfectly conducting boundaries. Suppose the two discs of a cavity are located at $z = 0$ and $z = d$. For the polarized electromagnetic fields in our toy model (with $\mathbf{E}$ in the x-z plane and $\mathbf{B}$ along y), the boundary conditions require

$$E_z(x, z)\big|_{x=0} = 0, \tag{6.121}$$

$$E_z(x, z)\big|_{x=a} = 0, \tag{6.122}$$

$$E_x(x, z)\big|_{z=0} = 0, \tag{6.123}$$

$$E_x(x, z)\big|_{z=d} = 0. \tag{6.124}$$

In addition, using Gauss's law in vacuum,

$$\nabla \cdot \mathbf{E} = 0, \tag{6.125}$$

and noting that the tangential electric field vanishes at $z = 0$ and $z = d$, one finds that E_z must satisfy the following boundary conditions at the disc surfaces:

$$\left.\frac{\partial E_z(x,z)}{\partial z}\right|_{z=0} = 0, \tag{6.126}$$

$$\left.\frac{\partial E_z(x,z)}{\partial z}\right|_{z=d} = 0 \tag{6.127}$$

We now solve for E_z using separation of variables. Let $E_z(x,z) = u(x)\,v(z)$. Substituting into the Helmholtz equation (6.34) gives

$$\frac{\mathrm{d}^2 u(x)}{\mathrm{d}x^2} + k_x^2 u(x) = 0, \tag{6.128}$$

$$\frac{\mathrm{d}^2 v(z)}{\mathrm{d}z^2} + k_z^2 v(z) = 0, \tag{6.129}$$

with the constraint

$$k_x^2 + k_z^2 = \frac{\omega^2}{c^2}. \tag{6.130}$$

Applying the boundary conditions in Eqs. (6.124)–(6.127), we obtain

$$E_z(x,z) = E_{z0}\sin\left(\frac{m_x \pi x}{a}\right)\cos\left(\frac{m_z \pi z}{d}\right) \tag{6.131}$$

or

$$E_z(x,z,t) = E_{z0}\sin\left(\frac{m_x \pi x}{a}\right)\cos\left(\frac{m_z \pi z}{d}\right)\mathrm{e}^{-\mathrm{i}\omega t}, \tag{6.132}$$

where m_x and $m_z = 0, 1, 2, \ldots$.

According to Eq. (6.130), the resonance condition is

$$\frac{\omega^2}{c^2} = \left(\frac{m_x \pi}{a}\right)^2 + \left(\frac{m_z \pi}{d}\right)^2. \tag{6.133}$$

Each pair (m_x, m_z) corresponds to a distinct standing-wave mode of the cavity. For fixed cavity dimensions a and d, only discrete frequencies satisfying Eq. (6.133) are allowed. This discrete spectrum is the defining feature of a resonant cavity and underlies its role as a frequency-selective structure.

6.6.2 *From Resonant Cavities to Disc-Loaded Waveguides*

In Sect. 6.6.1, each cell was treated as an isolated resonant cavity, supporting standing-wave modes with no net axial propagation. We now relax this condition by introducing a finite iris aperture ($w > 0$). The apertures enable electromagnetic

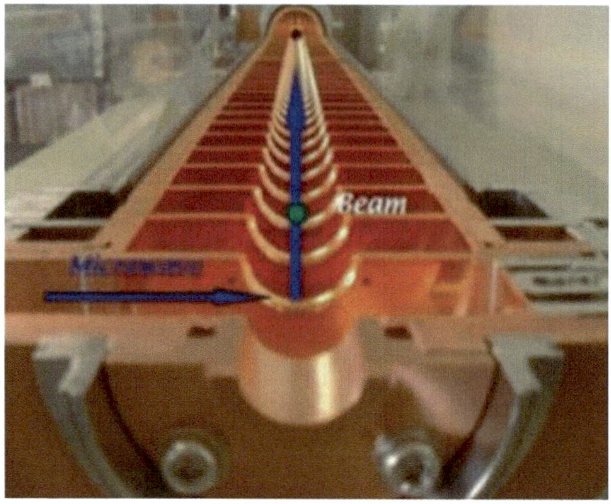

Fig. 6.16 A Cu disc-loaded acceleration waveguide designed for 3-GHz microwaves. Source of figure: Reference [10], licensed under Creative Commons Attribution 4.0 (CC BY 4.0), which permits redistribution and reuse of the material in any medium or format for any purpose when credit is given to the original source

energy to propagate along the waveguide, transforming the system from a set of isolated resonators into a periodically loaded waveguide (Fig. 6.16).

Physically, the apertures allow the fields in neighboring cells to overlap and exchange energy. As a result, the fields are no longer purely standing waves; instead, they form **traveling waves with a periodic spatial modulation**. A key feature of a disc-loaded waveguide is that the wave is repeatedly "delayed" as it traverses each cell: energy is temporarily stored in the cavity-like regions and then coupled to the next cell through the aperture. These repeated storage and release slow the phase progression of the wave. Consequently, the dispersion relation $\omega(k_z)$ differs significantly from that of a hollow waveguide, lying below the line $\omega = ck_z$, and the phase velocity, $v_p = \frac{\omega}{k_z}$, can be reduced below c.

By adjusting the aperture size w, one controls the coupling strength between cells and hence the dispersion relation, allowing the phase velocity to be tuned to match the velocity of electrons.

The resulting periodic structure can be analyzed using Floquet's theorem, which is discussed in the following subsection.

6.6.3 Floquet's Theorem

The analysis of the electric field in a disc-loaded waveguide can be greatly simplified by applying Floquet's theorem, which describes the solutions of the following differential equations with periodic coefficients:

$$\frac{\mathrm{d}^2 f}{\mathrm{d}z^2} + Q(z) f = 0, \quad \text{where } Q(z+d) = Q(z) \tag{6.134}$$

The theorem states that the solution can always be expressed in the form

$$f(z) = p(z)\,\mathrm{e}^{\mathrm{i}k_0 z}, \tag{6.135}$$

where $p(z)$ is periodic with the same period d as $Q(z)$ and k_0 is a constant known as the Floquet exponent (or characteristic exponent). Equation (6.135) simplifies the solution of the differential equation by separating it into a periodic part $p(z)$ and a phase factor $\exp(ik_0 z)$, which describes the cumulative phase advance along z.

Regardless of whether the waveguide is disc-loaded or not, E_z satisfies the Helmholtz equation (6.34):

$$\nabla^2 E_z + \frac{\omega^2}{c^2} E_z = 0 \tag{6.136}$$

Substituting $E_z = u(x)v(z)$ into Eq. (6.136) yields

$$\frac{\mathrm{d}^2 u(x)}{\mathrm{d}x^2} + k_x^2\, u(x) = 0, \tag{6.137}$$

$$\frac{\mathrm{d}^2 v(z)}{\mathrm{d}z^2} + k_z^2\, v(z) = 0, \tag{6.138}$$

where

$$k_z^2 + k_x^2 = \frac{\omega^2}{c^2}. \tag{6.139}$$

In a disc-loaded waveguide,

$$\text{structure}(z + d) = \text{structure}(z). \tag{6.140}$$

As a result, the effective longitudinal wavenumber k_z varies periodically with the same period d as the iris spacing:

$$\boxed{k_z^2(z + d) = k_z^2(z)} \tag{6.141}$$

Therefore, Eq. (6.138) can be expressed in the same general form as the differential equation in (6.134). According to Floquet's theorem, the solution of $v(z)$ must have the form of a periodic function multiplied by an exponential phase factor, so is the product of $v(z)$ and $u(x)$, namely, $E_z(x, z)$. Consequently, the longitudinal electric field can be written as

$$E_z(x, z) = F(x, z)\,\mathrm{e}^{\mathrm{i}k_0 z}, \tag{6.142}$$

or

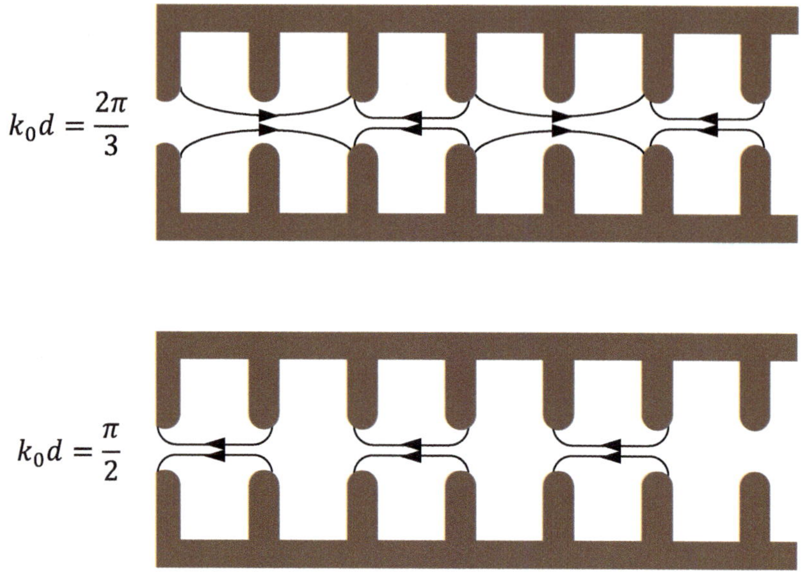

Fig. 6.17 A snapshot of the accelerating electric field in disc-loaded waveguides with $k_0 d = \frac{2\pi}{3}$ (top) and $k_0 d = \frac{\pi}{2}$ (bottom)

$$E_z(x, z, t) = F(x, z)\, e^{i(k_0 z - \omega t)}, \tag{6.143}$$

where $F(x, z)$ is periodic along z with period d equal to the inter-disc spacing and k_0 is a **global propagation constant along the z-direction**. The value of k_0 depends on the waveguide design, for example, in many clinical traveling-wave Linacs,

$$k_0 d = \frac{2\pi}{3}. \tag{6.144}$$

This choice corresponds to a phase shift of 120° between adjacent cells (Fig. 6.17). It represents a compromise between several competing requirements: sufficient cell-to-cell coupling, stable field distribution, and efficient power transfer along the structure [8, 12].

6.6.4 Space Harmonics in Disc-Loaded Waveguides

Because $F(x, z)$ is periodic in z with period d, it can be expanded as a Fourier series,

$$F(x,z) = \sum_{n=-\infty}^{\infty} A_n(x)\, e^{i\frac{2\pi n}{d}z}, \tag{6.145}$$

where n is an integer. The Fourier coefficients $A_n(x)$ are determined by the geometry of the waveguide and in practice decrease with increasing $|n|$.

Substituting this expansion into Eq. (6.143) gives

$$\begin{aligned} E_z(x,z,t) &= \sum_n A_n(x)\, e^{i\frac{2\pi n}{d}z}\, e^{i(k_0 z-\omega t)} \\ &= \sum_n A_n(x)\, e^{i[(k_0+\frac{2\pi n}{d})z-\omega t]} \\ &= \sum_n A_n(x)\, e^{i(k_z^{(n)}z-\omega t)}, \end{aligned} \tag{6.146}$$

where

$$k_z^{(n)} = k_0 + \frac{2\pi n}{d} \tag{6.147}$$

is the axial propagation constant of the n-th **space harmonic**.

Each space harmonic behaves like a plane wave with its own phase velocity:

$$v_p^{(n)} = \frac{\omega}{k_z^{(n)}} = \frac{\omega}{k_0 + \frac{2\pi n}{d}} \tag{6.148}$$

The global propagation constant k_0 is determined by the waveguide structure. Once k_0 is specified, all space harmonics are fixed through Eq. (6.147).

Importantly, although the overall field is a superposition of all space harmonics, only one harmonic denoted as n_s can have a phase velocity close to the electron velocity v_e. By properly choosing the structure period d and operating frequency, this harmonic can satisfy

$$v_p^{(n_s)} \approx v_e, \tag{6.149}$$

enabling **synchronous acceleration**.

For nonsynchronous harmonics, their effect on electrons averages to zero over time, so they do not contribute to net acceleration. However, they can still contribute to RF power loss and field distortion. An optimal Linac design therefore aims to minimize these nonsynchronous harmonics by appropriately choosing the aperture size w, the cell period d, and the cell geometry.

As the electrons accelerate along z, according to Eqs. (6.148)–(6.149), the iris spacing d must increase to maintain phase synchronism. However, once the electrons reach relativistic velocity ($v_e \approx c$), the spacing d can be stabilized.

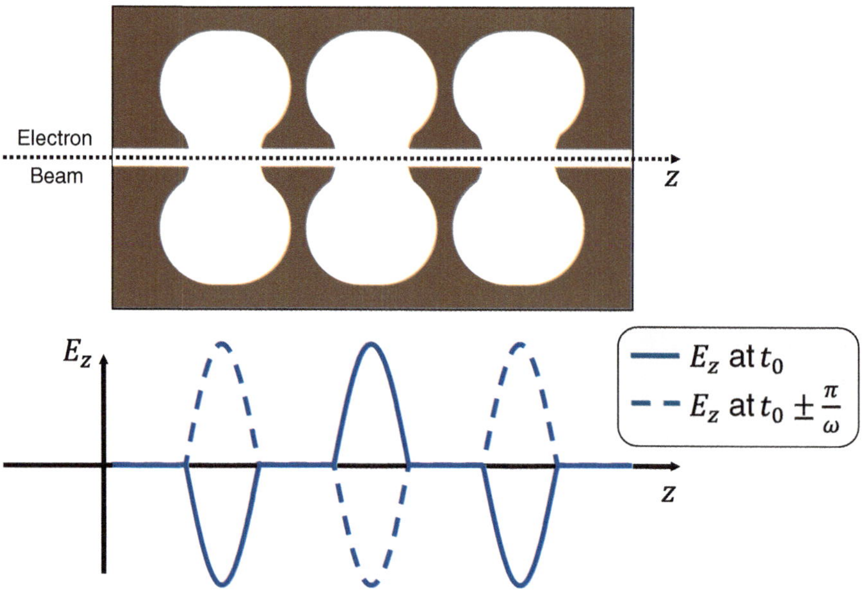

Fig. 6.18 (top) A schematic diagram of a π-mode standing-wave accelerating waveguide. (bottom) The accelerating electric field distributions at two time points separated by half a microwave period

6.6.5 Standing-Wave Accelerators

When a chain of identical resonant cavities is coupled through irises and arranged along the electron beam direction (Fig. 6.18), the structure supports a set of collective eigenmodes. These modes are standing-wave patterns established by the superposition of forward and backward waves, **resulting in no net power flow along the structure.**

If the dominant mode is engineered to be the π-mode, defined by a phase advance of π per cell, i.e.,

$$k_0 d = \pi, \tag{6.150}$$

according to Eq. 6.143, the longitudinal electric fields E_z in neighboring cavities are 180° out of phase (Fig. 6.18). This phase shift describes a spatially alternating field pattern that is fixed in space, rather than a propagating wave.

In this mode, an electron entering the first cavity is accelerated by the axial electric field. As it travels to the next cavity, two effects occur simultaneously: the field in the adjacent cavity has the opposite sign due to the π phase shift, and the electromagnetic field continues to oscillate in time. If the time of flight between cavities satisfies

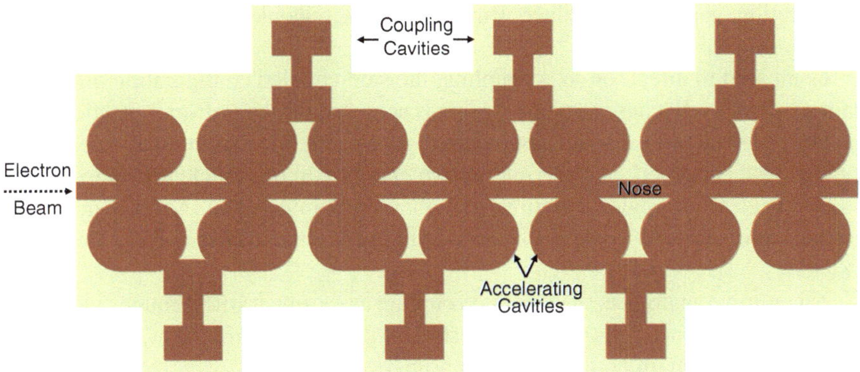

Fig. 6.19 A schematic diagram of a $\pi/2$-mode standing-wave accelerating waveguide

$$\Delta t = \frac{\pi}{\omega}, \tag{6.151}$$

the temporal phase advance compensates for the spatial phase reversal. As a result, the electron again experiences an accelerating field upon entering the next cavity, enabling continuous acceleration along the z direction.

Thus, despite the π phase difference between neighboring cells, the overall field pattern remains stationary in space, and the structure operates as a **π-mode standing-wave accelerating structure.**

In addition to the π-mode, the $\pi/2$-mode is commonly employed in standing-wave accelerators. It is characterized by a phase advance of

$$k_0 d = \frac{\pi}{2}. \tag{6.152}$$

A shapshot of electric field in $\pi/2$-mode standing-wave waveguide is shown in Fig. 6.17. As shown, the cells separate into accelerating cells and coupling cells. The accelerating cells support a strong longitudinal electric field along the beam axis, while the coupling cells ideally exhibit little or no on-axis accelerating field and primarily serve to transfer microwave power and stabilize the field distribution. These coupling cells can be positioned off-axis to minimize their interaction with the electron beam (Fig. 6.19).

In the ideal $\pi/2$-mode, the accelerating cavities sustain strong E_z fields, whereas the coupling cavities have negligible longitudinal electric fields but support magnetic fields that mediate electromagnetic energy transfer between adjacent accelerating cells.

To suppress direct coupling between adjacent accelerating cavities, the irises between them are made small, thereby reducing electric (capacitive) coupling. In addition, a nose cone is introduced to concentrate the electric field within each accelerating gap and to reduce fringe fields that would otherwise extend into neighboring cavities.

Compared to direct, on-axis coupling, the side-coupled configuration offers certain advantages. Because strong coupling is achieved through off-axis cavities, large apertures on the beam axis are not required. As a result, the longitudinal accelerating field is less perturbed, and the microwave energy is more efficiently concentrated in the accelerating regions, leading to higher shunt impedance. Another important advantage of the $\pi/2$-mode is its superior field stability, as discussed in Reference [8].

An intuitive analogy is to view side coupling as routing traffic through side streets surrounding a central plaza: the side streets handle the traffic flow, allowing the central area to remain optimized for its primary function.

6.7 Summary of Chap. 6

1. Medical Linacs use microwaves with frequencies of approximately 3 GHz to accelerate electrons to MeV energies, corresponding to speeds close to c. In free space, the electric field of a microwave is transverse to the direction of propagation and therefore cannot provide sustained acceleration along the beam direction. However, when confined within a waveguide, reflections from the conducting walls can produce a time-varying electric field component along the electron beam (z) direction.
2. Given a microwave frequency, knowledge of the waveguide geometry and wall material allows us to determine the electric and magnetic fields throughout the waveguide, including within its walls. This establishes a direct link between the input microwave power and the accelerating field. For waveguides with finite conductivity, it also enables the calculation of induced currents and power losses, connecting energy consumption to acceleration efficiency. Correspondingly, quantities such as the Q factor and shunt impedance are introduced to characterize the performance of an acceleration waveguide.
3. For normal conducting waveguides in the skin-effect regime, both the Q factor and the shunt impedance scale approximately with the square root of the electrical conductivity of the wall material. This scaling arises because the surface resistance decreases with increasing conductivity.
4. In this chapter, a simplified toy-model waveguide geometry is used to enable analytical solutions of the electromagnetic fields. In practice, these fields are typically obtained numerically by solving Maxwell's equations with appropriate boundary conditions.
5. To achieve sustained acceleration in a traveling-wave structure, the electron velocity must be matched to the phase velocity of the electromagnetic wave along z. In a hollow waveguide, the phase velocity always exceeds c. By introducing periodic structures such as irises, the phase velocity can be reduced to match the electron velocity.

6. Standing-wave structures with side-coupled cavities provide an alternative to traveling-wave waveguides for electron acceleration. These can be viewed as a series of coupled resonant cavities that support a time-varying but spatially stationary electric field pattern.

References

1. Anderson R, Lamey M, MacPherson M, Carlone M. Simulation of a medical linear accelerator for teaching purposes. J. Appl. Clin. Med. Phys. 2015;16(3):359–77. https://doi.org/10.1120/jacmp.v16i3.5139.
2. Gibbons JP. Khan's the physics of radiation therapy. 6th ed. Philadelphia: LWW; 2019.
3. Gilmour AS, Jr. Microwave and millimeter wave vacuum electron devices: inductive output tubes, klystrons, traveling-wave tubes, magnetrons, crossed-field amplifiers, and gyrotrons. Norwood: Artech House; 2020.
4. Greene D, Williams PC. Linear accelerators for radiation therapy. 2nd ed. Boca Raton: CRC Press; 1997.
5. Griffiths DJ. Introduction to electrodynamics. 5th ed. Cambridge: Cambridge University Press; 2023.
6. Karzmark CJ, Nunan CS, Tanabe E. Medical electron accelerators. New York: McGraw-Hill; 1992.
7. Kesari V, Basu BN. High power microwave tubes: basics and trends. Bristol: IOP Concise Physics; 2018.
8. Loew GA, Talman R. Elementary principles of linear accelerators. SLAC-PUB-3221. 1983.
9. Podgorsak EB. Radiation physics for medical physicists. 3rd ed. Cham: Springer; 2016.
10. Seeman J, Schulte D, Delahaye JP, Stapnes S, Ross M, Grudiev A, Yamamoto A, Latina A, Seryi A, Garcia RT, Guiducci S, Papaphilippou Y, Bogacz SA, Krafft GA. Design and principles of linear accelerators and colliders. Cham: Springer; 2020.
11. Wang Z. Modern electrodynamics (in Chinese). Beijing: Higher Education Press; 2023.
12. Warner DJ. Fundamentals of electron linacs. Geneva: CERN Document Server; 1996.

Index

K. Li, *Essential Physics of Medical X-ray Sources and Detectors*, https://doi.org/10.1007/978-3-032-27800-5

Q

R

GPSR Compliance

The European Union's (EU) General Product Safety Regulation (GPSR) is a set of rules that requires consumer products to be safe and our obligations to ensure this.

If you have any concerns about our products, you can contact us on ProductSafety@springernature.com

In case Publisher is established outside the EU, the EU authorized representative is:

Springer Nature Customer Service Center GmbH
Europaplatz 3
69115 Heidelberg, Germany

Batch number: 10682282

Printed by Printforce, the Netherlands